Infrared Astronomy

Infrared Astronomy

Proceedings of the NATO Advanced Study Institute held at Erice, Sicily, 9–20 July, 1977

edited by

GIANCARLO SETTI

Laboratorio di Radioastronomia CNR, University of Bologna, Italy

and

GIOVANNI G. FAZIO

Harvard-Smithsonian Center for Astrophysics, Cambridge, Mass., U.S.A.

D. Reidel Publishing Company

Dordrecht : Holland / Boston : U.S.A. / London : England

Published in cooperation with NATO Scientific Affairs Division

NATO ADVANCED STUDY INSTITUTES SERIES

*Proceedings of the Advanced Study Institute Programme, which aims
at the dissemination of advanced knowledge and
the formation of contacts among scientists from different countries*

The series is published by an international board of publishers in conjunction
with NATO Scientific Affairs Division

A	Life Sciences	Plenum Publishing Corporation
B	Physics	London and New York
C	Mathematical and Physical Sciences	D. Reidel Publishing Company Dordrecht, Boston, and London
D	Behavioral and Social Sciences	Sijthoff International Publishing Company Leiden
E	Applied Sciences	Noordhoff International Publishing Leiden

Series C – Mathematical and Physical Sciences

Volume 38 – Infrared Astronomy

Library of Congress Cataloging in Publication Data

International School of Astrophysics, 4th, Erice, Italy, 1977.
    Infrared astronomy.

    (NATO advanced study institutes series: Series C, Mathematical and physical sciences ; v. 38)
    "Lectures presented at the 4th course of the International School of Astrophysics, held in Erice
(Sicily) from July 9–July 20, 1977, at the 'E. Majorana' Centre for Scientific Culture."
    Bibliography: p.
    Includes index.
    1. Infra-red astronomy – Congresses. I. Setti, Giancarlo, 1935–   II. Fazio, Giovanni G.,
1933–   III. Title.   IV. Series.
QB470.A1I54   1977   523.01′5′012   78–22030
ISBN-13: 978-94-009-9817-9   e-ISBN-13: 978-94-009-9815-5
DOI: 10.1007/978-94-009-9815-5

Published by D. Reidel Publishing Company
P.O. Box 17, Dordrecht, Holland

Sold and distributed in the U.S.A., Canada, and Mexico
by D. Reidel Publishing Company, Inc.
Lincoln Building, 160 Old Derby Street, Hingham, Mass. 02043, U.S.A.

CONTENTS

SEMINARS

FOREWORD

This volume contains a series of lectures presented at the 4th
Course of the International School of Astrophysics, held in Erice
(Sicily) from July 9 - July 20, 1977 at the "E. Majorana" Centre
for Scientific Culture. The course was fully supported by a grant
from the NATO Advanced Study Institute Programme. It was attended
by 82 participants from 15 countries.

Even though the infrared portion of the electromagnetic spectrum
covers an extensive interval from the red region of the optical
spectrum (10,000 Å) to the microwave radio region (1 mm), its
role in astronomy has been minimal until the last two decades.
Until very recently, the only objects observed were the sun, the
moon and the planets. A primary reason for this late development
was the lack of sensitive detectors and the necessary cryogenic
technology that must accompany their use. Recent progress in this
technology has been paralleled by an ever increasing interest of
astronomers in infrared observations, leading to a number of extremely important results in different branches of astronomy.
This becomes evident when one realizes that in many astrophysical
conditions most of the energy is found to be channeled into the
infrared portion of the electromagnetic spectrum.

Stars were detected that yield most of their radiation in the
infrared; these objects present a new view of stellar evolution,
both in the birth and death stages.

Observations in the infrared of HII regions and dense dark
dust clouds, together with the studies of the molecular clouds,
are providing completely new insight into the physical conditions
and distribution of the interstellar gas in the Galaxy and into
the processes of star formation. Intense infrared radiation has
been discovered in the centre of our galaxy, numerous galactic
nuclei and other extra-galactic objects, including the very prominent emission from Seyfert's nuclei and quasars. Therefore,
observations in the near and far infrared are yielding new insights
into the most powerful objects in the universe, and, in particular,
one hopes for a better understanding of the physics and evolution
of galactic nuclei.

Other important achievements in infrared astronomy are in the
field of cosmology, where observations in the submillimetre

and far-infrared regions have helped establish the shape of the
3 K diffuse background radiation spectrum, consistent with its
origin in the initial explosion of the universe (hot big-bang).

The earth's atmosphere has limited most observations to that
short wavelength which is observable from the ground. Efforts
have been made, using aircraft, rocket and high-altitude balloon-
borne telescopes, to extend the observations to the far-infrared
and submillimetre regions. Now, however, we are on the threshold
of a new and exciting era in infrared astronomy, where large and
ambitious programmes will be undertaken from space platforms.
Detector and cryogenic technology continue to advance rapidly to
match this new capability. The scope of infrared observations and
their impact on astronomy will correspondingly increase.

In this context, it was a most opportune time to hold a NATO
Advanced Study Institute on this subject to acquaint students and
researchers, not only in infrared astronomy but in related fields,
with the many advances and future prospects of this most exciting
branch of astronomy.

The various aspects of this field were covered in a series of
lectures and topical seminars totalling 50 hours. They were divided into two general sections: the first dealt primarily with
galactic infrared sources and their role in star formation, the
nature of the interstellar medium and galactic structure; the
second with the interpretation of infrared, optical and radio
observations of extra-galactic sources and their role in the origin
and structure of the universe. In addition, a number of lectures
were dedicated to instrumental techniques and to a review of future
space observations.

We wish to express our gratitude to the Scientific Affairs
Division of the North Atlantic Treaty Organization for the generous
support given to the Institute. Sincere thanks are also due to
Mrs. B. Mandel and Mr. L. Baldeschi for their help with the organ-
ization of the meeting and for the patient typing and drawing of
a number of figures contained in the manuscript; to Mrs. D.
Weksberg and Mrs. S. Publicker for partial typing and to Mr. R.
Primavera for the photographic reproduction of the figures.

Special thanks are due, of course, to all lecturers and participants who contributed so much to the success of the course,
and to Prof. A. Zichichi for his hospitality at the "Ettore
Majorana" Centre for Scientific Culture.

Giancarlo Setti   Giovanni G. Fazio
University of Bologna   Harvard-Smithsonian
Laboratorio di Radioastronomia   Center for Astrophysics
Via Irnerio 46   60 Garden Street
40126 Bologna, Italy   Cambridge, Mass. 02138, USA

INTERSTELLAR MATTER

P.G. Mezger

Max-Planck-Institut für Radioastronomie
Bonn, West Germany

1. OUR GALAXY - AN OVERVIEW

1.1 Formation of Galaxies, Primordial Elements and the 3 K Background

Most relevant observations agree with the standard "big bang" model in which the universe started to expand about 1.6E10 yrs. ago from a state of high density and high temperature. ^{1}H and ^{4}He and (in very minor quantities) their isotopes ^{2}H = D and ^{3}He formed at an expansion time $t \sim 300$ s. However, no elements with atomic number $A \geq 4$ could form by neutron capture, the process by which D through ^{4}He formed. In Table 1 I have compiled what I consider to be the best estimates of the primordial abundances (see also Section 2.5 and Wagoner, 1973). At $t \sim 3E5$ yrs. and temperatures of ~ 3000 K, free electrons combined with protons and He-nuclei and the radiation field decoupled from the matter. Today the radiation field is observed as an isotropic black-body background radiation of radiation temperature 2.7 K (see Fig. 5). Density fluctuations present in the primordial matter increased as a result of the expansion of the universe and protogalactic clouds formed with masses ranging from 1E12 M_0 to 1E8 M_0 (and possibly less).

Gas clouds lose internal energy (and in this way decrease their kinetic gas temperature) by radiation. Once a critical set of values of temperature and density are reached, the protogalactic cloud starts to collapse. If it has angular momentum, the free-fall collapse occurs preferentially parallel to the axis of rotation - the protogalactic cloud collapses into a disk. An upper limit for the age of our Galaxy (i.e. when the protogalaxy started to collapse) is $T_G \sim 1.4E10$ yrs., corresponding to an expansion time of $t \sim 2E9$ yrs. At that time the bulk of galaxies may have

1

G. Setti and G. G. Fazio (eds.), Infrared Astronomy, 1-24.
All Rights Reserved. Copyright © 1978 by D. Reidel Publishing Company, Dordrecht, Holland.

Table 1

Abundances in the Galaxy

Element	Primordial Abundance	Primordial Abundance (D_G = 10 kpc)
$X(^1H)$	$0.78 - 0.76$	0.70
$X(D)$	$6\ 10^{-4} - 2.5\ 10^{-5}$	$2.5\ 10^{-5}$
$Y(^3He)$	$7\ 10^{-5} - 1.5\ 10^{-5}$	$\lesssim (4-8)\ 10^{-5}$
$Y(^4He)$	$0.219 - 0.242$	0.28
$Z(A>4)$	$<< 10^{-11}$	0.02

formed.

How efficient was the formation of galaxies out of the primordial gas? In Table 2 I compare the present mean density of the
universe ρ_b as derived for primoridal ^4He abundances of 0.07 -
- 0.08 (by number, corresponding to mass fraction Y_{prim} (^4He)
given in Table 1) with the mean density ρ_G of matter contained in
visible galaxies. Since $\rho_G/\rho_b \gtrsim 1$ (i.e. practically all available
mass went into galaxies), the efficiency of galaxy formation must
have been very high. Present observations, in fact, support only
the existence of intergalactic matter in galaxy clusters. There
also exists a critical density ρ_c; if the mean density of the
universe $\leq \rho_c$, the universe is "open" and will expand forever. This
latter situation is favored by present observations.

1.2 Star Formation and Chemical Evolution in Galaxies

Stars observed today have masses ranging from 0.08 M_Θ to $\sim$100 M_Θ.
Stars less massive than 0.08 M_Θ never reach central temperatures

Table 2

Mean Densities of the Universe

$\rho_c = 5.7\ 10^{-30}$ g cm^{-3}	(for H = 55 km s^{-1} Mpc^{-1})
Mean densities derived for $Y_{prim} = 0.219 - 0.242$	$\rho_b/\rho_c = 0.012 - 0.088$
Mean density of "visible" matter	$\rho_G/\rho_c = 0.08$
Matter contained in Galaxies	$\rho_G/\rho_b = 6.7 - 0.91$

of some 1E7K required for H-burning. However, there is no reason why fragments down to at least 0.007 M_Θ should not form together with other protostars. The upper mass limit appears to be determined by radiation pressure (see Section 2.2). Stars spend most of their lifetime on the main sequence (MS), where they produce energy by burning H into ^{4}He in their core region.

Fig. 1 is a schematic representation of the observed mass spectrum (curves a, b) and the Initial Mass Function (IMF) of stars derived by Salpeter (1955). I have extended Salpeter's IMF to the theoretical limit for protostars 0.007 M_Θ (Low and Lynden-Bell, 1976). Salpeter's IMF represents the present stellar mass spectrum. It represents the stellar birth rate function only if star formation in the Galaxy was independent on time (see Section 2.1). The true IMF may be steeper than $M_*^{-1.35}$ for $M_* > 10$ M_Θ.

Galaxies as observed today can be classified according to the Hubble sequence, which ranges from elliptical (E) through spiral (S) to irregular (I) galaxies. The bulk of the galaxies are spirals. Radiospectroscopy of the 21 cm line of atomic H has revealed that the Hubble classification is primarily a classification according to an increasing gas content of the galaxies, ranging from $\mu = M_{gas}/M_{tot} \ll 1\%$ (E) through some % (S) to some 10% (I). If galaxies don't lose gas to intergalactic space, they can only decrease their gas content by transforming gas into stars. However, the final stages of stellar evolution, white dwarfs and neutron stars have upper limits of 1.5 M_Θ and 2 M_Θ, respectively. One therefore assumes that all stars more massive than a certain lower limit return mass to the interstellar gas in their final evolutionary stages. This affects primarily the massive stars with short MS lifetimes; one therefore considers an "immediate return" of matter and a return rate r which refers to the total mass of gas transformed into stars. Integrated over the stellar

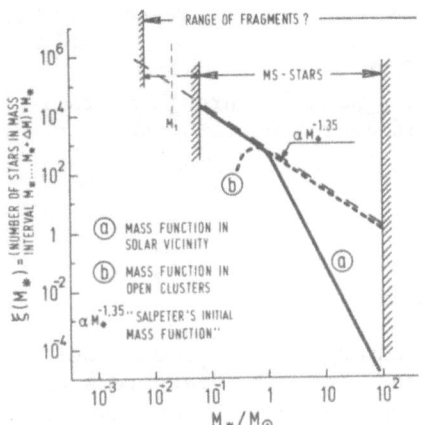

Fig. 1. Schematic representation of the stellar mass spectrum.

mass function r $\sim$ 0.3. With S = M_*/M_{tot}, the total fraction of the protogalaxy transformed into stars is $(1-r)S + \mu = 1$. One usually assumes that the rate at which gas is transformed into stars depends on the total amount of gas available, i.e.

$$dS/dt = \mu^k/(1-r)\tau \text{ with } (k \neq 0) \tag{1.1}$$

Combination of the two equations yields a differential equation with solutions

$$\mu = \begin{cases} \exp\{-t/\tau\} & \text{for } (k=1) \\ \{(k-1)t/\tau+1\}^{1/(1-k)} & \text{for } (k \neq 1) \end{cases} \tag{1.2}$$

τ is the time in which a fraction $0.63/(1-r)$ of the protogalaxy has been transformed into stars. Thus τ^{-1} is a measure of the efficiency of star formation, which appears to be one of the prime physical parameters underlying the Hubble classification, and which decreases from E through S to I galaxies.

   If this simple picture is correct, the gas content of a galaxy should also be related to its chemical evolution, i.e. the chemical composition of its gas as compared to the chemical composition of the primordial gas. The following model together with a number of more sophisticated analytical models are discussed by Tinsley (1974). With P_Z the mass fraction of gas ejected from evolved stars in the form of newly synthesized metals[1], the metal abundance of the interstellar gas changes with time according to $d(Z\mu) = -(1-r)ZdS/dt + P_ZdS/dt$. This equation has the solution

$$Z = P_Z(1-r)^{-1}\ln(1/\mu) \tag{1.3}$$

While the interstellar gas passing through stars is enriched in metals and ^{4}He, one assumes $Y = Y_{prim} + \Delta Y$ with

$$\Delta Y(^4He) = \alpha Z , \tag{1.4}$$

its primordial D is destroyed during a passage through a star. Therefore, the present D-abundance is decreased relative to the primordial value according to

$$X(D) = X(D)_{prim}\mu^{r/(1-r)} \tag{1.5}$$

1.3 Large-Scale Structure of the Galaxy

Fig. 2 is a schematic cross section through the center of our Galaxy and perpendicular to its plane. The disk stars form a bulge ("nuclear bulge") around the center of the Galaxy and taper

1) In the following we refer to $Z(A > 4)$ as the "metal abundance".

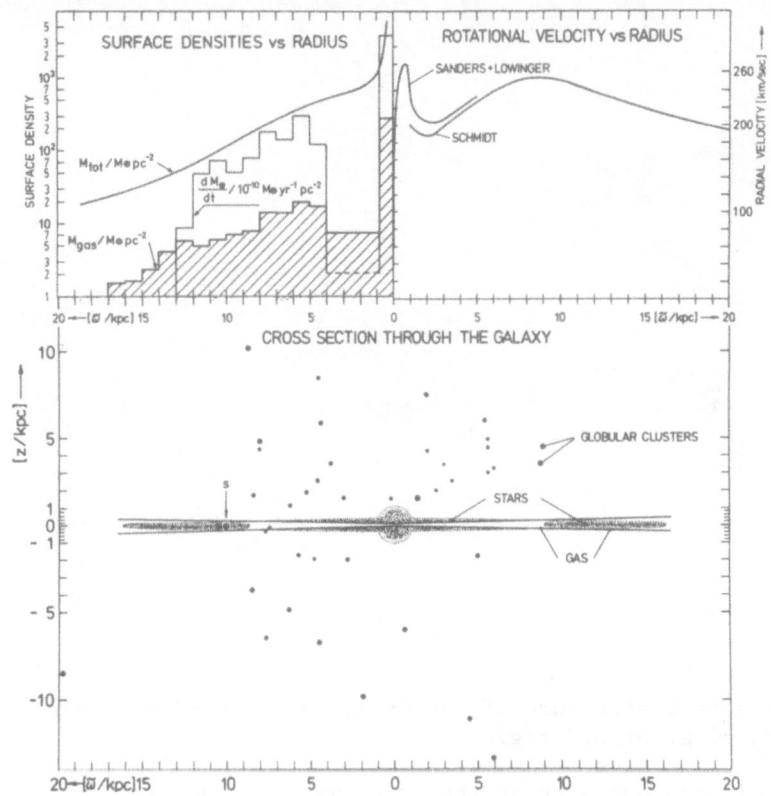

Fig. 2. A schematic cross section through the Galaxy.

off towards its edge. The globular clusters are supposed to be the oldest stellar systems which outline the size of the spherical protogalaxy. The existence and possible mass of a galactic halo is still a matter of debate. The interstellar gas as outlined by atomic H forms a flat layer with widths of $\sim$80 pc in the central region, $\sim$200 pc at D_G = 10 kpc and widening up to 600 pc at D_G = = 20 kpc. The upper right part of the diagram gives the rotation curve; the upper left diagram gives the surface density M_{gas} (including H and He), $M_{tot} = M_* + M_{gas}$ and the star formation rate dM_*/dt (for this diagram and the lower part of Table 1 see also Section 1.4).

Our galaxy is an S-galaxy, classified between Sb and Sc. Fig. 3 shows its spiral structure as outlined by giant HII regions (Georgelin and Georgelin, 1976). In the "spiral arm region", i.e. the range 4 < D_G/kpc < 13, all observed giant HII regions can be accommodated in four main spiral arms which wind about 3/4 of a revolution around the center. In the interarm region, one observes

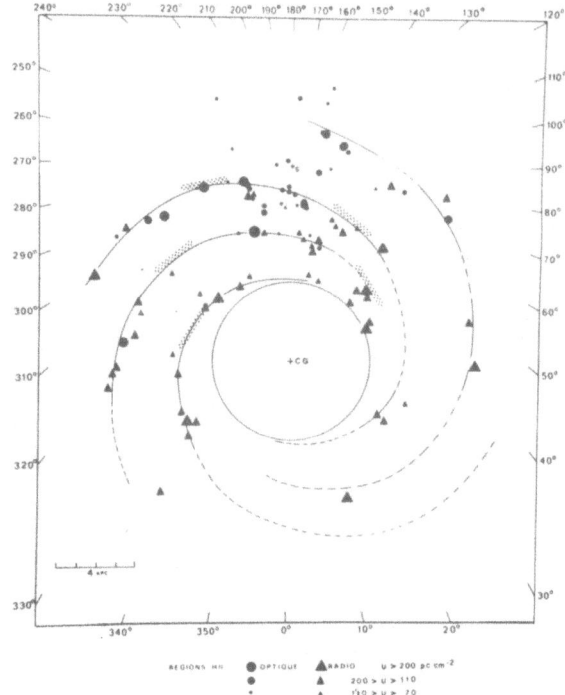

Fig. 3. The spiral structure of the Galaxy as deduced from the distribution of giant HII regions.

only small HII regions such as the Orion Nebula. Outside the main spiral arms, giant HII regions are only found in a small region (D_G < 200 pc) surrounding the galactic center. On the basis of this distribution of HII regions we estimate that of all O-stars formed during the past 10^6 yrs., 74% formed in main spiral arms and 13% formed in the interarm region and close to the galactic center. Proceeding along these lines Smith, Biermann and Mezger (1977) estimated a present total star formation rate in the Galaxy $dM_*/dt \sim 5 M_\odot$ yrs.$^{-1}$ and an exponent $k \sim 0.5 - 1$ in eq. 1.1. Being aware of the oversimplification of the model, we evaluated eqs. 1.2 through 1.5 with values compiled in Table 1 and Table 3 referring to the solar vicinity. We obtained P_z = 0.005 and α = 2-3 in eqs. 1.3 and 1.4 in reasonable agreement with detailed computations based on stellar structure and evolution. Eq. 1.5 yields $X(D)/X(D)_{prim}$ = 0.29; the corresponding primordial value is $X(D)_{prim}$ = 7.3E-5, well within the limits imposed by the estimated primordial He-abundance. Eq. 1.2 yields τ = 9.2E9 yrs. for k = 1/2 and the appropriate substitution of the age of the sun, 4.6E9 yrs., yields $\mu(t_\odot)$ = 0.24 and $Z(t_\odot)$ = 0.010 which would be the fractional gas content and metal abundance of the interstellar gas at the time when the solar system was formed from the surrounding interstellar gas. The evolution of the ^{3}He-abundance is much more complicated since it is both destroyed and generated by stellar evolution. The quantitative results obtained from this rather crude model indicate that the picture of chemical evolution of the Galaxy as a consequence of stellar evolution is basically correct.

1.4 The Physical State of Interstellar Space and Matter

I refer to Table 3. About 4% of the total mass of the Galaxy is in the form of interstellar gas. This number based on recent work by Gordon and Burton (1976) is lower than previous estimates. $M_{gas}(D_G)$ in Fig. 1 also relates to the data by Gordon and Burton. However, for the nuclear disk (D_G < 0.8 kpc) I have used a mass estimate by Bania (1977) of $\sim$4E8 $M_\odot$;

Our understanding of the physical state of the interstellar matter (ISM) has considerably changed during the past few years. The once favored two-phase model, in which cosmic and/or X-rays partially ionized and heated the gas, was not supported by recent observations, especially those made in the UV by means of the Copernicus satellite. Today one recognizes three major constituents: i) a fully ionized component; ii) clouds and cloudlets composed primarily of atomic H ($n_H \sim$ 1E1 - 1E2 cm^{-3}; $T_k \sim$ 20 - - 100 K); iii) molecular clouds ($2n_H >$ 1E3 cm^{-3}; Tk $\sim$ 10 K) which contain H primarily in molecular form. Atomic H extends from the center to the outermost regions of the Galaxy. Ionized gas and molecular clouds are concentrated in the central region (D_G < 200 pc) and in the spiral arm region (4 < D_G/kpc < 13) and are both

Table 3

Masses of the Galactic Disk

$\dfrac{M_{tot}}{M_\odot} = \dfrac{M_{gas}+M}{M_\odot}$ = 1.45 10^{11}	($D_G \leq$ 20 kpc)	
$M_{gas}/M_\odot$ = 6.1 10^9	clouds cont. H$_2$: 52%	
	clouds cont. H : 30% (40%)	
	hot neutral in- : 15% (5%)	
	tercloud gas	
	fully ionized in- : 3%	
	tercloud gas	
$\mu = M_{gas}/M_{tot}$ = 0.042 for $D_G \leq$ 20 kpc		
0.057 for $D_G \leq$ 9-11 kpc		
0.057 for $D_G \leq$ 10 kpc		

generically related to star formation. The layer containing mole-
cular H extends to $|z| \sim 80$ pc.

In the ionized gas one discriminates at least three subcomponents: 1) Radio HII regions with emission measures E > some 1E4 pc
cm^{-6} and densities > 1E2 cm^{-3}; 2) A fully ionized low density gas
with an rms electron density of ~0.4 cm^{-3}; 3)A very hot and tenuous
plasma ($n_e \sim$ 1E-3 cm^{-3}; $T_e \sim$ 1E6 K). Subcomponents 1) and 2) have
electron temperatures < 1E4 K. Subcomponents 2) and 3) together
fill probably most of the interstellar space. A hot, neutral low-
density gas may also be present, however its relative importance
until recently appears to have been highly overrated. The fully
ionized gas extends at least to 1000 pc above the plane with a mean
electron density of ~0.03 cm^{-3}. The mass of ionized gas in Table
1 refers only to the ionized gas within $|z|$ < 100 pc.

In the solar vicinity a fraction m_d/m_g < 0.5% of the interstellar matter (especially of "metals") is locked up in dust grains,
whose sizes range from 1 μm to 1E-3 μm. The grain cores appear to
be graphite or silicates; in dense clouds it is usually assumed
that $m_d/m_g \propto Z$; however it is likely that m_d/m_g also varies with
gas density.

The temperature of gas and dust is established as an equilibrium between energy gain and loss. The gas gains energy by absorption of ionizing radiation and by mechanical heat input (e.g.
SN explosions, stellar winds) and loses energy by radiation (e.g.
of collisionally excited lines of ions, atoms and molecules). Dust
grains gain energy by absorption of radiation (especially UV) and
lose energy via a quasi-blackbody radiation. This latter process
is very effective. In the general ISM grain temperatures are confined to a range 5 - 20 K. They only rise close to luminous stars.
Ice mantles sublimate for T_d > 150 K, the most refractory graphite
cores for T_d > 3650 K.

The cosmic ray particles consist of atomic nuclei (primarily
of ^{1}H and ^{4}He) and relativistic electrons; the former component is
usually referred to as "cosmic radiation" proper. The main effect
of cosmic radiation on the IMS is to maintain a low degree of ionization even in dense clouds, which in part accounts for the gas
temperature $T_K \sim$ 10 K and which initiates chemical gas-phase reactions. The cosmic electrons, together with a galactic magnetic
field of strength ~2E-6 Gauss, are responsible for the diffuse
galactic synchrotron radiation. Most of the cosmic radiation
supposedly originates in SN.

More important than cosmic particles for the energy balance
of the interstellar gas is the radiation field. I have plotted
in Fig. 4 the energy density of photons U_ν in erg cm^{-3} Hz multiplied by the frequency ν. This quantity is the energy density per
frequency interval. Note that this picture relates to a region
~ 10 kpc from the galactic center. Closer to the center all contributions except the 3 K microwave background may increase. This
updated version of a previous diagram contains the following new
entries: 1) A theoretical upper limit for the isotropic IR-radiation,

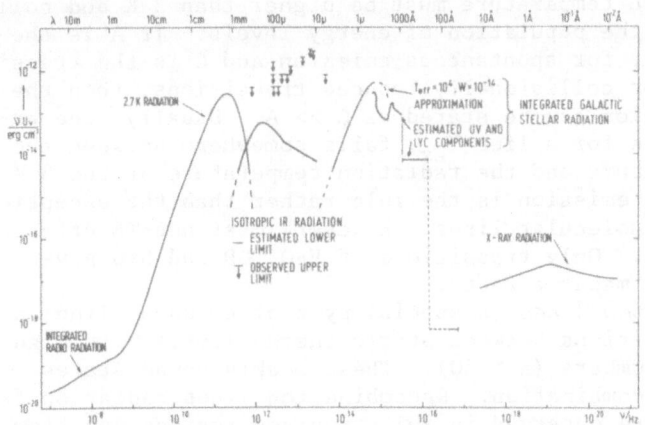

Fig. 4. Radiation densities in the solar vicinity.

which considers both galactic and extragalactic contributions
(dashed curve, Stecker et al. 1977) and observed upper limits as
compiled by Harwit (private communication) from various sources;
2) The UV points for λ > 912 Å are from Witt and Johnson, 1973;
for the energy density of Lyman continuum (Lyc) photons 912 ≥
> λ/Å ≥ 227 (see Section 3.2). It is important to realize that
these estimates for the IR- and far UV energy density relate
strictly to the solar vicinity, i.e. a typical interarm region.
The radiation density close to the sites of star formation (e.g.
in main spiral arms) should be considerably higher. The IR-radiation could affect the level populations of molecules and result in
non-LTE ("Thermodynamical Equilibrium") line emission. The Lyc-
radiation appears to be responsible for the ionization of the low-
density interstellar gas.

1.5 Radiation Mechanisms in the Radio and IR Wavelength Range

a) *Magneto-Bremsstrahlung or synchrotron radiation* arises from
cosmic electrons decelerated in a magnetic field. The intensity
of the diffuse galactic synchrotron radiation decreases rapidly with
wavelength. However, some supernovae (such as Taurus A) and a
compact non-thermal source in the center of our Galaxy have very
flat spectra.
b) *Bremsstrahlung* or free-free and free-bound continuum radiation
is due to free electrons decelerated in the Coulomb field of or
recombining with ions, respectively. Free-bound radiation is of
importance only at very short IR-wavelengths.
c) *Thermal radiation from dust grains*. Here the maximum emission
occurs at $\lambda_m T_d = 2400$ μm K. At present it is observed only at
wavelengths λ ≤ 1000 μm.
d) *Emission lines:* For a line emitted under TE conditions to be

seen the gas kinetic temperature must be higher than 3 K and collision must dominate the population of energy levels. If A is the
Einstein coefficient for spontaneous emission and C is the transition probability for collisionally induced transitions, then the
condition for emission can be stated as C >> A. Usually, the excitation temperature for a line, T_{ex}, falls somewhere between the
kinetic gas temperature and the radiation temperature of the 3 K
background. Non-TE emission is the rule rather than the exception
in the emission of molecular lines. However, most non-TE effects
are relatively weak. Only transitions of H_2O, OH and SiO give
rise to spectacular maser effects.
e) *Radio recombination lines* (a special type of emission line) are
the result of transitions between atomic energy levels with high
principal quantum numbers (n > 50). These weakly bound states
are populated by recombination. Recombination lines radiation from
Hydrogen + Helium are observed in fully ionized regions and from
carbon in partially ionized regions.
f) *Absorption lines* are observed against a background continuum
source (of radiation temperature T_c) or - in some special cases -
against the general microwave background. For such a line to be
observed in absorption, $T_{ex} - (T_c + 3 K) < 0$.

2. STAR FORMATION AND RELATED TOPICS

2.1 A Summary of Recent Work

This topic was recently reviewed in two invited papers (Mezger
and Smith, 1975; 1976). Two further papers (Smith, Biermann and
Mezger, 1977; Churchwell et al. 1977) consider especially the large-
scale star formation rate and its effect on the chemical evolution
of the Galaxy. I will briefly summarize these papers.
    According to the density-wave theory of spiral structure
developed by Lin and associates, the spiral pattern rotates around
the galactic center in the same sense as the general galactic
rotation (see Fig. 2, upper right diagram), but with about half
the velocity. Interstellar gas streaming into these density wave
spiral arms (DWA; thought to be represented by the main spiral arms
in Fig. 3) is compressed and star formation is initiated. However,
as shown in Section 2.4, only a small fraction of the compressed
gas is directly transformed into stars, while most of the gas is
transformed into dense clouds. These clouds travel into the interarm region, where star formation proceeds, although on a much
smaller scale. This picture qualitatively explains the large
difference in star formation rate (74% in DWA, 13% in the interarm
region) mentioned in Section 1.
    Star counts in the solar vicinity yield the mass spectrum
shown in Fig. 1. The kink in the spectrum at $M_* \sim 1 M_\odot$ is due to the
fact that the MS lifetime of more massive stars is shorter than the
age of the Galaxy. Correction for this evolutionary effect yields

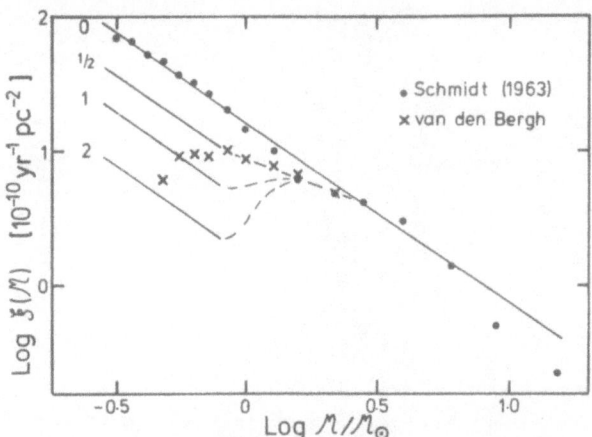

Fig. 5. Stellar birthrate functions compatible with the stellar mass spectrum (corrected for evolutionary effects) observed for the vicinity of the sun.

Salpeter's Initial Mass Function (IMF) where $\xi(M_*) \propto M^{-1.35}$. The birthrate function is identical with Salpeter's IMF only if the star formation rate in the Galaxy in the past was constant (corresponding to k = 0 in eq. (1.1)). For k > 0 one obtains birthrate functions as shown in Fig. 5, which are compatible with Salpeter's IMF representing the mass distribution of stars. Exponents k = = 0.5 - 1 give the best fit to observations.

   Birthrate functions computed for K > 0 show a discontinuity in the mass range around ~ 1 $M_\odot$. Observations support this result. In open clusters and subgroups of OB-associations, stars of different mass are supposed to have formed at approximately the same time. Curve ⓑ in Fig. 1 shows that the cluster mass function is in fact identical with Salpeter's IMF up to about 1 $M_\odot$. The observed turnover, however, is due to a true deficiency of low-mass cluster stars. We interpret this as follows: The basic physical process for the formation of protostars leads to a mass spectrum $\propto M^{-1.35}$. However, stars with M < 1 $M_\odot$[†]) only form under rather special conditions, such as the sudden compression of gas in DWA or as occurs in dense cloud complexes in the interarm region. Low-mass stars, M < 1 $M_\odot$[†]), on the other hand, appear to form continuously in cloud complexes. T-Tauri associations are examples of interstellar clouds in which only low-mass stars form.

2.2 Radio Astronomical and IR-Observations Related to Star Formation

[†])This limit between massive and low-mass stars is rather arbitrary and may in fact vary over a considerable mass range.

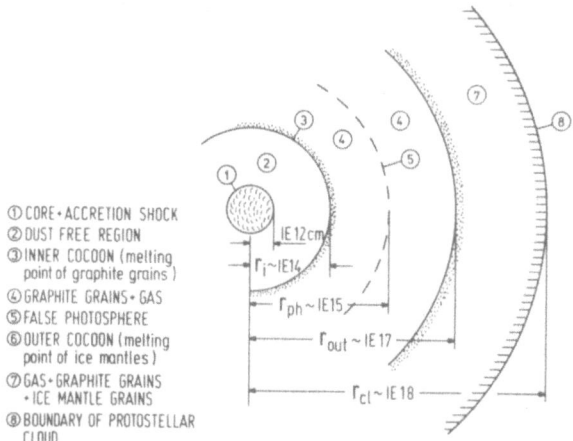

① CORE · ACCRETION SHOCK
② DUST FREE REGION
③ INNER COCOON (melting
   point of graphite grains)
④ GRAPHITE GRAINS · GAS
⑤ FALSE PHOTOSPHERE
⑥ OUTER COCOON (melting
   point of ice mantles)
⑦ GAS · GRAPHITE GRAINS
   · ICE MANTLE GRAINS
⑧ BOUNDARY OF PROTOSTELLAR
   CLOUD

Fig. 6. Geometry of a protostellar shell (see text).

In the radio and far IR range, stars are primarily observed through
their interaction with the surrounding interstellar matter. Examples of such interaction are: 1) HII regions (H and most He fully
ionized); 2) C^+-regions (only C is ionized); 3) far IR sources
(dust heated by absorption); 4) near IR sources (heated dust close
to the star and free-bound emission); 5) "Hot spots" in molecular
clouds (molecules heated by collisions with dust grains); 6) H_2O,
OH and SiO maser sources (pumping mechanism not yet understood).
Dynamical computations of massive protostellar clouds have been
made by Yorke and Krügel (1977), Yorke (private communication).
I refer to Fig. 6 and Table 4 for a discussion of their results.
t_1 yrs after the collapse of a dense ($n_H \sim$ 1E6 cm^{-3}) protostellar
cloud (size ∿1E18 cm) has started, a star-like nucleus forms. All
grains sublimate inside the radius r_i, at which distance an inner
dust cocoon forms which absorbs all stellar radiation and reemits
it in the near IR. Radiation pressure acts on the grains outside
r_i, and at t_2 yrs an outer cocoon is formed of size r_{out} cm. Infalling gas is held up by the outer cocoon which limits the final
mass of the accreting star. This process probably determines the
upper limit of the stellar spectrum. At t_3 yrs the inner cocoon becomes transparent to Lyc-radiation, a compact HII region forms and
expands as ionization bound (ib) HII region. At t_4 yrs the ionization front reaches the edge of the protostellar cloud; the compact (ib) HII region then becomes density-bound (db). It will be
observable as a strong, compact radio HII region on the average
for about t_4 yrs. The stage of an optically observable HII region
(primarily through H_α and forbidden lines) depends on the mass of
the surrounding dense cloud out of which the protostellar cloud is
formed. It may last through most of the MS lifetime t_6 yrs of the
ionizing star.

Table 4

Parameters Related to the Evolution of OB-Stars,
IR-Sources and HII-Regions

$M_{cl}/M_\odot$	$M/M_\odot$	sp.T.	M/M_{cl}	t_1/yr	r_i/cm	r_{ph}/cm	T_{eff}/K	t_2/yr	r_{out}/cm	T_3/yr	t_4/yr	t_5/yr	t_6/yr
150	36	06-07	0.24	1.5 E5	5 E13	1.6 E15	400	1.6 E5	3E 17	1.9 E5			6.0 E6
50	17	08-09	0.34	3.2 E5	4.5E13	5.8 E14	540	3.6 E5	1E 17	3.7 E5	3E4	5E5	7.6 E6
20	12	BO	0.60	4.7 E5	2.9E13	3.1 E14	460	6.1 E5	4E 16	6.4 E5			8.4 E6

t_1: star-like core and inner cocoon form;

t_2: outer cocoon form;

t_3: inner cocoon becomes transparent to near IR and subsequently to UV-and Lyc-radiation;

t_4: life-time of compact ionization bound HII region ⎱ "Radio HII Regions"

t_5: life-time of compact density bound HII region ⎰

t_6: MS life-time of stars.

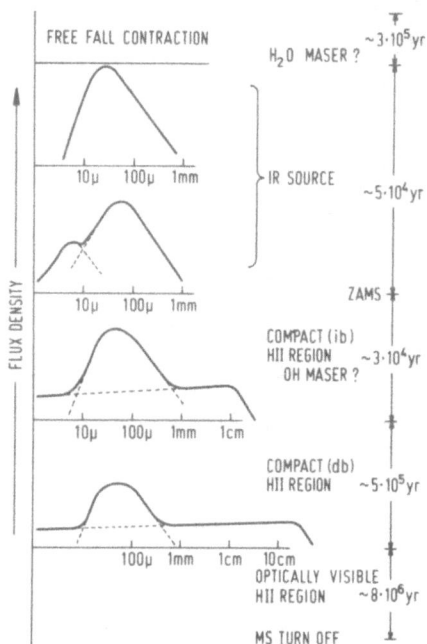

Fig. 7. Spectra of evolving protostellar shells (see text).

Fig. 7 shows observable stages related to the evolutionary
stages of massive pre-MS and MS stars. For the IR stages see
Yorke (1977). Time scales refer to an MS star of 17 $M_\odot$.
No dynamical calculations are available for protostellar
clouds with masses <20 $M_\odot$, out of which late B-stars should form.
Qualitatively, one sees that the fraction of the protostellar
cloud which ends up in the MS star increases with decreasing M_{cl}.
Early-type B-stars are still visible as strong far IR sources,
associated with C^+-regions. I presume that for late-type B-stars
the outer cocoon will become less and less important; these stars,
therefore, may only be seen as near IR sources. The association
of H_2O and OH masers with evolutionary stages in Fig. 7 is very
tentative and should be considered with caution. The predicted IR
and radio sources are actually observed. While it is certainly
possible to construct many models which could fit these observations, the predicted observational stages shown in Fig. 7 are consistent with the computed evolutionary tracks of early-type stars.
Yorke and Krügel did not include in their model a (highly
probable) rotation of the contracting protostellar cloud. However,
recent observations by Thompson et al. (1977) of the highly reddened
emission line object MWC 349 indicate how rotation will modify the
cocoon star model. IR and optical observations of MWC 349 yield
the following picture: A central O6.5 star is surrounded by a

luminous accretion disk of radius 1E13 cm, thickness $\sim$1/20 of the radius and mass 2E-2 $M_\odot$. The star is surrounded by an ionization bound HII region of radius 4E16 cm and density 3E5 cm^{-3}. Dust in the HII region must be depleted and the observed reddening therefore should be due primarily to dust located in the neutral shell surrounding the HII region.

Comparison with the Yorke and Krügel model shows that (at least the inner part of) the contracting protostellar cloud has formed a rotating disk, from which the star accretes matter. The disk (which may extend beyond its visible part) loses angular momentum through viscous dissipation, which gives rise to a visual luminosity of 5E38 erg s^{-1} with a spectrum $F_\nu \propto \nu^{1/3}$. As expected rotation disturbs the spherical symmetry of the Yorke and Krügel model in the distance range <1E14 cm which includes accretion shock and inner cocoon. Accretion times will be longer than for models with no rotation.

2.3 The Evolution of OB-associations in DWA and in the Interarm Region

Blaauw (for this and following references not given in these lecture notes, see the review by Mezger and Smith) recognized subgroups as the units of star formation in OB-associations. Fig. 8 shows the four subgroups which Blaauw could discern in the Orion OB association. Each subgroup has a stellar content (extrapolated from observed early-type to late-type stars) of $\sim$2000 $M_\odot$. On optical photographs of this region, one recognizes the presence of

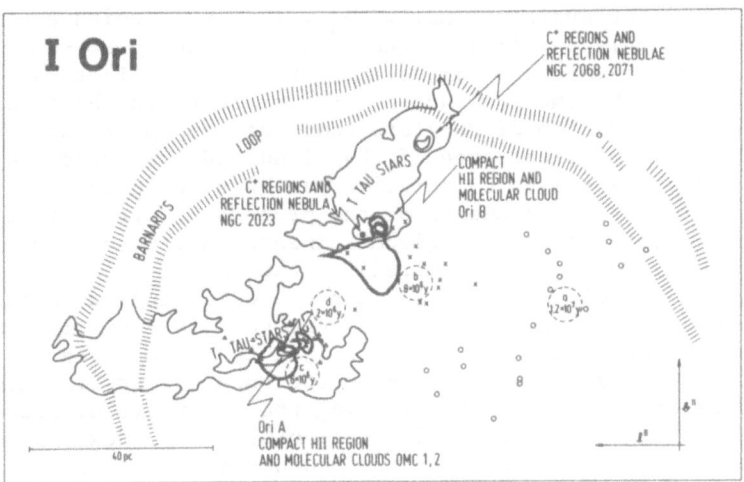

Fig. 8. The Orion OB association.

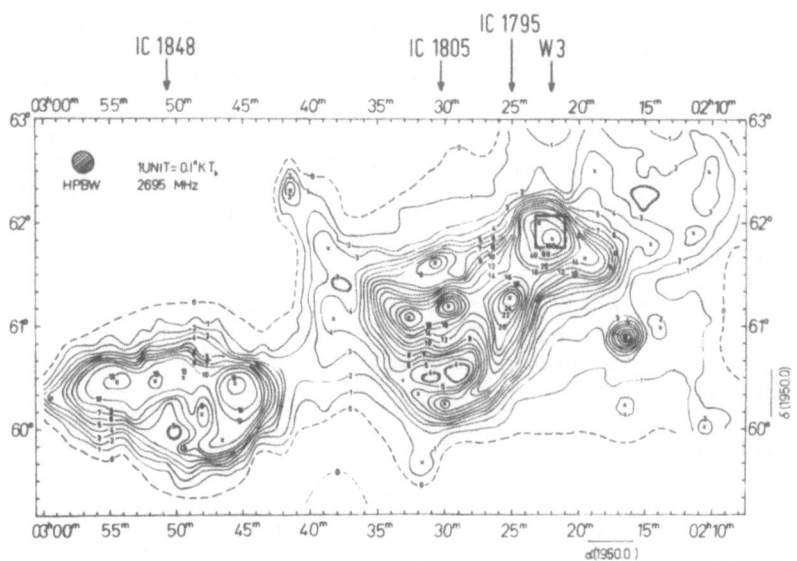

Fig. 9. The giant HII regions IC 1848, IC 1805, 1C 1795 and W3, shown by their free-free radio emission.

extended dust clouds. Kutner et al. (1977) have mapped the CO $\sim$ $\sim$ 2.6 mm emission from these clouds. The clouds contain two cen- ters of active star formation, Orion A and B, and have total gas masses of the order of 1E5 $M_\odot$.

Originally, the Orion OB association appears to have been a com- plex of gas clouds of average density ($n_H \sim$ 1E3 cm^{-3}) which ex- tended $\sim$130 pc roughly parallel to the galactic plane. Low-mass stars have formed in (and probably still form in part of) this cloud (as indicated by the presence of T-Tauri stars in the extended CO-clouds). The formation of OB-stars and stars >1 $M_\odot$ in subgroups started at the low-longitude edge of the cloud and proceeded towards higher longitudes. Since no CO-emission has been found to be associated with the visible subgroups the process of star formation and subsequent dispersion of the cloud remnants must have been rather efficient. Conditions required for the formation of massive stars are those derived, for example, for the well-studied condensa- tions OMC1 and 2: Masses of $\sim$1E3 $M_\odot$, on average densities $\sim$1E4 cm^{-3} which increase towards the core of the clouds to values as high as 1E6 cm^{-3}. That massive stars form inside these clouds is shown by the presence of IR-sources representing the early evolutionary stages of protostars as described in Section 2.2. What exactly creates these conditions is not yet fully understood.

I doubt if the Trapezium star cluster is a separate subgroup. The evolutionary ages of the O-stars ionizing Orion A and those

which heat the IR-sources in OMC1 and 2 differ probably by less
than 1E5 yr. If all these stars have evolved to the MS and dispersed their surrounding gas, they will form a subgroup with stars
of roughly equal age. It appears to me that the true units of
star formation form out of gaseous condensations such as OMC1 and
2 and that subgroups as defined by Blaauw are composed of several
of these subunits, which evolve at about the same time.

The Orion OB association should be representative of star
formation in the interarm region. The best investigated giant HII
region - which represents an evolutionary stage of an OB-associa-
tion in a main (DWA) spiral arm - is W3 and its associated optically
visible HII regions NGC 1795, 1805 and 1848, located in the Perseus arm at a distance of ∿3 kpc. The free-free radio emission
from these four HII regions is shown in Fig. 9 (Wendker and Altenhoff, 1977). The projected size of these four giant HII regions is
290 pc or more than twice the size of the Orion OB association.
But, while the ages of the four discernable subgroups in the Orion
OB association differ by more than 1E7 yr. the ages of the various
subgroups in the giant HII regions in Fig. 9 (as represented by
the various peaks in the free-free emission) differ by less than
1E6 yr. (as estimated from the evolutionary stages of the different compact radio components). The fact that all O-stars in
this huge association reach the MS at about the same time shows that
O-star formation there was initiated on a large scale (as predicted for DWA). This has to be compared to the much more localized
initiation of star formation observed in cloud complexes in the
interarm region, such as, e.g. the Orion OB association.

2.4 The Efficiency of Star Formation

For (ib) compact HII regions all Lyc-photons are absorbed by the
surrounding gas and dust. If appropriate corrections for the absorption by dust are made, the radio flux density multiplied by
the square of its distance from the sun yields directly the number of Lyc-photons emitted by the ionizing stars. For a given
birthrate function one can derive a mass-to-Lyc luminosity ratio,
which for $k = 1/2$, $M_1 = 0.01$ (the lower mass limit) attains the
value

$$<M_*>/<N_c> = 4.6E\text{-}47\ M_\odot/\text{Lyc-photons s}^{-1} \qquad (2.1)$$

Thus, with an integrated flux density of 400 Jy and a distance of
3 kpc, the total stellar content associated with the giant HII
regions shown in Fig. 9 becomes $M_* > 1.3E4\ M_\odot$. This stellar mass
is only a lower limit since most of the compact components are
already density bound so that a certain fraction of the Lyc-photons escape into the low-density regions of the interstellar space.
On the other hand, if only stars $>1.4\ M_\odot$ form in OB associations
eq. (2.1) must be multiplied by 0.25.

In Table 3 I consider the efficiency of star formation by compression of gas entering DWA's. Star formation rates are from

Table 5

Efficiency of Star Formation

R kpc	dM$_*$/dt M$_\odot$ yr^{-1}		M$_{gas}$ 10^6 M$_\odot$	T$_R$ 10^6 yr	M$_{gas}$(T$_R$/2)$^{-1}$ M$_\odot$ yr^{-1}	M$_*^\cdot$ /M$^\cdot$ gas
4–5	0.34	0.085	475	1.26	7.5	0.01
5–6	1.04	0.26	681	1.45	9.4	0.03
6–7	0.57	0.14	572	1.65	6.9	0.02
7–8	0.87	0.22	664	1.85	7.2	0.03
8–9	0.42	0.11	417	2.07	4.0	0.03
9–10	0.30	0.08	418	2.32	3.6	0.02
10–11	0.48	0.12	396	2.61	3.0	0.04
11–12	0.35	0.088	354	2.93	2.4	0.04
12–13	0.068	0.017	456	3.28	2.8	0.006
(1)	(2)	(3)	(4)	(5)	(6)	(7)

(1) Galactic radii

(2) Star formation rate (M$_*$ = 100–0.01 M$_\odot$)

(3) Star formation rate (M$_*$ = 100–1.4 M$_\odot$)

(4) M$_{gas}$ = 1.4 [M(H) + M(H$_2$)]

(5) T$_R$ = π(R$_{in}$ + R$_{out}$)θ_c^{-1}

(6) Gas compressed by shock in DWA's

(7) Efficiency of star formation in DWA's

Smith et al. (1977). Note that in DWA's only high mass stars
supposedly form. The spiral arm pattern is supposed to rotate
with 0.5 Θ_c, the orbital velocity of the interstellar matter.
Since there are four spiral arms, a certain volume of gas enters
a DWA every 0.5 T$_R$ yr where T$_R$ is the time required for the ISM
to complete one rotation around the galactic center. Column (7)
gives the ratio of gas which is transformed directly into stars
to the amount of compressed gas. Only a few % of the compressed
gas is directly transformed into stars. The main effect of the
compression of gas in DWA's, thus, appears to be the formation of
molecular clouds, which subsequently travel into the interarm region
where they are the sites of low-mass star formation and the occa-
sional formation of O-stars. Stabilization of these clouds against
further collapse could be provided by rotation and magnetic fields
(see e.g. Gold, 1976; von Hoerner, 1968).

2.5 Radio Observations Related to Abundance Gradients in the Galaxy

Optical observations of HII regions in external galaxies have shown
the existence of abundance gradients. For the interpretation of
a number of observations, it is of importance to know if similar
abundance gradients exist in our Galaxy. With the simple theory
for chemical evolution of the interstellar gas outlined in Section
1.2, one expects an increase in $Y(^4He)$ and $Z(A > 4)$ with decreasing
$\mu = M_{gas}/M_{tot}$ and a corresponding decrease of $X(D)$. Since μ does
not change very strongly with galactic radius D_G (see Fig. 2, upper
left diagram), one cannot expect very strong abundance variations
in the Galaxy.

$\underline{Y(^4He)}$: Churchwell et al. (1977) measured the He^+/H^+-abundance
in galactic HII regions which increases from $\sim$7.5% ($D_G > 10$ kpc)
to values between 9 - 10% ($D_G \sim 10 - 8$ kpc) and subsequently drops
monotonically to very low values close to the galactic center. The
low He^+-abundance at large galactic radii is considered as representing the primordial He-abundance. The decrease in He^+/H^+ from
8 to 0 kpc is interpreted as an ionization effect where dust grains
absorb preferentially He-ionization Lyc-photons (see Section 3.1).

$\underline{Z(A > 4)}$: Since the above interpretation requires that m_d/m_g
increases with decreasing D_G and since one usually assumes m_d/m_g
$\propto Z$, an increase in Z with decreasing D_G is implied. This suggestion is supported by the determination of electron temperatures
of galactic HII regions. Churchwell et al. (1977) find a decrease
of T_e between $D_G = 10$ kmc and 5 kpc of $\sim$1500 K. Model calculations
show that this decrease can be explained by an increase in Z by a
factor of 2.

$\underline{X(D)}$: In principle, D/H gradients could be deduced from observations of isotopic molecular lines. Penzias et al. (1977)
observed DCN and $H^{13}CN$ lines in molecular clouds throughout the
Galaxy and find for the Sgr B2 molecular cloud intensity ratios
which are markedly lower than in the spiral arm region. However,
due to chemical fractionation, a quantitative interpretation of a
possible depletion of D in the galactic center region is not yet
possible.

$\underline{In\ summary}$: There are strong observational indications that
^4He, heavier elements and the dust-to-gas ratio increase from the
sun towards the center of the Galaxy. This would affect the physical state of both neutral and ionized interstellar gas.

3. THE INTERSTELLAR MATTER: SELECTED TOPICS

3.1 Absorption Characteristics of Dust

The UV radiation field and the ionization structure of HII regions
are to a large extent determined by the absorption characteristics
of dust grains. Absorption cross sections for the wavelength range
$\lambda > 1500$ Å and the diffuse ISM have first been determined by Witt

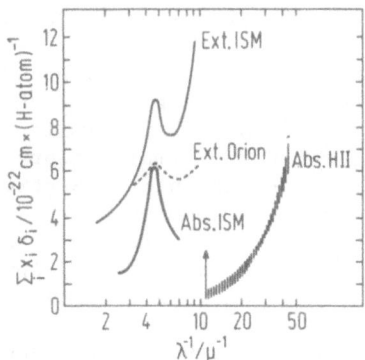

Fig. 10. Absorption and extinction cross sections of dust.

and Lillie (1973); and for the Lyc-range 912 $\geq \lambda/\text{Å} \geq$ 227 and the
ionized gas of compact HII regions by Mezger et al. (1974). Both
results were questioned because they implied extremely high values
of the albedo $\Gamma = \sigma_{sca}/\sigma_{ext}$ and a very strong increase of the absorption cross section in the Lyc-region. In the meantime, Witt
and Lillie (1976) and Panagia and Smith (1977) redetermined the
above mentioned absorption cross sections, using the same methods
but improved observations. I have arranged their results in Fig.
10.

All cross sections are normalized to the cross section for
visual extinction in the solar vicinity $\sigma_v = 3.7 \ 10^{-22}$ cm^2/H-atom.
Extinction cross sections for the diffuse ISM are from Witt and
Lillie (1976) and Bless and Savage (1970) and for the Orion Nebula
from a compilation by Wickramasinghe and Nandy (1972; Fig. 12).
Absorption cross sections for $\lambda > 1500$ Å and the diffuse ISM are
from Witt and Lillie (1976). Panagia and Smith (1977) determined
the ratio of absorption cross section for He- and H-photons in the
Lyc-range $a_0 = \sigma_{He}/\sigma_H = 4 \pm 1$ (as compared to 7^{+4}_{-3} by Mezger et al.
1974). To convert this ratio into a smooth curve, we adopt a power
law for the Lyc absorption cross section

$$\sigma_{Lyc} = \sigma(\nu_1)(\nu/\nu_1)^{\alpha} \tag{3.1}$$

with ν_1 the Rydberg frequency. For a not too steep photon spectrum,
$a_0 = 4$ corresponds to an exponent $\alpha = 1.9$. $<\sigma_{He}>/\sigma_v \simeq 1$ for the
solar vicinity (Churchwell et al. 1977) yields $\sigma(\nu_1)/\sigma_v = 0.13$.
These values together with the estimated uncertainties $\Delta a_0 = \pm 1$
correspond to the hatched area in Fig. 10. There are two ways of
explaining the very low absorption cross sections derived for the
Lyc-region: i) The different extinction curves observed for the
diffuse ISM and the Orion Nebula (Fig. 10) suggest different grains
(both in size and chemical composition) in diffuse ISM and HII regions; ii) Absorption cross sections for the Lyc-region are derived
on the assumption of a homogeneous gas and dust distribution. If

the ionized gas clumps (clumping factor $\Phi > 1$) the Lyc-absorption
cross sections in Fig. 10 are underestimated by a factor $\Phi^{1/2}$.
A value of $\Phi = 30$, which may be typical for HII regions, thus
would increase the Lyc-absorption cross section by a factor of
five, as indicated by the arrow. However, the strong increase of
the absorption cross section in the Lyc range still remains a puzzle.

3.2 The Fully Ionized Component of the ISM and the Energy Density
of Lyc-Photons in the Solar Neighborhood

The existence of a low-density ($<n_e> \simeq 0.03$ cm^{-3}) ionized gas of
scale height $\sim$1000 pc is inferred from the dispersion measure of
pulsars (see e.g. Terzian and Davidson, 1976). The existence of
a denser component with smaller scale height is inferred from both
a separation of thermal and non-thermal diffuse galactic continuum radiation (Westerhout, 1968), surveys of radio recombination
line emission from the galactic plane (e.g. Lockman, 1976; Hart
and Pedlar, 1976) and diffuse Hα emission from a region with radius 2 - 3 kpc from the sun (Reynolds et al. 1974). It can be
shown (Mezger, in preparation) that the ionization of both fully
ionized components can be accounted for by Lyc-radiation from O-
stars which have dispersed their compact radio HII regions (see
Section 2.2). Here I will limit myself to the solar neighborhood
which appears to be typical for an interarm region. Some characteristics of the Lyc-radiation are compiled in Table 6.

Table 6

Lyc-Photons in Solar Neighborhood

Ring extending from - to	ΣN_c giant HII	ΣN_c interarm	number of O6*	$\dfrac{d}{pc}$	$\dfrac{n_e}{cm^{-3}}$
9 - 10 kpc	2.7E51	4.3E51	179	326	≤ 0.15
10 - 11 kpc	4.4E51	6.9E51	288	270	≤ 0.20
Remarks	(1)	(2)	(3)	(3)	(4)

(1) Number of Lyc-photons absorbed per sec in giant radio HII
    regions (Smith et al. 1977);
(2) Small radio HII regions in interarm region = 0.175 x ΣN_c (giant
    HII), which intercept $\sim$10% of the total Lyc-flux from O-stars.
    Number of Lyc-photons available for ionization of low-density
    interarm region ΣN_c (interarm) = 9 x 0.175 ΣN_c (giant HII);
(3) O6-stars with N_c = 2.4E49 s^{-1} (or $r_H + n_e^{2/3}$ = 92.6 pc) will
    contribute most to Lyc-flux. Number of O6*πd^2 = $(R_{out}^2 - R_{in}^2)\pi$;
(4) HII region overlap if $dn_e^{2/3} \leq 92.6$ pc.

Column (2) gives the total number of Lyc-photons emitted by those O-stars in the interarm region which are not embedded in radio HII regions. For the sake of simplicity I represent this Lyc-radiation as being emitted by O6-stars (which contribute most to the Lyc-radiation from O-stars). The required number of O6-stars and their average separation $\sim 2d$ are given in column (3). If the intercloud gas in the interarm region has densities less than those given in column (4), it can be fully ionized by the combined radiation of the O6-stars.

For a gas density of $n_H \sim 0.1$ cm^{-3} in an HII region, the average ratio of protons to recombinated H-atoms $\langle n_p/n_1 \rangle \sim \langle n_e/n_1 \rangle \sim$ ~ 1000. The absorption depth for Lyc-photons within such a low density HII region, $\tau(\nu_1) = n_H \dot{a}_1(\nu_1)r = (n_H/\langle n_e/n_1 \rangle)$ (6.3E -18 cm^2) (r/cm) with $a(\nu)$ the absorption cross section for Lyc-photons (here approximated by $\nu_{Lyc} = \nu_1$) is <1 for $\lesssim 300$ pc. One can therefore assume that the intercloud gas is transparent to Lyc-photons if the line of sight to the O-star does not intersect a cloud. For an estimate of the radiation density of Lyc-photons close to the sun, I assume that the luminosity of an O6-star in the Lyc-region varies with frequency according to $L_\nu = L(\nu_1)\nu/\nu_1$ for $\nu_1 \leq \nu \leq 4\nu_1$ and is practically zero for $\nu > 4\nu_1$. Then $N_c \langle E_{Lyc} \rangle$ $= \nu_1 \int^{4\nu_1} L_\nu d\nu$ and $L(\nu_1) = N_c \langle E_{Lyc} \rangle/\nu_1 \ln 4$, with $\langle E_{Lyc} \rangle = 3.8E{-}11$ erg the average energy of a Lyc-photon. For $d \gg R$ the relationship between radiation intensity L_ν and radiation density $u_\nu = = L_\nu/c4\pi d^2$ and

$$u(\nu_1)\nu_1 = u_\nu \nu \quad = \frac{1}{c} \frac{N_c}{4\pi} \frac{\langle E_{Lyc} \rangle}{d^2 \ln 4} \tag{3.2}$$

for $\nu_1 \leq \nu \leq 4\nu_1$. Substitution of N_c and $\langle E_{Lyc} \rangle$ related to an O6-star, $d = 9E20$ cm ~ 300 pc and summation over all O6-stars located within a circle of 1 kpc around the point in the galactic plane yields

$$\Sigma u_\nu \nu = 3.33 u(\nu_1)\nu_1 = 7.2E{-}15 \text{ erg cm}^{-3} \tag{3.3}$$

the radiation density shown in Fig. 4. Beyond the ionization limit of He$^+$ the UV-flux of an O6-star (and of O-stars in general) drops by about a factor E-4; thus MS-stars should not contribute much to the soft X-ray radiation.

It has been suggested that observed $\lambda 21$ cm radiation from hot atomic H originates in the transition layer between HII and HI. One can easily show that the contribution from this gas is (for a line width of 21.4 km s^{-1} corresponding to a Doppler temperature of 10^4 K) only of the order of a few hundredths K, while the observed line temperatures are of the order of several degrees K. Therefore, the hot atomic H must be located somewhere else. It appears, however, as mentioned in Section 1.4, that the fraction of hot atomic gas has been previously overestimated and that the fraction given in Table 3, 15%, is decreased to 5% or even less if the stray radiation of radio telescopes is properly taken into account (Mebold, private communication).

3.3 The diffuse Galactic IR Radiation

Recently Low et al. (1977) detected the existence of an extended
IR emission in the longitude range $|\ell| < 33°$ which can be observed
out to $|b| \lesssim 1°$. Its surface brightness, observed with a 15' beam,
is $I_{IR} \lesssim 7 \ 10^{-9} \ W \ cm^{-2} \ ster^{-1}$. Ryter and Puget (1977) suggest that
this IR radiation comes from molecular clouds which can be traced
by their CO emission. For dense molecular clouds ("Hot spots"
as described in Section 2.2) associated with well-known compact
HII regions, they derive an IR-luminosity per H-atom $L_{IR}^H = 2 \ 10^{-30}$
W (H-atom)$^{-1}$ which - when multiplied with the estimated number of
H-atoms tied up in galactic H_2 molecules - yields a total IR luminosity of the Galaxy of $L_{IR} = 7.5 \ 10^{36}$ W. I doubt if IR luminosities derived for molecular clouds associated with sites of formation of OB-stars can be extrapolated to the majority of molecular
clouds of medium density in which - if at all- only stars of low
masses form. Alternatively, I suggest (Mezger, in preparation)
that the diffuse IR radiation comes from dust grains embedded in
the diffuse thermal component. One, in fact, gets agreement between observed radio continuum temperature and IR surface brightness if each dust grain emits about eight times as much energy as
is available in Lyman alpha photons. (Remember that each Lyc-photon absorbed by the gas eventually gets degraded into a Lyman
alpha photon which can only be destroyed by absorption by dust
grains). An IR excess of 8 is a reasonable value. Integration
over all ionized gas in the Galaxy (excluding, however, the very
complex galactic center region) yields therefore a total IR luminosity of $L_{IR} = 2.2 \ 10^{36}$ W or $6 \ 10^9 \ L_O$.

ACKNOWLEDGEMENTS

It is my pleasure to acknowledge helpful criticism and stimulating
discussions with many colleagues at the MPI and at the Erice Summer
School. I also wish to thank T. Pauls for his critical reading of
this manuscript. Discussions with J. Schmid-Burgk were essential
for Section 3.2. C. Thum compiled Fig. 8.

REFERENCES

Bania, T.M., 1977, Astrophys. J., in press.
Bless, R.C., Savage, B.D., 1970, "Ultraviolet Stellar Spectra and
Ground-Based Observations", edited by L. Houziaux and H.E. Butler,
D. Reidel, Dordrecht, Holland.
Churchwell, E., Smith, L.F., Mathis, J., Mezger, P.G. and Huchtmeier,
W., 1977, Astron. & Astrohys., submitted for publication.

Georgelin, Y.M. and Georgelin, Y.P., 1976, Astron. & Astrophys. 49, 57.

Gold, T., 1976, Accademia Nazionale Dei Linci, Contribution Nr. 31.

Gordon, M.A. and Burton, W.B., 1976, Astrophys. J. 208, 346.

Hart, L. and Pedlar, A., 1976, Mon. Not. Roy. astr. Soc. 176, 547.

Hoerner, V.S., 1968, "Interstellar Ionized Hydrogen", edited by Y. Terzian, W.A. Benjamin, Inc., New York, p. 101.

Kutner, M.L., Tucker, K.D., Chin, G. and Thaddeus, P., 1977, Astrophys. J., submitted for publication.

Lockman, F.J., 1976, Astrophys. J. 209, 429.

Low, C., and Lynden-Bell, D., 1976, Mon. Not. Roy. Astr. Soc. 176, 367.

Low, F.J., Kurtz, R.F., Poteet, W.M. and Nishimura, T., 1977, Astrophys. J. 214, L115.

Mezger, P.G., Smith, L.F. and Churchwell, E., 1974, Astron. & Astrophys. 32, 269.

Mezger, P.G. and Smith, L.F., 1975, IAU Proc. 3rd European Astr. Meeting, "Stars and Galaxies from Observational Point of View", edited by E.K. Kharadze, Tbilisi, p. 369.

Mezger, P.G. and Smith, L.F., 1976, Proc. IAU Symp. No. 75,"Star Formation", in press.

Panagia, N. and Smith, L.F., 1977, Astron. & Astrophys., submitted for publication.

Penzias, A.A., Wannier, P.G., Wilson, R.W. and Linke, R.A., 1977, Astrophys. J. 211, 108.

Reynolds, R.J., Roesler, F.L. and Scherb, F., 1974, Astrophys. J. 192, L53.

Ryter, C.E. and Puget, J.L., 1977, Astrophys. J. 215, 775.

Salpeter, E.E., 1955, Astrophys. J. 121, 161.

Smith, L.F., Biermann, P. and Mezger, P.G., 1977, Astron. & Astrophys., submitted for publication. (SBM)

Stecker, F.W., Puget, J.L. and Fazio, G.G., 1977, Astrophys. J. 214, L51.

Terzian, Y. and Davidson, K., 1976, Astrophys. and Space Science 44, 479.

Thompson, R.I., Strittmatter, P.A., Erickson, E.F., Witteborn, F.C. and Strecker, P.W., 1977, preprint.

Tinsley, B.M., 1974, Astrophys. J. 192, 629.

Wagoner, R.V., 1973, Astrophys. J. 179, 343.

Wendker, H.J. and Altenhoff, W.J., 1977, Astron. & Astrophys. 54, 301.

Westerhout, G., 1958, BAN 488, 215.

Wickramasinghe, N.C. and Nandy, K., 1972, Rep. Progr. Phys. 35, 157.

Witt, A.N. and Lillie, C.F., 1973, Astron. & Astrophys. 25, 397.

Witt, A.N. and Lillie, C.F., 1976, Astron. & Astrophys. 208, 64.

Witt, A.N. and Johnson, M.W., 1973, Astrophys. J. 181, 363.

Yorke, H.W. and Krügel, E., 1977, Astron. & Astrophys. 54, 183.

Yorke, H.W., 1977, Astron. & Astrophys., submitted for publication.

INFRARED OBSERVATIONS OF HII REGIONS

G.G. Fazio

Harvard-Smithsonian Center for Astrophysics
Cambridge, Massachusetts 02138, U.S.A.

1. INTRODUCTION

An HII region is an extended gaseous nebula, which is optically
bright, and excited by an O- or early B-type star or cluster of
such stars. Often these regions are associated with dense clouds
of dust and gas, and there is strong evidence that star formation
is also occurring. In the early 1970's observations indicated that
many HII regions radiate most of their energy at far-infrared wavelengths; therefore detailed infrared observations of these regions
could yield important new information on the evolution of pre-main
sequence objects. Indeed, infrared observations obtained over the
last seven years have confirmed these ideas. In these two lectures
I will review what infrared observations have been made and what we
have learned from them.

2. GENERAL PROPERTIES OF AN HII REGION

Fig. 1 shows a schematic diagram of a simple HII region. It consists of a volume of gas ionized by an O- or early B-type star.
The surface temperature of the star (T_*) is $\gtrsim 3 \times 10^4$ K, and its
luminosity (L_*) is $\gtrsim 10^4$ $L_\odot$. Fig. 2 shows the spectra of a typical
05 star. The ultraviolet radiation emitted by the star photoion-
izes the surrounding gas, which is predominately hydrogen. Photons
with $\lambda < 912$ Å ionize the hydrogen gas and they are absorbed in the
process, with the excess energy transferred to the kinetic energy
of the emitted photoelectron. Helium, the next abundant gas is
ionized at $\lambda < 504$ Å, and He^+ at 204 Å. The energetic electrons in
this process collide with other electrons and ions to redistribute
their energy to maintain a Maxwellian energy distribution, with T_e

G. Setti and G. G. Fazio (eds.), Infrared Astronomy, 25–49.

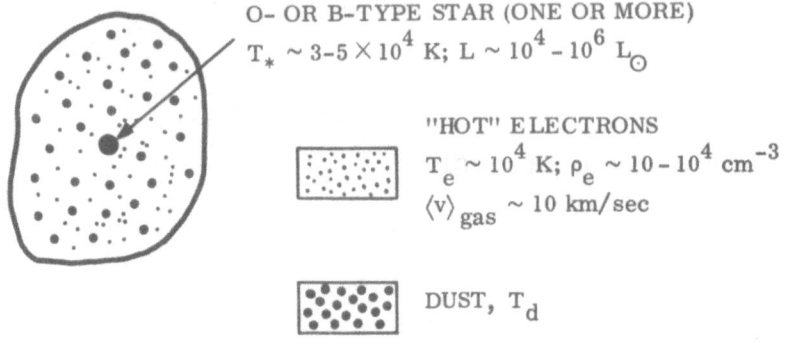

O- OR B-TYPE STAR (ONE OR MORE)

$T_* \sim 3\text{-}5 \times 10^4$ K; $L \sim 10^4 - 10^6$ $L_\odot$

"HOT" ELECTRONS

$T_e \sim 10^4$ K; $\rho_e \sim 10 - 10^4$ cm^{-3}

$\langle v \rangle_{gas} \sim 10$ km/sec

DUST, T_d

Fig.1.   Schematic diagram of a simple HII region.

~ 10 K.   Thermal electron-ion collisions also excite low-lying energy levels of the ions, causing the emission of a forbidden line spectrum.   Recombination of thermal electrons by ions results in emission of the line spectra of H I, He I, and He II.   Throughout the nebula hydrogen is almost completely ionized, helium is predominately singly ionized and other elements are singly or doubly ionized.   Gas densitites in the nebula can range from 10 to 10^4 cm^{-3}, with internal gas velocities of the order of 10 km sec^{-1}.   The hot ionized gas tends to expand into the surrounding colder neutral gas.   It is also known that considerable amounts of dust are present in and around the HII region.   This is evident from comparison of radio and optical maps as well as the observation of stellar reddening.   We shall show later that this dust plays an important role in the distribution of energy in the nebula.   Typical masses of HII regions are of the order of 10^2 to 10^4 $M_\odot$ and typical sizes are 1 to 10 l.y. in diameter.

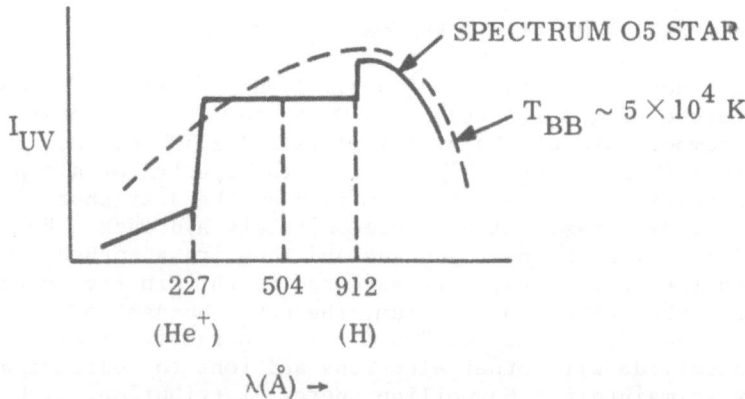

Fig. 2.   Spectrum of typical O5 star.

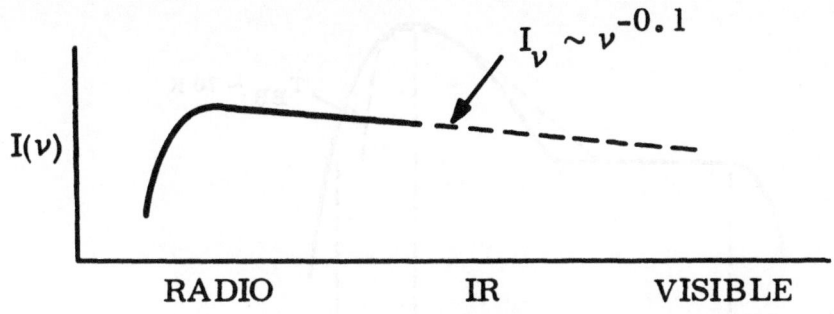

Fig.3. Thermal bremsstrahlung spectrum of an HII region.

2.1 Radiation Spectrum of HII Regions

We would expect the radiation from an HII region in the radio and infrared spectral range to be dominated by thermal bremsstrahlung emitted by the hot electron gas; i.e., radiation produced by free-free transitions of electrons decelerated in the Coulomb field of positive ions. The resulting spectrum can be calculated (e.g. Chaisson, 1976) and is shown in Fig.3. The spectrum is rather flat with $F_\nu \propto \nu^{-0.1}$, over most of the spectral range. At the low frequency end the spectrum turns over in the region where the source becomes optically thick to the radiation. Measurement of the radio continuum spectrum of a source can be used to identify a thermal source, such as a gaseous nebula, as distinguished from a non-thermal source such as a supernova, external galaxy or quasar, for which in a typical case, $F_\nu \propto \nu^{-0.7}$.

However it has been observed that in many HII regions, most of the radiation appears at wavelengths of the order of 100 μm, and that the infrared radiation was in excess of the electron free-free emission observed over the wavelength interval 5 μm to 1 mm. This is shown in Fig.4. The current interpretation of the infrared excess is that it is energy reradiated from cool (~100 K) dust in and around the HII region. In Fig.4 a 70 K blackbody spectrum is shown for comparison with the infrared spectrum: Note that the observed spectrum exceeds the blackbody curve in the 5-20 μm region and falls more steeply than the blackbody curve in the 200 μm to 1 mm range.

2.2 Origin of the Infrared Radiation

An O-type star with a surface temperature of 5×10^4 K emits ~ 2/3 of its energy in the Lyman continuum (Lyc), i.e. at wavelengths less than 912 Å. Part of this radiation will be absorbed by the gas and part by the dust. Each of the photons absorbed in

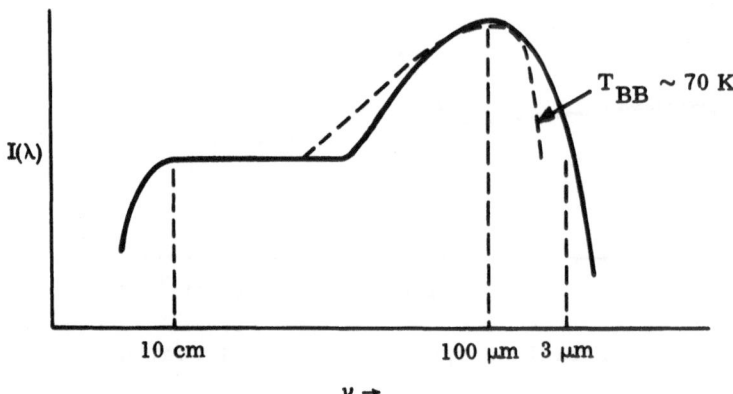

Fig.4. Total spectrum of HII regions with 70 K blackbody curve
for comparison.

the hydrogen gas around the star produces a Lyman α (Lyα) photon
(1216 Å), a Balmer photon plus several lower energy photons.[†] About
one-fourth of the energy of the O5 star is converted into Lyα photons, which usually remain trapped in the ionized region and are
destroyed only by absorption on dust grains in the ionized volume.
The amount of dust in the HII region can vary considerably. If the
gas cloud in which the source is embedded is large enough, there
will exist an outer edge to the ionized gas. Such an HII region is
said to be ionization bounded.

At first, it was thought that absorption of Lyα photons was
the primary source for heating the dust inside an HII region. However the observations by Harper and Low (1971) showed that although
the infrared luminosity was proportional to the Lyα luminosity, the
former exceeded the latter by a large factor. Therefore some other
source of heat had to exist to explain the infrared luminosity.
One possibility was that Lyc radiation was being absorbed directly
by the dust; other possibilities were that the radiation longward
of the Lyc limit, i.e. > 912 Å, or shortward of the He ionization
limit, i.e. < 504 Å, were also absorbed by the dust.

3. WHAT CAN BE LEARNED FROM INFRARED AND RADIO FLUX AND SPECTRAL
DISTRIBUTIONS?

In this lecture we shall limit our discussions to radio and
infrared observations. We shall first discuss the flux and spectral distributions, then the brightness distribution of HII regions.

† This statement is not exactly true, but we shall discuss it in
detail later.

3.1 Radio Observations

The observed total radio continuum flux, $I_R(\nu)$, at a frequency ν, is proportional to the number of Lyc photons absorbed per sec, N_c', by the gas in the HII region. The number of Lyc photons absorbed per sec to maintain H^+ ions equals the rate of recombination of protons and electrons. The total radio flux is proportional to the number of recombinations. This is true only at equilibrium, and where the region is optically thin at the frequency ν. If we consider only hydrogen gas then N_c' is given by (Mezger et al.,1974):

$$N_c' = 4.761 \times 10^{48} \alpha(\nu_1 T_e)^{-1} \nu^{0.1} T_e^{-0.45} I_R(\nu) D^2 \ \mathrm{sec}^{-1}, \quad (3.1)$$

where the function $\alpha(\nu_1 T_e)^{-1}$, tabulated by Mezger and Henderson (1967), is ~ 1 for $T_e \sim 10^4$ K; ν is given in GHz, T_e in K, $I_R(\nu)$ in Janskys (10^{-26} W m^{-2} Hz^{-1}); and D, the distance to the source in kpc.

The number of Lyα photons, N_α, produced per Lyc photon depends somewhat on the electron density in the HII region: $N_\alpha = \beta N_c'$ where $0.68 < \beta < 1$, and $\beta \to 1$ for electron density $\gg 10^3$ cm^{-3}. The competing process is two photon decay of the 2^2S level of HI. Therefore from the total radio continuum flux we can derive the number of Lyα photons produced per sec, and the Lyα luminosity.

$$I_R(\nu) \to N_c' \to N_\alpha \to L_\alpha \qquad\qquad (3.2)$$

3.2 Infrared Observations

From observation of the total infrared flux, $I_{IR}(\nu)$, as a function of ν the value of the infrared optical depth, $\tau_{IR}(\nu)$; the dust temperature, T_d; and the total luminosity of the source, L , can be derived.

From measurement of $I_{IR}(\nu)$ at two frequencies, ν_1 and ν_2, the color temperature of the dust, T_d, can be determined. T_d is the temperature of the Planck function having the same wavelength dependence as the observed distribution in the interval (ν_1,ν_2). We shall assume that the dust is at this temperature and that it radiates as a blackbody, $B_\nu(T_d)$. If the source has an infrared optical depth, $\tau_{IR}(\nu)$, then the observed infrared flux is given by (Fazio, 1976):

$$I_{IR}(\nu) = (1-e^{-\tau_{IR}})B_\nu(T_d) \qquad\qquad (3.3)$$

For the optically thick case, i.e. $\tau_{IR} \to \infty$:

$$I_{IR}(\nu) = B_\nu(T_d) \qquad\qquad (3.4)$$

and for the optically thin case, $\tau_{IR} \to 0$:

$$I_{IR}(\nu) = \tau_{IR}(\nu) B_\nu(T_d) \tag{3.5}$$

For most HII regions the dust is optically thin to far-infrared radiation and the latter case applies. Typical values of $\tau_{IR}(100\ \mu m)$
are $\sim 10^{-2}$.

Assuming that most, if not all the energy emitted by the source
is absorbed by dust and reradiated in the infrared, then the total
infrared luminosity, L_{IR}, is a lower limit to the source luminosity, and is proportional to the total number of Lyc photons emitted
per sec, i.e.

$$L_{IR} \lesssim L_* \propto N_c \tag{3.6}$$

Measured values of L_{IR} range from approximately 4×10^4 to 4
$\times 10^7 L_\odot$.

3.3 Combined Results of Infrared and Radio Data

Petrosian, Silk, and Field (1972) have shown that in an HII
region the total infrared luminosity is given by

$$L_{IR} = L_\alpha + (1-f)\, \langle h\nu \rangle_{Lyc} \cdot N_c + L_{\nu < Lyc}(1 - e^{-\tau_0'}), \tag{3.7}$$

where $L_\alpha = N_\alpha(h\nu_\alpha)$ is the Lyman α luminosity, $\langle h\nu \rangle_{Lyc}$ is the average energy of the stellar continuum photons, f is the fraction of
photons < 912 Å absorbed by the gas in the ionized region, $L_{\nu < Lyc}$
is the luminosity below the Lyman continuum, τ_0' is the effective
absorption optical depth of the dust in the HII region for photons
> 912 Å.

The first term is the contribution due to Lyα heating of the
dust, the second term is the Lyc absorbed by the dust, and the
third term the absorbed radiation longward of Lyc, corrected for
internal absorption.

They have also shown that f can be approximated by:

$$f \sim \exp\{-\tau_{Lyc}\}, \tag{3.8}$$

where τ_{Lyc} is the optical depth of the dust at Lyman continuum
wavelengths.

Since $f = N_c'/N_c$, its value can be computed from the experimental data:

$$\left. \begin{array}{l} L_{IR} \to L_* \to N_c \\[2mm] L_R \to N_c' \to L_\alpha \end{array} \right| \quad \frac{N_c'}{N_c} = f \to \tau_{Lyc} \tag{3.9}$$

which in turn permits the determination of τ_{Lyc}. The value of τ_{Lyc}

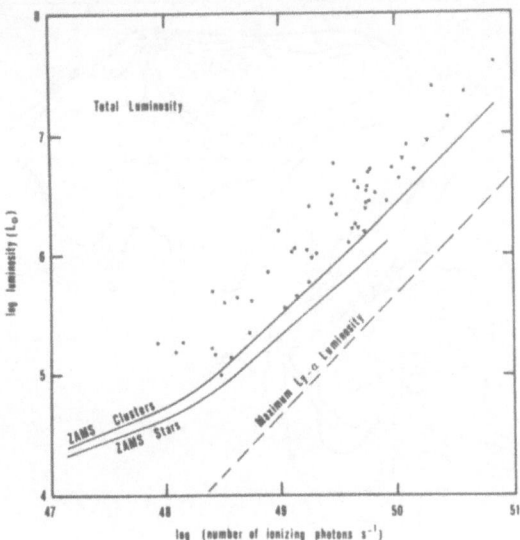

Fig.5. The total infrared luminosity plotted against the Lyman continuum photon flux for 46 HII regions (Panagia, 1976).

is important because it can yield some new insights into the physical properties of the dust.

The University College, London, infrared astronomy group using a 40 cm balloon-borne telescope has measured the value of L_{IR} for numerous HII regions (e.g., Furniss, Jennings, and Moorwood, 1975; Emerson, 1977). In Fig.5 (Panagia, 1976), for 46 of these HII regions, the total infrared luminosity is plotted against the Lyc photon flux (N_c'). For reference the curves of Zero Age Main Sequence (ZAMS) stars, ZAMS clusters of stars, and the line L_{IR} = $(L_\alpha)_{max}$ are plotted. Note the following: (1) as expected, $L_{IR} \propto$ Lyc for the sources; (2) all the sources lie above the line defined by the ZAMS stars and clusters; (3) the infrared excess, defined as $L_{IR}/N_c'(h\nu_\alpha) = L_{IR}/L_\alpha$ has a value of ~ 14, and is fairly independent of the luminosity (Panagia, 1976).

The displacement of the data points from the theoretical ZAMS curves is to be expected, because each point should move upward if $L_{IR} \leq L_*$, and to the right if some Lyc radiation is absorbed directly by the dust. In fact, the horizontal displacement gives the value of $f = N_c'/N_c$. The values of f derived are ~ 0.3 to 0.5.

4. WHAT CAN BE LEARNED FROM THE BRIGHTNESS DISTRIBUTIONS?

By comparing the radio continuum and infrared brightness distributions of HII regions, e.g., Figures 6, 7, and 8, several conclusions can be drawn:

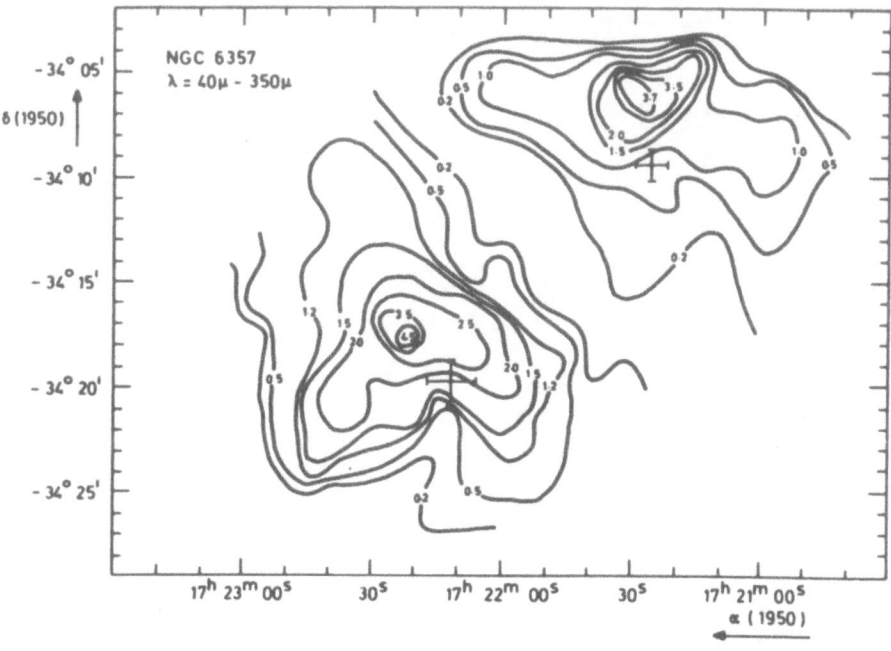

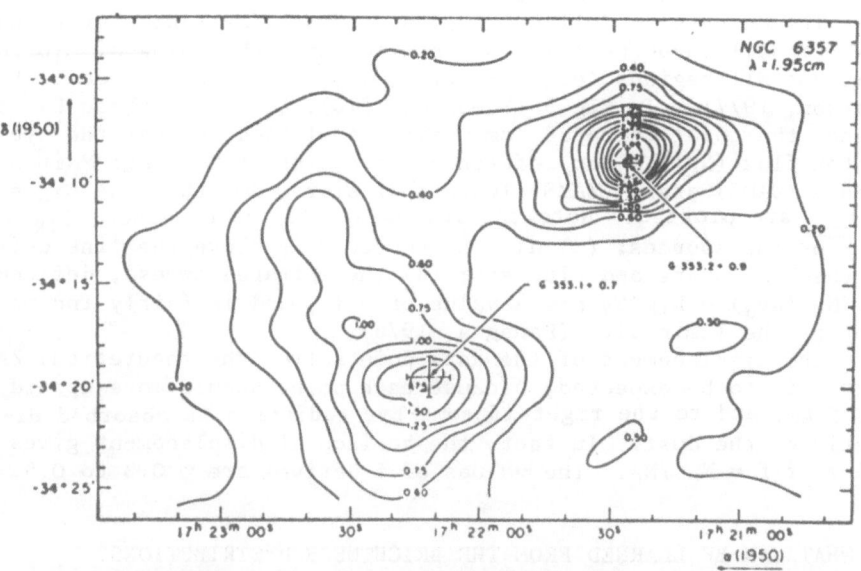

Fig.6. Comparison of the radio continuum (1.95 cm wavelength) map
with the far-infrared (40-350 μm) map of NGC 6357 (Emerson, Jennings
and Moorwood, 1973).

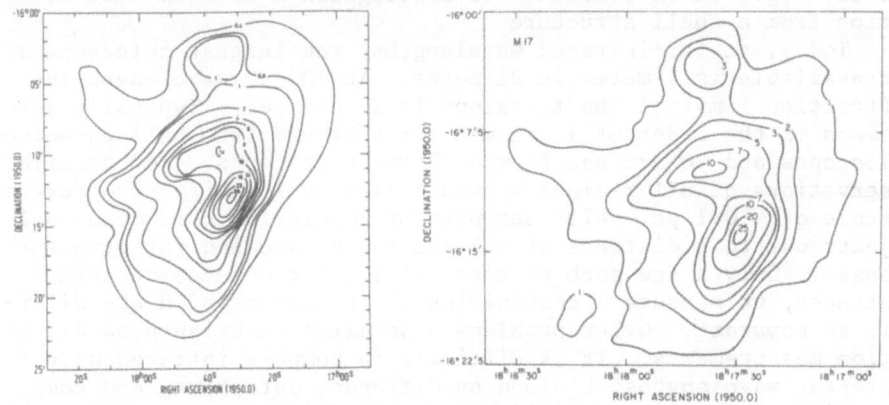

Fig.7. Comparison of the radio continuum (8.6 GHz) map with the
far-infrared (69 μm) map of M17. (Radio map courtesy of T. Wilson,
Max-Planck Institut für Radioastronomie.)

(1) There is a general coincidence between infrared and radio iso-
photes. Some exceptions do exist. This implies the dust is mixed
well with the ionized gas.
(2) The far-infrared intensity distribution is usually centered on
the corresponding radio peak.
(3) The radio and infrared source sizes are approximately equal.
(4) Limited angular resolution at far-infrared wavelengths prevents
details of the dust distribution in the HII region from being ob-

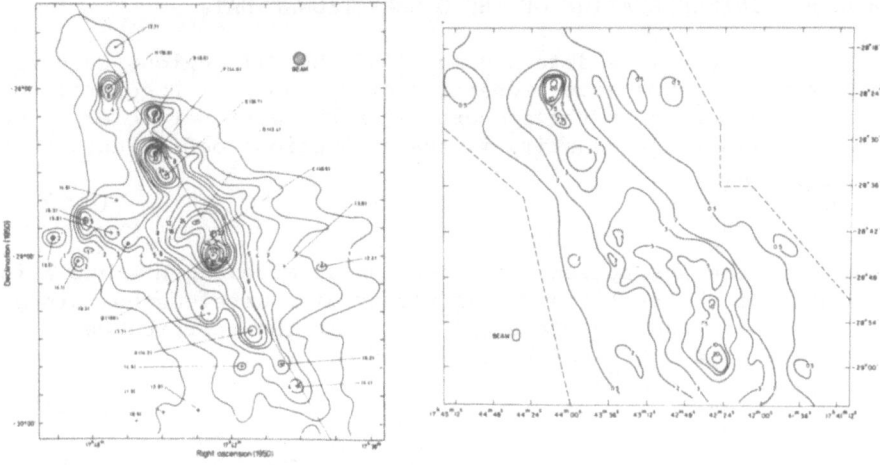

Fig. 8. Comparison of the radio continuum (5 GHz) map with the far
infrared (69 μm) map of the Galactic Center region. (Radio map
from Whiteoak and Gardner, 1973).

served, e.g., it is difficult to distinguish a uniform dust distribution from a shell structure.

Today, at far-infrared wavelengths, the largest telescope mirror available is 1 meter in diameter. At 100 μm wavelength the
diffraction limit of the telescope is 25 arc sec. Typically reso-
lutions of the order of 1 arc min are achieved from balloon-borne
telescopes and 30 arc sec from airborne telescopes. For detailed
observations of HII regions we would like to identify structure on
a scale of ~ 0.1 pc. With our present resolution this means only
objects out to a distance of ~ 1 kpc can be studied for structural
changes. Only a few such objects exist. For objects at larger
distances, we measure a combination of processes which are difficult to separate. Other problems also exist in brightness distribution measurements. It is difficult to compare infrared data at
different wavelengths obtained by different detectors, and comparison with radio data is difficult due to different beam sizes.

Recently L. Rodriguez (1977) mapped a series of HII regions
at high frequency (23 GHz) with a single radio dish antenna. This
permitted an accurate measurement of extended emission with high
resolution ($\sim 1'$). This extended emission is often lost in radio
interferometric observations. In comparison of 8 HII regions with
maps made with the 102 cm balloon-borne infrared telescope (1'), he
concluded: (1) radio and infrared peak intensities are approximately ($\lesssim 1'$) at the same position; (2) infrared size is greater than
or equal to radio size with the ratio varying from 1 to 2; (3)
shape of contour maps are similar; (4) there was no evidence of a
radio shell structure.

5. FURTHER INTERPRETATION OF THE OBSERVATIONAL DATA

To produce a more detailed model of the HII region, we need to
know, or must assume some properties of the dust. In general our
present information on the properties of the dust particles is very
poor. One important quantity is the absorption coefficient, Q_λ^{abs}:

$$Q_\lambda^{abs} = \frac{\sigma_\lambda}{\pi a^2} \qquad (5.1)$$

where σ_λ is the absorption cross section at wavelength λ and a is
the radius of the dust grain. At ultraviolet wavelengths, since
$\lambda << a$, $\sigma \sim \pi a^2$ and therefore $Q_{Lyc}^{abs} \sim 1$. At the infrared wavelengths $\lambda >> a$ and Q_{IR}^{abs} can be approximated by

$$Q_{IR}^{abs} \sim A_n/\lambda^n \qquad (5.2)$$

where n is a constant in the range 0 to 2. The optical depth, τ_λ,
is related to Q_λ by

$$\tau_\lambda = \int_R n_d Q_\lambda \pi a^2 dr \qquad (5.3)$$

where n_d is the number of dust particles per unit volume, and R is the thickness of the source.

Combining infrared and ultraviolet wavelength data, we can deduce some interesting properties of the dust in HII regions.

a. The far-infrared absorption coefficient. We have shown that observational data permits the determination of τ_{IR} and τ_{Lyc} or τ_{UV}. We can then determine Q_{IR}:

$$Q_{IR} = Q_{UV}(\tau_{IR}/\tau_{UV}) \sim \tau_{IR}/\tau_{UV} \ . \qquad (5.4)$$

For example, in a typical case Q (50 μm) = 0.01. From this value a minimum grain size can be deduced (>0.05 μm).

b. Theoretical dust temperature in an HII region. By equating the energy absorbed by the dust (mostly ultraviolet) to the energy emitted (mostly infrared), we can estimate the dust temperatures in the ionized volume:

$$\int_{o}^{\infty} F(\lambda)Q_{abs}(\lambda)d\lambda = \int_{o}^{\infty} \pi B_{\lambda}(T_d)Q_{abs}(\lambda)d\lambda. \qquad (5.5)$$

Inserting $Q_{UV}^{abs} \sim 1$ and $Q_{IR}^{abs} \sim A_n/\lambda^n$, the equation can be solved for T_d. In the ionized gas region the theoretical results are that $T_d \sim 100\text{-}200$ K. However the measured spectra indicate a combination of temperatures: 50-70 K from peak radiation near 100 μm and 100-200 K from 5-20 μm excess radiation. There have been two explanations proposed to explain this phenomena: i) A shell of cold dust (50-70 K) exists around the outside of the HII region, and the dust is depleted inside the HII region (Wright, 1973). The dust shell is heated by stellar radiation escaping from the HII region. ii) Two or more types of dust grains exist in and around HII regions (Natta and Panagia, 1976). This latter theory is currently favored for it explains most of the observed spectral and brightness distributions.

c. Submillimeter spectral distribution. The infrared spectral distribution can be written as:

$$I_{\lambda} \propto Q(\lambda)B_{\lambda}(T_d) \propto \frac{1}{\lambda^n} B_{\lambda}(T_d) \qquad (5.6)$$

In the submillimeter region of the spectrum the blackbody radiation is given by Rayleigh-Jeans' relation:

$$B_{\lambda}(T_d) \propto 1/\lambda^4 \ . \qquad (5.7)$$

Therefore $I_{\lambda} \propto \lambda^{-(4+n)}$ or $I_{\nu} \propto \nu^{2+n}$.
We have previously noted that at these wavelengths the observed spectrum is steeper than the blackbody curve. The data is best fitted by values of n = 1 to 2, which yields additional information on the absorption properties of the dust at submillimeter wavelengths.

d. Dust depletion in HII regions. Assuming normal interstel-

lar values of the mass of dust to gas in an HII region the Lyc optical depth can be computed. The extinction,A_v, at optical wavelengths is given by

$$A_v = 1.086\tau_v = 1.086\pi a^2 Q_v^{ext} n_d R. \tag{5.8}$$

At Lyc wavelengths

$$\tau_{Lyc} = \pi a^2 Q_{Lyc}^{abs} n_d R. \tag{5.9}$$

Combining the above:

$$\tau_{Lyc} = \frac{N_H}{1.086}\left(\frac{Q_{Lyc}^{abs}}{Q_v^{ext}}\right)\left(\frac{A_v}{N_H}\right), \tag{5.10}$$

where N_H is the column density of hydrogen atoms. The quantity
A_v/N_H is basically a measure of the mass of dust to gas and has
been experimentally determined for the interstellar medium to be
4.5×10^{-22} mag cm^2.

In the ionized region the value of N_H is given by $N_e R$, and the
ratio of Q values can be determined from the extinction curves of
Yorke et al. (1973), correcting for the albedo (Lillie and Witt,
1977). The value of τ_{Lyc} computed in this way can be compared to
the value computed previously from radio and infrared observations.
If the latter value is less than the former, the dust-to-gas ratio
is less in the HII region than it is in interstellar space, and the
dust is said to be depleted.

For the HII regions in the University College, London, survey
we would expect from the above equations a linear relationship between τ_{Lyc} and $n_e R$. In fact, Emerson (1977) found an anticorrela-
tion, with τ_{Lyc} decreasing with increasing values of $n_e R$. To investigate further Emerson plotted $(Q_{Lyc}/Q_v)(A_v/N_H)$ versus $n_e R$ and
found $(Q_{Lyc}/Q_v)(A_v/N_H) \propto (n_e R)^{-2}$. Since $N_c' \propto n_e^2 R^3$, (Q_{Lyc}/Q_v)
$(A_v/N_H) \propto (N_c/R)^{-1}$. Indeed, the observations indicate (Q_{Lyc}/Q_v)
$(A_v/N_H) \propto (N_c')^{-1}$, although the scatter in the data points is large.
Possible explanations of this result, given by Emerson, are:
(1) gas and/or dust clumping;
(2) τ_{Lyc} calculated from observational data could be in error due
to failure of stellar atmosphere models;
(3) dust-to-gas mass ratio decreases in HII regions excited by earliest type stars;
(4) modification of grain properties (Q_{Lyc}/Q_v);
(5) Q_{Lyc} may decrease with N_c'.
In general, the UCL survey indicates HII regions are not dust
depleted, i.e. $M_d/M_g \sim 10^{-2}$; however there is evidence (Thronson,
1977; Fazio et al., 1978) that in very compact HII regions heated
by the hottest O-stars there is significant dust depletion, as suggested in explanation (3) above.

e.<u>Another method of determining the dust-to-gas mass ratio.</u>
The dust-to-gas mass ratio (M_d/M_g) in the ionized volume of an HII

region can also be calculated by another method if certain assumptions are made about the dust properties:

$$M_d = (\frac{4}{3}\pi R^3)n_d\rho(\frac{4}{3}\pi a^3)$$

$$\tau_{UV} = n_d\pi a^2 Q_{UV}R$$

$$\therefore M_d = \frac{16}{9}\pi R^2 \rho a(\tau_{UV}/Q_{UV}) \tag{5.11}$$

These formulas assume a spherical volume and uniform mixing of gas and dust. Likewise the mass of gas is given by:

$$M_g = (\frac{4}{3}\pi R^3)n_e M_H(X+4Y), \tag{5.12}$$

where M_H is the mass of a hydrogen atom and X and Y are the fractional abundances by number of H and He, respectively, i.e. X = 0.9 and Y = 0.1. Therefore

$$\frac{M_d}{M_g} = \frac{4\rho a\tau_{UV}}{3Q_{UV}n_e R M_H(X+4Y)} \tag{5.13}$$

This method, however, is not as accurate as that given in (d) above.

6. THE RELATIONSHIP BETWEEN IONIZED HELIUM ABUNDANCE AND INFRARED EXCESS RADIATION

Churchwell, et al. (1974), from radio recombination line measurements determined the ratio $N(He^+)/N(H^+)$ in approximately 40 HII regions. In general they observed the expected value of ~ 0.1. However, in a small number of cases they observed the ratio to be much less than 0.1, and interpreted this as due to the non-coincidence of the He^+ and H^+ Stromgren spheres, resulting from selective absorption of ionizing photons by dust. They based this conclusion on another observed fact, a correlation of the He abundance with excess infrared radiation $IR_{exc} = (L_{IR} - L_\alpha)/L_\alpha$, i.e., He^+/H^+ ratio decreases as IR_{exc} increases. One explanation of this result is that the dust opacity is much higher in the wavelength interval 504 Å > λ > 228 Å. A factor of 6 to 8 increase is needed. Recent work by the UCL group tends to verify this relationship, but the relationship is still somewhat weak. More data is needed.

7. COMPACT HII REGIONS

Compact HII regions are the smallest gaseous nebulae known, and probably represent a class of very young objects. Their diameters are usually ≤ 0.5 pc, and in the ionized region the electron density is $\geq 10^4$ cm^{-3}. They are often found associated with OH and H_2O masers; their estimated lifetime is $\leq 10^4$ years. The "diffuse" HII regions we have previously discussed are considerably more tenuous and an order of magnitude larger in size. Compact HII regions are often found nearby or in a more extended source containing evolved objects or associated with molecular clouds.

Typical examples of compact HII regions include the infrared sources W3OH, W51 IRS-2, K3-50, ON-3 W75(S), DR-21, and NGC 7538(N) which have recently been discussed by Thronson (1977) and G10.6-0.4 (near W31) observed by Fazio et al. (1978).

From his infrared observations Thronson described these objects as having luminosities in the range of 9 x 10^4 to 2 x $10^6 L_o$, with the source of this energy being very luminous main sequence or pre-main sequence stars. The energy output is dominated by infrared radiation, with a source size unresolved with a 30 arcsec beam; in W3OH this implies a source size ≤ 0.2 pc. The infrared sources are coincident in position with the compact radio continuum sources. From infrared measurements of the amount of the dust depletion in the ionized region there is evidence that the bulk of the dust that radiates is outside the HII region. The dust radiates in emission, with a temperature in the range of 30 to 50 K, with relatively high optical depths at infrared wavelengths, among the highest known ($\tau_{IR} \sim 0.1$ to 0.9). Heavy visual extinction arises within 0.5 pc of the source, with $A_V \sim 100\tau_{IR}$. Likewise the value of $\tau_{UV} \sim 0$ to 2. Within the ionized volume the dust is depleted relative to the expected interstellar dust-to-gas ratio by factors varying from 1 to 20.

In an attempt to interpret the evolutionary state of these compact HII regions Thronson (1977) noted that of the sources he observed four were excited by main sequence objects and four by pre-main sequence objects. The latter objects had the highest values of infrared excess (L_{IR}/L_α) observed, e.g., W75(S) has $L_{IR}/L_\alpha \geq 950$.

8. EVOLUTION OF HII REGIONS

In summary, let us now consider a possible evolutionary sequence for HII regions, in which we will try to order the various types of objects we have described. The following description is taken from Mezger (1976) and is based on the work of Davidson and Harwit (1967) and Krügel (1974).

When a massive proto-star approaches the main sequence, its surface temperature is low and it cannot ionize the ambient gas, but it has already acquired its full luminosity. The radiation pressure acts on the dust and drives it at high speed (~ 10 km s^{-1}) outwards, without dragging along the neutral gas, because friction between the gas and dust by elastic collisions is weak. The dust piles up in a front until its density there is so high that coupling between the gas and dust becomes effective. The dust front

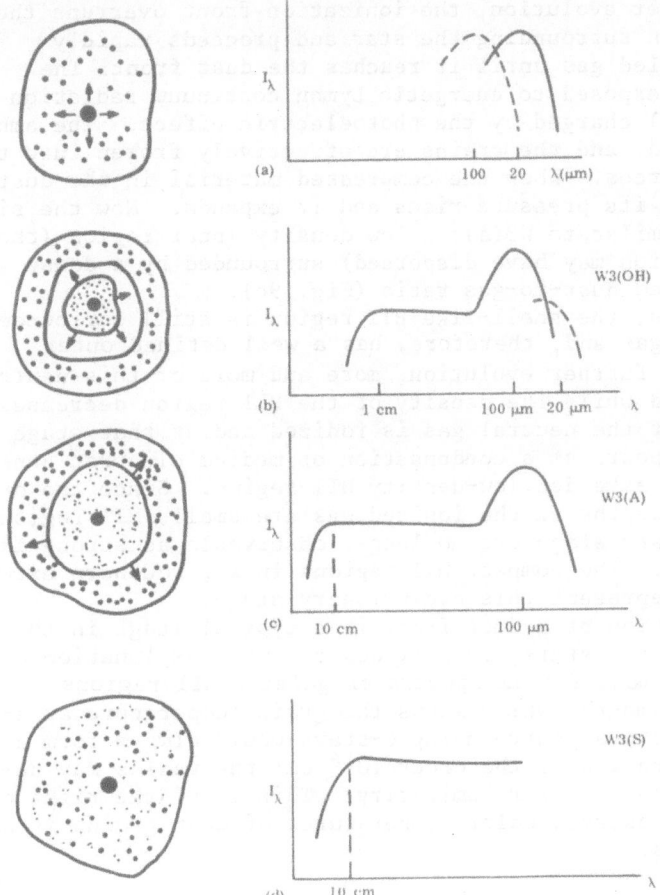

Fig. 9. Possible evolutionary sequence of an HII region; see text for description.

is optically thick at visible wavelengths and hides the star from the optical observer, but it is a strong infrared source (Fig. 9a). The dust front takes up the whole momentum of the stellar radiation. Its drift velocity through the gas is small (about 1 km s^{-1}) compared to the speed at which it is driven away from the star. Therefore, the dust front acts like a supersonic piston on the surrounding gas leaving in its wake a highly rarified region.

When the star becomes hot enough, it ionizes (part of) its surrounding gas from which the dust has been expelled, thus forming a dust-depleted HII region, which is surrounded by a low-density region (which has been cleared by the dust front), and further out by the dust front itself (Fig. 9b). This may be the stage at which we observe W3(OH).

In the further evolution, the ionization front overruns the
compact HII region surrounding the star and proceeds rapidly
through the rarified gas until it reaches the dust front. The
grains there are exposed to energetic Lyman continuum radiation
and are (probably) charged by the photoelectric effect. The ambient gas is ionized, and the grains are effectively frozen into the
gas by Coulomb forces. When the compressed material in the dust
front is ionized, its pressure rises and it expands. Now the situation is very similar to W3(A): a low density inner region (the
central condensation may have dispersed) surrounded by a dense
shell with enhanced dust-to-gas ratio (Fig. 9c).

At this stage, the shell-like HII region is still surrounded
by dense neutral gas and, therefore, has a well defined outer
boundary. In its further evolution, more and more of this neutral
shell gets ionized while the density of the HII region decreases.
Eventually, all of the neutral gas is ionized and at that stage
the HII region appears as a condensation of medium electron density embedded in an extended low-density HII region. Since the optical absorption depths in the ionized gas are small, HII regions
at this evolutionary stage are no longer observable as strong IR
sources (Fig. 9d). The compact HII regions in W3, southern exten-
sion, appear to represent this evolutionary stage.

If the formation of a dust front is a typical stage in the
pre-MS evolution of O-stars, it provides a simple explanation of
the similarity of most far IR spectra of galatic HII regions,
providing the thermostat which keeps the grain temperature at a-
bout 70-80 K. The dust surrounding O-stars would always form a
shell of typical radius of the order 10^{18} cm; the radius may depend somewhat on the stellar luminosity. This is of the right order to yield the observed color temperatures of dust grains in and
around HII regions.

9. EXAMPLES OF HII REGIONS

We shall now consider for detailed study two HII regions, the
first is M20, the Trifid Nebula (Fig. 10), a simple, near ideal
object; the second is the more complex region, M42, the Orion Nebula (Fig. 11).

9.1 Trifid Nebula (M20)

This gaseous nebula exhibits a distribution of derived nebular properties that is in closer agreement to what is expected
theoretically of galactic nebulae than any other HII region. It
has considerable symmetry about the exciting star (HD 164492, an
O7.5 star), normal helium abundance, subsonic turbulence, an electron temperature that is in agreement with both radio and optical
observations, and a positive thermal gradient as a function of neb-

Fig. 10. Optical photograph of M20 (Trifid Nebula); courtesy of the Hale Observatories.

ular radius.

Chaisson and Willson (1975) summarized the following proper-ties of the nebula from both radio recombination line and radio continuum observations: $n_e \sim 10^2$ cm^{-3}, $T_e \sim 8150$ K, EM $\sim 10^{4 \cdot 7}$ pc cm^{-6}, M $\sim 210 M_\odot$, M(H$^+$) = $160 M_\odot$, $N_c = 10^{48.8}$ photons sec^{-1}, n(He)/n(H) = 0.095, $v_t \sim 8.5$ km sec^{-1}. Its distance is estimated to be 2.1 kpc and its diameter is 5.3 pc. The radio data also indicates that the dust lanes present physically interact with the ionized gas.

Molecular line observations indicate an extensive molecular cloud engulfing the nebula from the south and west. Visual pho-tography also indicates greater obscuration toward the south and west, with little or no obscuration in the northeast. Approxi-mately 5 arcmin southwest of the exciting star, and enclosed in the larger cloud, is a small concentration of molecular gas (called M20SW) of considerable mass ($\geq 1300 M_\odot$) and density (≥ 400 cm^{-3}), coincident with a globule of total obscuration. Kinematic properties of this mass suggest it is in a state of gravitational

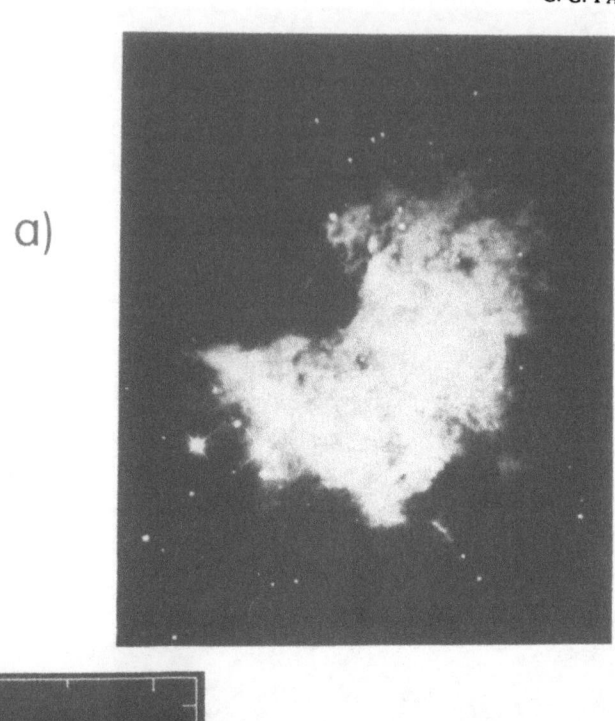

a)

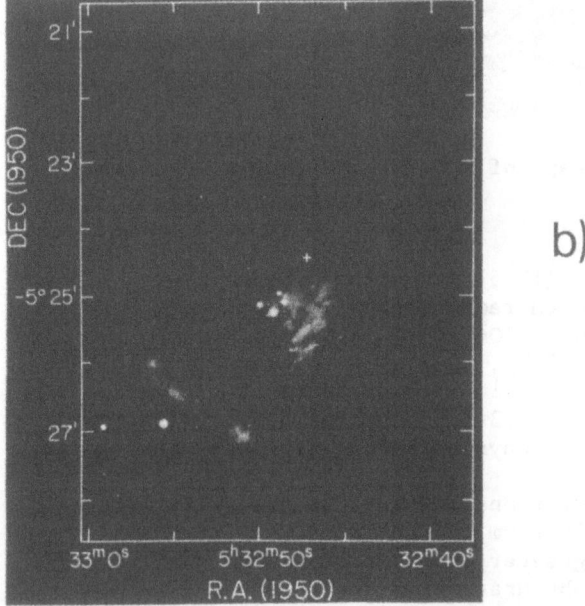

b)

Fig. 11. (a) Optical photograph of M42 (Orion Nebula); courtesy of
Lick Observatory; (b) Photograph of Orion Nebula in Hα recombination line, showing more clearly Trapezium stars and bright bar;
cross marks infrared cluster sources (Werner et al., 1977).

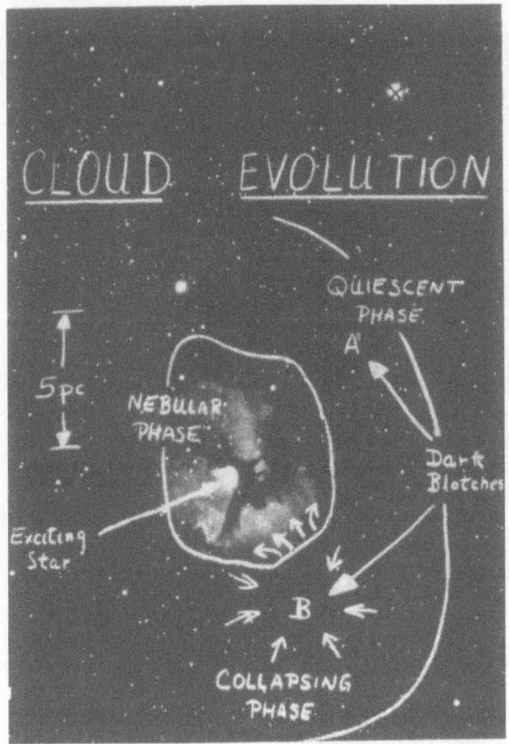

Fig. 12. **Suggestive observational** evidence for an evolutionary scenario in M2O depicting a cold dense cloud proceeding to a young hot star (Chaisson, 1976).

collapse.

Chaisson and Willson (1975) interpret their data as evidence for three principal phases of star formation in this region (Fig. 12). In the northwest is the initial or quiescent phase of the molecular cloud. At M2OSW the molecular cloud has probably reached sufficiently high density and mass to initiate gravitational collapse, the intermediate state of star formation. The HII region represents the final phase of star formation.

Because infrared observations of these regions would be of considerable interest in determining the dust properties and temperatures and in searching for the existence of protostars, the Harvard-Smithsonian Center for Astrophysics/University of Arizona 102-cm balloon borne telescope was used to observe this region at far-infrared wavelengths, using a broadband photometer (40-250 microns). The total infrared luminosity was found to be $\geq 2 \times 10^5 L_\odot$, and the infrared size was comparable to the radio size, except that the infrared source was more extended in the northwest-southeast direction. The infrared peak intensity appeared to be shifted about 0.5 arcmin north of the radio peak. The dust temperature was

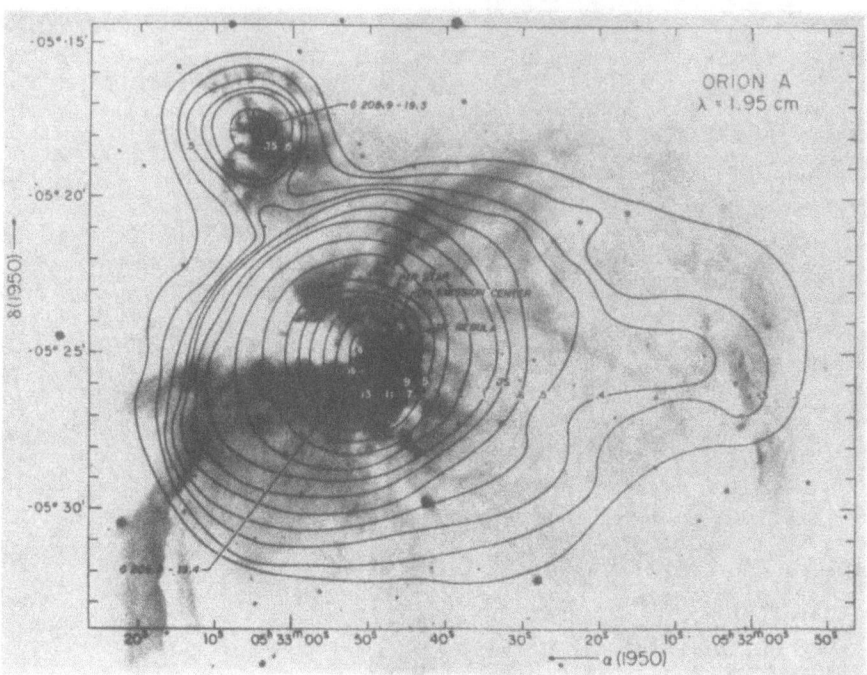

Fig. 13. Radio continuum map at 1.95-cm wavelength of the Orion
Nebula (Schraml and Mezger, 1969).

determined to be 49 ± 14K. Surprisingly, no infrared source was
observed in M2OSW, the collapsing molecular cloud. This means that
no stars with a luminosity greater than 3000$L_\odot$ have yet formed.
The infrared observations also indicate that the Lyα power available for heating the dust is only 1.5 x $10^4 L_\odot$, and therefore most
of the dust is heated by absorption of stellar continuum radiation.

9.2 Orion Nebula (M42)

     An example of a relatively complex HII region is the Orion
Nebula, which is associated with the Orion Molecular Cloud 1 (OMC-
1). The latter object is part of a complex region of giant molecular clouds in the Orion constellation. It is relatively nearby,
only 1500 light years from the sun and one of the most studied objects in the sky. In the central part of OMC-1, about 2 light
years wide, is found a rather fascinating collection of sources.
Optically one notices the famous Orion Nebula (Fig. 11), a diffuse,
luminous HII region, centered on four, hot, luminous main sequence
stars called the Trapezium cluster. Large dust obscurations are
evident as well as a bright "bar" to the southeast. Radio continu-

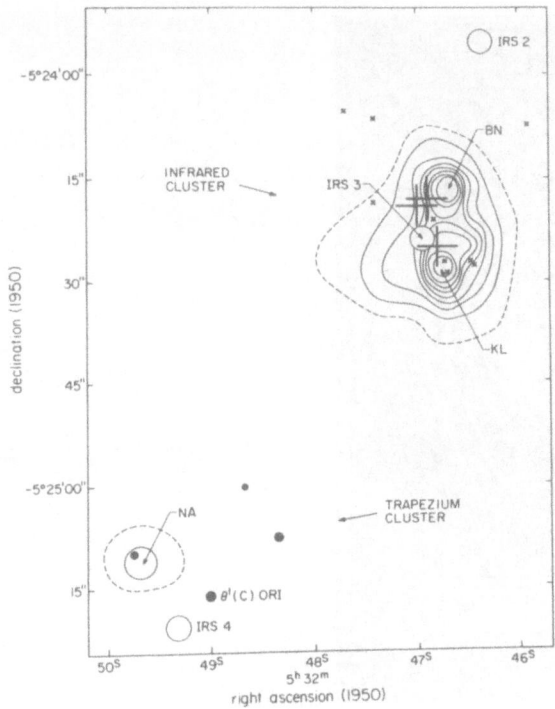

Fig. 14. Schematic representation of the relative positions of
the Trapezium stars and the infrared cluster in the Orion Nebula
--OMC-1 complex (Wynn-Williams and Becklin, 1974; Rieke, et al.,
1973).

um measurements (Fig. 13) show the Trapezium stars excite and ionize the nebula, producing the radio and optical radiation. Their
total luminosity is about $3 \times 10^5 L_\odot$. The ionized gas has a temperature of about 10^4 K, a density of 10^4 atoms/cm^3, and a total mass
of 10 M$_\odot$. Centered on the Trapezium star furthest to the east is
a strong 20 μm infrared source, called the Ney-Allen (NA) source
(Fig. 14). A 10 μm emission feature due to silicates indicates
the source of infrared emission is a heated dust shell around the
star. Displaced by 1 arcmin to the northwest of the Trapezium
stars is the center of a dense neutral cloud of molecular gas (OMC
-1). Fig. 15 shows the 2-cm formaldehyde emission line map of the
central region of the cloud. The molecular cloud has a temperature
of about 100 K, a density greater than 10^6 molecules/cm^3, and a total mass of 500 M$_\odot$. Optical and radio continuum measurements centered on this cloud show no prominent emission. However, at infrared wavelengths, it is a very bright source, with a luminosity
greater than $10^5 L_\odot$. Fig. 16 is a far-infrared (100 μm) map of the
nebula which shows the radiation centered on the molecular cloud.
Higher resolution measurements at 10 μm indicate the center of the

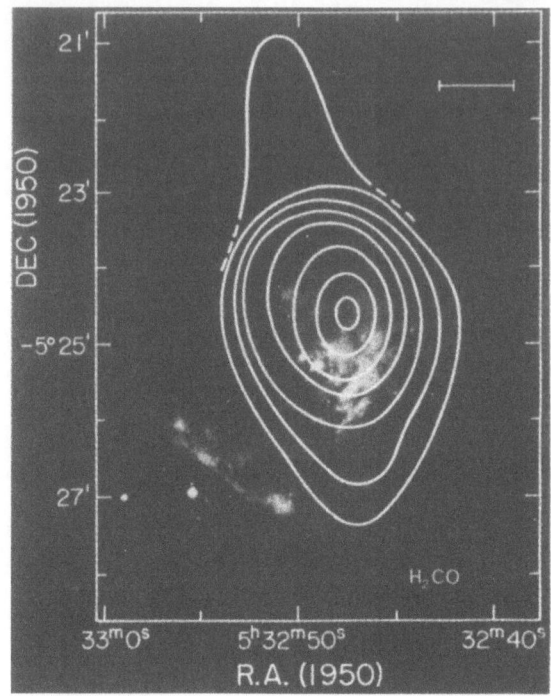

Fig. 15. Contour map of the 2-cm wavelength emission line of the
formaldehyde molecule from Orion Molecular Cloud 1 (Werner, Becklin, and Neugebauer, 1977).

cloud consists of a complex of small infrared sources. Fig. 14
shows the relationship of the infrared cluster to the Trapezium
stars. In the 3 to 10 μm region the brightest infrared source is
the Becklin-Neugebauer (BN) object, which has a spectrum corresponding to a 600 K blackbody, with absorption features at 3.1 μm
due to ice and 10 μm due to silicate dust. The object is strongly
polarized at 2.2 and 10 μm. Its diameter is less than 300 AU, and
its luminosity is $10^3 L_\odot$. These properties are all very similar to
those of a protostar in its pre-main sequence evolution. At far-
infrared wavelenghts the brightest member of the cluster is the
Kleinmann-Low (KL) Nebula, located 12 arcsec south of the BN objects. Its temperature is about 70 K; its size is approximately
2000 AU and its luminosity is $7 \times 10^4 L_\odot$. About 30 percent of the
total infrared luminosity of the total region comes from the cluster of sources centered on the molecular cloud, with the remaining
flux coming from dust mixed with the ionized gas region and the
dust heated at the transition zone in the bright bar. Another important property of this complex is the presence of sources of OH,
H_2O, SiO microwave maser emission which appear to be near the infrared cluster sources.

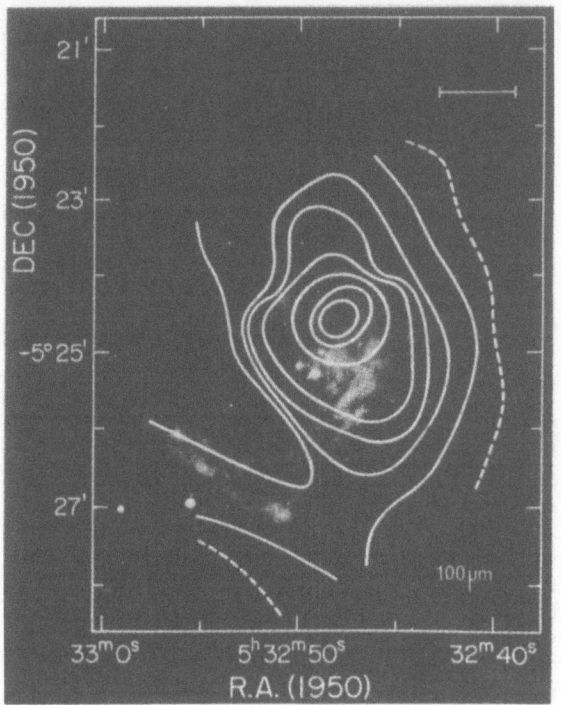

Fig. 16. Contours of the 100 μm wavelength infrared emission from
Orion Molecular Cloud 1 (Werner, Becklin, and Neugebauer, 1977).

    The Trapezium cluster of stars and the infrared sources at
the center of the molecular cloud have comparable total luminosities. The Trapezium stars appear to be newly formed stars, just
arriving on the main sequence. The density and mass of the surrounding ionized region is typical of such a stage of evolution.
The infrared cluster in the center of the molecular cloud is probably a group of protostars still in the high density region where
they were born. The infrared cluster of stars may be evolving
toward the stage the Trapezium stars are in. The molecular cloud
appears to be behind the ionized gas cloud and in contact with it.
A proposed model of the system is shown in Fig. 17. The Trapezium
stars probably condensed from the same molecular cloud. The bright
bar in the southeast appears to be a HI-HII transition zone where
the ionized gas meets the molecular cloud. This small scenario of
early stellar evolution is in turn part of a larger scale association in the Orion constellation which is filled with massive young
main sequence stars. In fact, it appears that the Orion association provides an outstanding example of a sequence of stellar
groups of decreasing age which culminates in the Orion Nebula
sources.

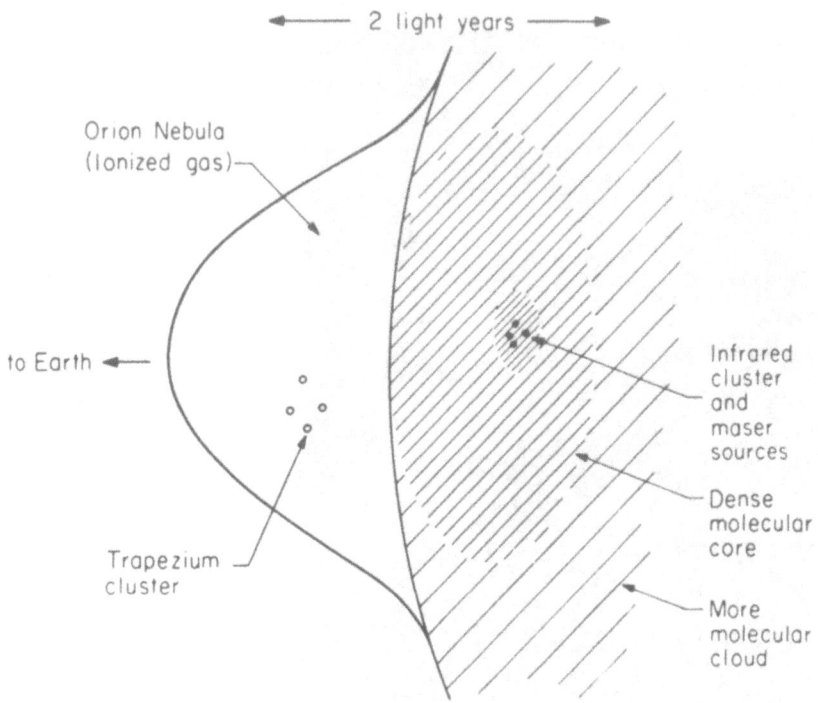

Fig. 17.   Schematic representation of the relative positions of
the sources in the Orion Nebula--OMC-1 complex, based on a model
by B. Zuckerman, University of Maryland (Zuckerman, 1973; Werner,
Becklin, and Neugebauer, 1977).

Not all infrared astronomers agree on the above protostar interpretation of the infrared cluster sources. It is still possible
that the BN source, for example, could be a luminous F supergiant
star obscured by cold dust, with 80 magnitudes of extinction, or
possibly an evolved star with a hot dust shell (Allen and Penston,
1974).

REFERENCES

Allen, D.A. and Penston, M.V., Nature 251, 110-112 (1974).
Chaisson, E.J., in "Frontiers of Astrophysics," E.H. Avrett, Ed.
(Harvard University Press, Cambridge, Massachusetts), pp. 259-351,
1976.
Chaisson, E.J. and Willson, R.F., Astrophys. J. 199, 647-659 (1975).
Churchwell, E., Mezger, P.G., Huchtmeier, W., Astron. Astrophys.
32, 283-308 (1974).
Davidson, K. and Harwit, M., Astrophys. J. 148, 443-448 (1967).
Emerson, J.P., Ph.D. Thesis, University College, London (1977).

Emerson,J.P., Jennings, K.E., and Moorwood, A.F.M., Astrophys. J. 184, 401-414 (1973).
Fazio, G.G., in "Frontiers of Astrophysics," E.H. Avrett, Ed. (Harvard University Press, Cambridge, Massachusetts), pp. 203-258, 1976.
Fazio, G.G., Lada, C.J., Kleinmann, D.E., Wright, E.L., Ho, P.T.P., and Low, F.J., to be published Astrophys. J. (Letters), 1978.
Furniss, I., Jennings, R.E., and Moorwood, A.F.M., Astrophys. J. 202, 400-406 (1975).
Harper, D.A., and Low, F.J., Astrophys. J. (Letters) 165, L9-13 (1971).
Krügel, E., Ph.D. Thesis, University of Göttingen (1974).
Lillie, C.F. and Witt, A.N., Astrophys. J. 208, 64-74 (1976).
Mezger, P.J., in "Far Infrared Astronomy," M. Rowan-Robinson, Ed., (Pergamon Press, Oxford), pp. 231-246, 1976.
Mezger, P.J., Smith, L.F., and Churchwell, E., Astron.Astrophys. 32, 269-282 (1974).
Mezger, P.J., and Henderson, A.P., Astrophys. J. 147, 471-489 (1967).
Natta, A. and Panagia, N., Astron. Astrophys. 50, 191-211 (1976).
Panagia, N., in "Infrared and Submillimeter Astronomy," G.G.Fazio, Ed. (D. Reidel Publishing Co., Dordrecht, Holland), pp. 43-54, 1977.
Petrosian, V., Silk, J., and Field, G.B., Astrophys. J. (Letters) 177, L69-73 (1972).
Rieke, G.H., Low, F.J., and Kleinmann,D.E., Astrophys. J. (Letters) 186, L7-11 (1973).
Rodriguez, L., preprint, to be published Astrophys. J., May, 1978.
Schraml, J. and Mezger, P.G., Astrophys. J. 156, 269-301 (1969).
Thronson, H., Jr., Ph.D. Thesis, University of Chicago (1977).
Werner, M.W., Becklin, E.E., and Neugebauer, G., Science 197, 723-732 (1977).
Whiteoak, J.B., and Gardner, F.F., Ap. Lett. 13, 205-207 (1973).
Wright, E.L., Astrophys. J., 185, 569-572 (1973).
Wynn-Williams, C.G. and Becklin, E.E., Proc. Astron. Soc. Pacific 86, 5-25 (1974).
Zuckerman, B., Astrophys. J. 183, 863-869 (1973).

Additional General References

"Frontiers of Astrophysics," E.H. Avrett, Ed. (Harvard University Press, Cambridge, Massachusetts), 1976.
"Infrared and Submillimeter Astronomy," G.G. Fazio, Ed. (D. Reidel Publishing Co., Dordrecht, Holland), 1977.
Proceedings of the Eighth ESLAB Symposium on HII Regions and the Galactic Center (ESRO SP-105), A.F.M. Moorwood, Ed., 1971.
"Lecture Notes in Physics," Vol. 42, Proceedings of the Symposium "HII Regions and Related Topics," T.L. Wilson and D. Downes, Eds. (Springer-Verlag), 1975.

PHYSICS AND ASTROPHYSICS OF INTERSTELLAR DUST

J.M. Greenberg

Laboratory of Astrophysics, Huygens Laboratorium,
University of Leiden, Leiden, The Netherlands

PREFACE

Although this exposition covers the topics discussed in the series of lectures presented at the Summer School on Infrared Astronomy, their order has been changed. This is largely a result of the experience gained during the course. A number of results not presented in Erice have been included for completeness because there was inadequate time to do so during the course of lectures. I have tried to preserve the character of the lectures by limiting the mathematical detail and using physical or heuristic arguments to derive some of the formulae.

1. INTRODUCTION

The overall absorption and emission properties of interstellar dust are among the basic ingredients in the interpretation of observations of infrared absorption and emission.

Since infrared observations relate to such varied objects as: (1) dark clouds, (2) H II regions, (3) compact H II regions, (4) circumstellar regions, (5) planetary nebulae, (6) the galactic center, (7) Seyfert galaxies (Markarian galaxies), it must be accepted that the physical and chemical and, consequently, the optical properties of the dust may also vary substantially. Indeed, it is not unlikely that only in several of the above instances can the dust be completely homogeneous and similar to the average of the interstellar dust. For example, the circumstellar dust associated with cool evolved stars is certainly not the sort of dust that causes interstellar extinction. On the other hand it is quite likely that in a wide variety of cases what we observe is some modification of

51

G. Setti and G. G. Fazio (eds.), Infrared Astronomy, 51-95.

the standard interstellar dust. Thus the starting point in most
theories is the assumption that the infrared absorption and/or e-
mission are produced by some constituent related either directly
or indirectly to the interstellar dust. In order to draw inferen-
ces about, say, visual extinction from infrared emission or absorption it is necessary to make assumptions about the dust model.

The first portion of these lectures will therefore be devoted
to the general scattering properties of small particles and the
problem of determining the best available generalized interstellar
dust model based as directly as possible on observations over the
full spectral range.

The subsequent discussions will consider the optical properties of the specific dust materials, the spectral absorption features, the general absorption characteristics, temperature under
varying conditions, temperature fluctuations, far infrared emission, physical interactions on small solid particles – evaporation,
sputtering, growth, irradiations, specific heat. Along the way the
questions of shape and size effects on the grain characteristics
will be examined.

The theoretical basis and assumptions used in deriving a num-
ber of basic parameters such as: $A(V)/A(\lambda)$, (λ in the near or far
infrared); ρ_d/Q_{ir}; $\rho_{sil}/\tau_{9.7}$ etc. will be presented so that one may
arrive at constraints on their values in differing circumstances
depending on whether or not the dust involved is related to the interstellar dust in some evolutionary way. The basic constraints
are size, amount, chemical composition, physical state, and shape.

2. BASIC SCATTERING RELATIONSHIPS

The optics of interstellar dust is an application of the theory of
scattering of electromagnetic radiation by small particles (see,
for example, van de Hulst, 1957). In terms of the dimensionless
parameter $2\pi a/\lambda$, where a is a measure of the particle size, and λ
is the wavelength of the radiation, we are led to consider the entire range $0 < 2\pi a/\lambda < \infty$. The higher end of the range is almost
trivial (for our purposes), the lower end is fairly simple and the
middle range $2\pi a/\lambda \sim 1$ presents the major calculational difficulties.

It turns out that for infrared wavelengths longward of about
3-5 μm all the interstellar dust as we know it may be treated as if
$2\pi a/\lambda \ll 1$. In this case we are permitted to use the simple Rayleigh approximation to obtain the scattering and absorption. Consideration of the visible and ultraviolet spectral regions requires
a detailed treatment of the middle range in order to understand
such observations as the extinction and polarization by interstellar dust.

2.1 Definitions

When light impinges on a particle it is either scattered (deflec-
ted) or absorbed. We define the extinction (total) cross section
of a particle as

$$C_{ext} = C_{sca} + C_{abs} \qquad\qquad (2.1)$$

$$= \frac{\text{Total radiant energy scattered or absorbed/unit time}}{\text{Incident radiant energy per unit area/unit time}}$$

These cross sections have the dimensions of area and in each case
represent the "effective" blocking area to the incident radiation,
effective in the sense that they are seldom equal to the geometrical blocking area.

We define the efficiency of each process (extinction efficiency, etc.) as the ratio of the respective cross section to the geometrical cross section. Thus

$$Q_{ext} = C_{ext}/\pi a^2$$

$$Q_{abs} = C_{abs}/\pi a^2 \quad \text{for a sphere} \qquad\qquad (2.2)$$

$$Q_{sca} = C_{sca}/\pi a^2$$

When the particle is very large compared with the wavelength
$(2\pi a/\lambda \gg 1)$ we expect that the electromagnetic radiation may be
treated by geometrical optics rather than physical optics and the
total cross section should approach the geometrical cross section;
i.e. the size of the shadow. Actually the limit is *twice* the geometrical cross section in the interstellar dust case and this can
be readily understood by a simple application of basic optics principles. Each ray which impinges on the particle is either scattered or absorbed. This gives rise to a contribution of 1 x G to
the cross section, G being the shadow or projected area. In addition *all* the rays in the field which do not hit the particle give
rise to a diffraction pattern whose light is contained within an
angular cone of the order of λ/a. This light is, according to the
Huygens principle, exactly the same as that produced by a hole in
a screen with the same area as the particle. If the detection *ex-
cludes* this diffracted light then an additional contribution of 1
x G is made to the total cross section. If the particle is at a
distance D from the telescope of size R the angle subtended by the
telescope is R/D. Even for nearby interstellar dust (D = 100 pc),
$R/D < 10^{-18}$ so that the ratio of the area subtended by the telescope to that containing the major portion of the diffracted radiation is $(R/D)^2/(\lambda/a)^2 < 10^{-34}$ (!) for $a/\lambda = 10$. Thus the diffracted light is excluded almost entirely by the telescope. This
clearly satisfies the required condition.

At the other extreme, where $a/\lambda \ll 1$, the particle cross sec-

tions (and efficiencies) approach zero. Exactly how this occurs
can be shown by dimensional arguments alone.

When the particle size is very small compared with the wavelength, the passing electromagnetic wave as seen by the particle
appears like a time varying but uniform (across the particle) electric field E(t). At any instant there is a dipole mement $p(t) =$
$\alpha E(t)$ induced in the particle. Dimensionally $p = qL$ and $E = q/L^2$
so that $\alpha \sim L^3$, where $q \equiv$ charge and $L \equiv$ length. Since the only
length in this problem is the particle size we see that α is at
least proportional to the particle volume. Using the fact (which
can also be shown dimensionally) that the rate at which energy is
radiated by an oscillating dipole is proportional to $|p|^2 c$ we see
that the ratio of this to the incident flux of energy is $\sim |p|^2 c/$
$|E|^2 c \sim \alpha^2$ where c is the velocity of light. The scattering efficiency is thus proportional to $\alpha^2/a^2 \sim a^4$. The only other length
in the problem being the wavelength λ we see that C_{sca} must be proportional to $(a/\lambda)^4$ which goes to zero as $a/\lambda \rightarrow 0$. This is called
Rayleigh *scattering*.

If the material of the particle is such as to absorb some of
the radiation we may think of this as occurring as the result of
an ohmic loss because as the dipole moment changes it is equivalent to a current passing through the particle. The current, de-
fined as $i = dq/dt \sim (1/a)(dp/dt) \sim a^{-1}\omega p$ where ω is the (circular)
frequency of the incident radiation. Multiplying this times the
potential difference, $V = Ea$, across the particle gives the ohmic
loss $iV \sim |\alpha\omega||E|^2$ which when divided by $|E|^2 c$ gives the absorption
cross section as proportional to $\alpha\omega/c$. The absorption efficiency
is then proportional to $(\alpha/a^2)(\omega/c)$. We have $Q_{abs} \sim a/\lambda$ which
again $\rightarrow 0$ as $a/\lambda \rightarrow 0$. The absorption cross section is proportional
to the volume of the particle.

The Rayleigh approximations for scattering and absorption may
be derived rigorously following the above methods and the results
for spheres are

$$Q_{sca} = \frac{8}{3} x^4 \left|\frac{\epsilon-1}{\epsilon+2}\right|^2 \tag{2.3}$$

$$Q_{abs} = -4x\,\text{Im}\left(\frac{\epsilon-1}{\epsilon+2}\right) \tag{2.4}$$

where $x = 2\pi a/\lambda$, ϵ = complex dielectric constant. The complex dielectric constant may be expressed as $\epsilon = m^2$ where m is the complex
index of refraction $m = m' - im''$.

Both absorption and scattering are affected by deviations from
sphericity but we shall only be concerned with absorption (and emission) by the particles. The absorption by spheroids may be readily
calculated in the Rayleigh approximation but it will be sufficient
for our purposes to present the result for axially symmetric particles.

Define the polarizability along the principal axes as $\alpha_{\parallel}$,
$\alpha_{\perp}$ where $\parallel$ and $\perp$ refer to the direction of the E field with respect to the particle axis. Then

$$C_{\parallel,\perp} = -4\pi k \, \mathrm{Im}\alpha_{\parallel,\perp} \tag{2.5}$$

where $\alpha_{\parallel}$ and $\alpha_{\perp}$ are defined for both oblate (flat) and prolate
(elongated) spheroids (van de Hulst, 1957).

We shall also have occasion later to use the absorption of
small inhomogeneous particles consisting of cores and mantles. The

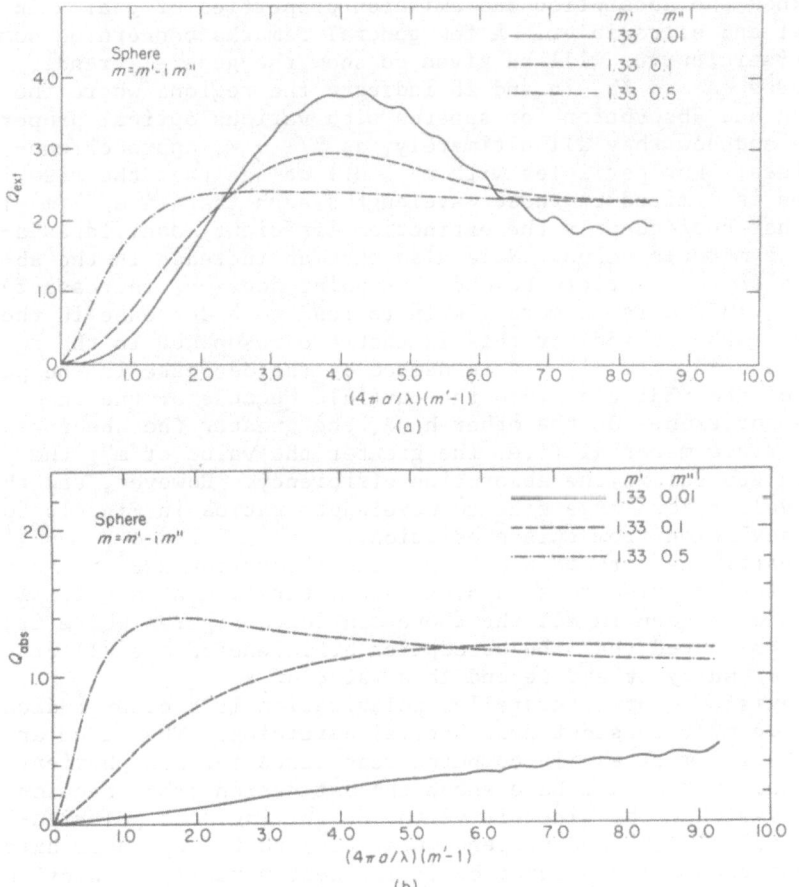

Fig. 1. a) Extinction efficiencies for spheres as a function of
absorptivity. The curves for m = 133 - 0.01 i and m = 1.33 - 0.1
i are characteristic of dielectric particles. The curve for m" =
0.5 is almost like that for a metal. b) Absorption efficiencies.
Note the asymptotic approach to a value close to one for short
wavelengths.

absorption of concentric spherical particles may be derived in the
Rayleigh approximation and the polarizability for such particles
is given here for completeness (van de Hulst, 1957).

$$\alpha = a^3 \frac{(\varepsilon_m-1)(\varepsilon_c+2\varepsilon_m) + (a_c/a_m)^3(2\varepsilon_m+1)(\varepsilon_c+\varepsilon_m)}{(\varepsilon_m+2)(\varepsilon_c+2\varepsilon_m) + (a_c/a_m)^3(2\varepsilon_m-2)(\varepsilon_c-\varepsilon_m)} \qquad (2.6)$$

where a_c and a_m are the inner outer radii and ε_c and ε_m are respectively the core and mantle dielectric constants.

The Rayleigh approximation may no longer be applied when we
wish to know the absorption and emission properties of grains in
the visual and ultraviolet. A few general remarks concerning some
characteristic curves will be given to show the general trends.

Figures 1a and 1b, 2a and 2b indicate the regions where the
extinction and absorption for spheres with various optical proper-
ties rise and how they all ultimately, as $a/\lambda \rightarrow \infty$, approach constant values. For particles with $m'' \lesssim 0.1$ we see that the rise in
extinction is limited to those wavelengths such that $(4\pi a/\lambda)(m'-1)$
$\lesssim 4$ and that subsequently the extinction efficiency oscillates a-
bout its asymptotic value. Note also that an increase in the absorptivity of the particle beyond this point does *not* increase the
extinction, indeed it is more likely to lead to a decrease in the
extinction. The reason for this is that the resonance in the region around $(4\pi a/\lambda)(m'-1) \simeq 4$ is damped by the decrease in the pen-
etration of the radiation into the particle because of the increased absorption. On the other hand, the greater the absorption
of the particle material (i.e. the greater the value of m'') the
greater in general is the absorption efficiency. However, see the
relative values for large size to wavelength ratios in Fig. 1b for
a slight deviation from this prediction.

For particles with $m'' \gtrsim 0.5$ *both* the absorption and extinction efficiencies rise to an approximate saturation at $x = 2\pi a/\lambda$
$= 1.5$. This is seen in all the above mentioned figures where it
is also to be noted that the absorption efficiencies are all of
the order of unity at and beyond this value of x.

The existence of interstellar polarization is a clear indication that we must consider nonspherical particles. The circular
cylinder is the most easily computed representative nonspherical
particle and in Fig. 3 I have shown the extinction cross section
for the cases when the electric vector of the incident polarization is parallel (Q^E) and perpendicular (Q^H) to the cylinder axis.
If incident unpolarized radiation is incident normally on a cylinder it appears, intuitively, that the E component should be more
effectively blocked than the H component. This turns out generally
to be the case as we see in Fig. 3 where the difference $Q^E - Q^H$ is
positive for a large range of $2\pi a/\lambda$. Since the E component of the
incident radiation is blocked more than the H component we deduce
that a set of such aligned cylinders would cause the radiation passing through to be partially linearly polarized *perpendicular* to
the alignment direction and would act like a sheet of polaroid

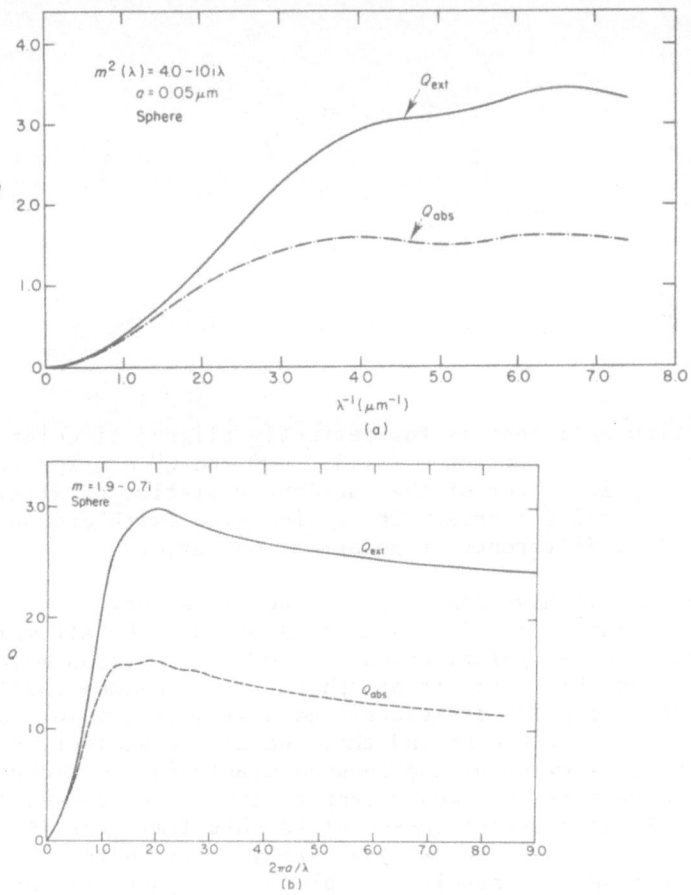

Fig. 2. a) Extinction and absorption efficiencies for a spherical particle of radius a = 0.05 μ with graphite-like index of refraction m^2 = 4-10 iλ. Both the extinction and absorption curves "saturate" at about x = 2πa/λ = 1. b) Extinction and absorption efficiencies for a silicate-type grain with index of refraction characteristic of its far ultraviolet values: m = 1.9-0.7 i. Note similarity of dielectric grains in the far ultraviolet to metallic grains in the visual.

(except not nearly so strongly). The effect is demonstrated schematically in Fig. 4.

Circular polarization is a somewhat more subtle phenomenon than linear polarization. It can, however, be made intuitively evident that, if one *starts* with polarized radiation, and sends it through a medium containing aligned particles whose axes are not exactly along or perpendicular to the incident polarization vector,

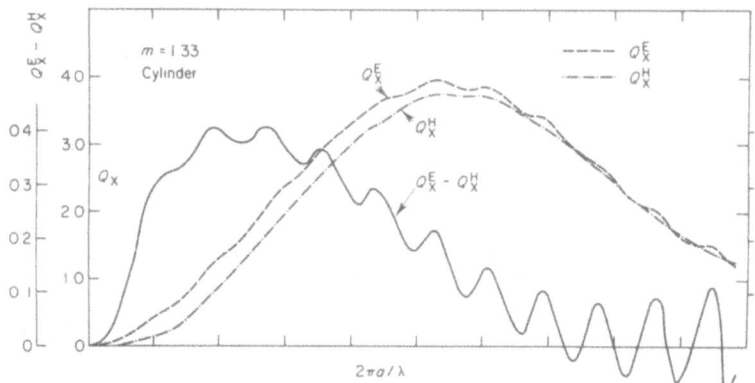

Fig. 3. Extinction efficiencies for perfectly aligned circular cylinders with index of refraction m = 1.33. Q^E and Q^H are for the electric and magnetic vector of the incident radiation parallel respectively to the cylinder axis. The solid curve (with ordinate to the left) is the difference or polarization curve.

the emerging beam will have some circular polarization.

First of all let us consider a tenuous medium like air which can *not* produce circular polarization. Since air contains molecules as scatterers which are either spherical or assumed unaligned the index of refraction (or dielelctric constant) of air must be isotropic; remember that $\epsilon = m^2$ and that the static dielectric constant, ϵ at $\lambda = \infty$, is given by the induced dipole in the medium. This is, as we have seen, the equivalent to the use of the Rayleigh approximation. If we restrict ourselves to this limit for illustrative purposes - which is actually entirely valid in the long wavelength limit - we see that if we replace the spherical particles by aligned particles whose dipole moment is different along orthogonal axes, the medium exhibits two different dielectric constants (and therefore indices of refraction) along orthogonal directions. Thus each plane polarized component of the transmitted light

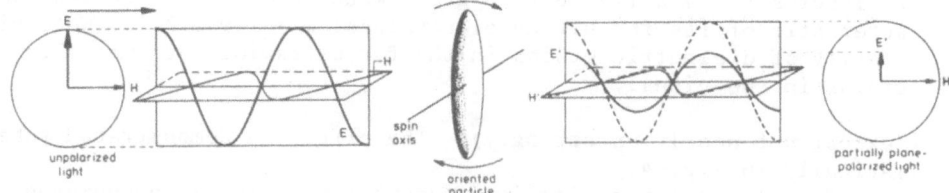

Fig. 4. The transformation of initially unpolarized light into partially plane polarized light by passage through a system of aligned particles. The E component is shown attenuated more than the H component by the spinning particles.

projected along and perpendicular to the alignment direction will undergo different phase shifts resulting, upon emergence and re-constitution, in circular polarization.

We shall now try to relate this circular polarization to the individual particle extinction cross sections. There is a funda-mental scattering theorem, known as the optical theorem, which states that the total cross section is given by the real part of the complex amplitude of the wave scattered by the particle in the forward direction (zero degree scattering angle). This implies that we can represent the full complex cross section efficiencies by $\bar{Q} = Q + iP$ where Q and P are the real and imaginary parts. We have already seen that the total cross section $C_{ext} = GQ$ (G = geo-metrical cross section) is the particle blocking area so that in terms of a plane wave it produces an attenuation for one particle per unit volume represented by

$$I = I_o \exp(-C_{ext}z) = |\psi|^2 \tag{2.7}$$

which implies a wave amplitude

$$\psi \sim \exp(-\tfrac{1}{2}C_{ext}z) \tag{2.8}$$

We know that there also exists a phase shift due to the index of refraction of the medium. This means that the wave must have the more general form

$$\psi = \exp(ikz + i\eta z)\exp(-\tfrac{1}{2}C_{ext}z) \tag{2.9}$$

where η is the phase shift per unit length or per particle imbed-ded in the medium. The implication of Eq. (2.9) is that if C_{ext} is the real part of the cross section then η must be the imaginary part of the cross section so that

$$\psi \sim \exp\{ik(1 + \frac{C_{Im}}{2k} + \frac{iC_R}{2k})z\} \tag{2.10}$$

The term in parentheses is explicitly displayed as an ordinary complex index of refraction of the medium whose real and imaginary parts are

$$m' - 1 = C_{Im}/2k \sim P$$
$$m'' = -(C_R/2k) \sim Q \tag{2.11}$$

The results shown in Eq. (2.11) are indeed precisely as derived on a more direct analytical basis (see van de Hulst, 1957).

The complex extinction efficiency for the sphere is shown in Fig. 5 just to demonstrate its general nature. However, in Fig. 6

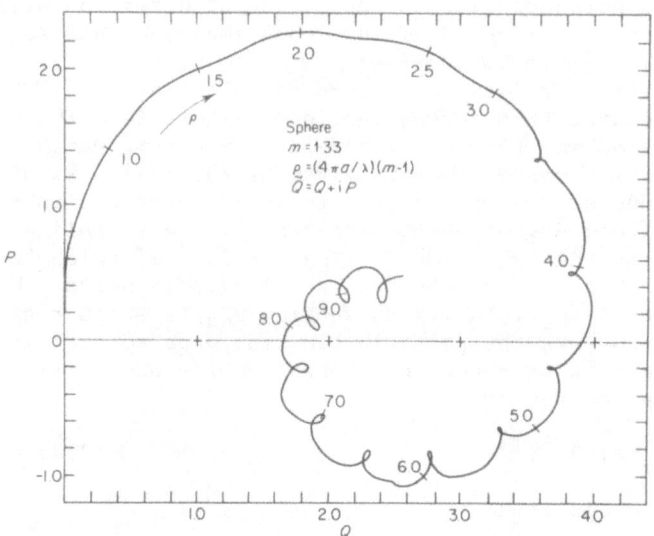

Fig. 5. The complex extinction efficiency $\bar{Q} = Q + iP$ calculated from Mie theory for a sphere with index of refraction m = 1.33. The running values of $\rho = 4\pi(a/\lambda)(m - 1)$ are shown as short lines along the spiral.

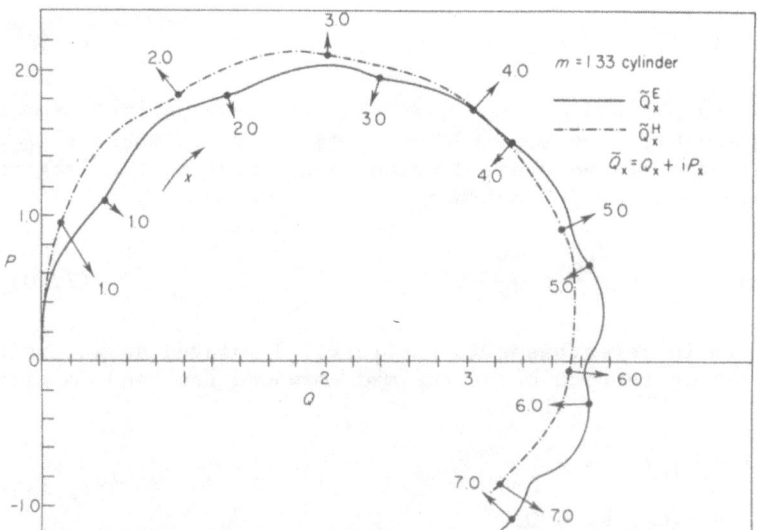

Fig. 6. Complex extinction efficiencies for E and H alignment of circular cylinders with m = 1.33. Running values of $x = 2\pi a/\lambda$ are indicated. Note that the x values advance more rapidly along the E spiral than along the H spiral.

the full story of extinction and linear and circular polarization is revealed by plotting the complex E and H extinction efficiencies for the infinite cylinder. For a given x (or a given λ for fixed size) the *average* of $\tilde{Q}$ projected on the real axis is the extinction. This becomes a maximum roughly at the right hand edge of the spiral. The linear polarization is the *difference* between the projected values of $\tilde{Q}^E$ and $\tilde{Q}^H$ on the real axis and we see that this difference tends to be a maximum for values of x at the top of the spiral. Finally, the difference in the phase shifts which is responsible for the circular polarization is the difference in the projections of $\tilde{Q}^E$ and $\tilde{Q}^H$ on the imaginary axis. It is evident that this difference tends to be zero for x values at the top of the spiral and is positive or negative on either side.

In a sense this completes the description of *how* the interrelationships of the three manifestations of nonspherical particles are produced. However, it is important to see whether it can also be demonstrated in such a way that we might have an idea of *why* the extinction and linear and circular polarization are interrelated as they are. The answer to this question turns out to be obtainable from a quite simple analytical scattering approximation. The starting point is the ray approximation for scalar wave scattering by spheres and a crude generalization which allows its application to the electromagnetic scattering by nonspherical particles.

The ray, or eikonal, approximation consists in following undeviated rays through the scatterer and coherently summing up these phase shifted rays to get their total effect. This assumes that the index of refraction of the particle differs but slightly from that for a vacuum; i.e. $(m-1) \ll 1$. In Fig. 7 we show that the initial wave front is modified by a spherical homogeneous particle to produce a non-planar wave front,

$$e^{ikz} \to e^{ikz + i\chi(t)} \tag{2.12}$$

where $\chi(t) = 2ka(m-1)\{1 - (t/a)^2\}^{1/2}$ is the phase shift at a dis-

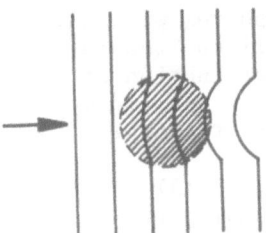

Fig. 7. Schematic of the distortion of a plane wave front passing through a spherical object with index of refraction close to unity. The depressions in the wave front are proportional to the phase advance inside the scatter.

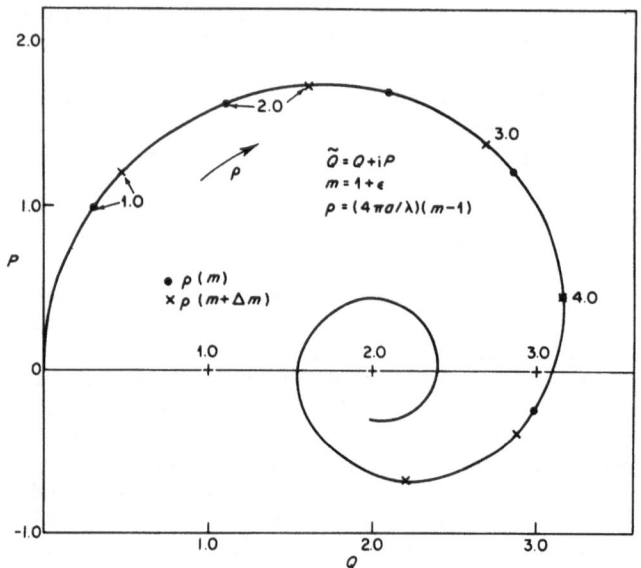

Fig. 8. Complex extinction efficiency computed from the ray approximation for the sphere (Eq. (2.13)). The dots and crosses along the curve are for spheres whose indices of refraction are m and m + Δm respectively.

tance t from the axis.

We present the final result of appropriate integration of the wave form Eq. (2.12) to give the complex extinction efficiency

$$Q = Q + iP = 4\{\tfrac{1}{2} + (i\rho)^{-1}e^{-i\rho} - \rho^{-2}(e^{-i\rho}-1)\}\qquad (2.13)$$

where, for simplicity, we have considered m to be pure real. Eq. (2.13) is shown in Fig. 8.

To demonstrate the well-known power of the approximation the values of Q from Eq. (2.13) are compared in Fig. 9 with those obtained by exact Mie theory computations for spheres with indices m = 1.33 and m = 1.66. The key to this comparison lies in the demonstration that to a good approximation the extinction efficiency of a particle depends on the *single* dimensionless parameter $\rho = (4\pi a/\lambda)(m-1)$ which is the phase shift of a ray traversing the diameter of the sphere.

We now generalize the result given in Eq. (2.13) to the electromagnetic scattering by a slightly nonspherical particle by letting the sphere be made of a slightly birefringent material, as illustrated in Fig. 10, such that the indices of refraction corresponding to the electric vector E of the incident light along and perpendicular to the symmetry axis are m_1 and m_2 respectively with $m_1 > m_2$. It is as if we were to build up the sphere with very small

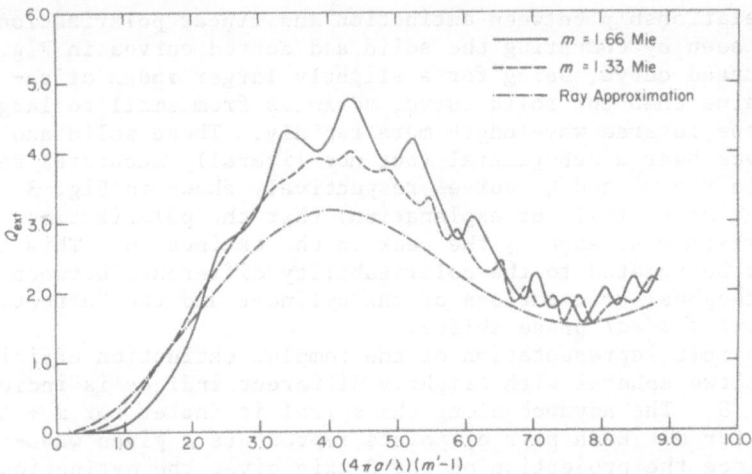

Fig. 9. Extinction efficiencies $Q = C/\pi a^2$ for spheres with indices of refraction m = 1.33, m = 1.66 and m = 1 + ε (ε << 1). The first two are calculated by Mie theory. The last is calculated from the ray approximation Eq.(2.13).

particles whose polarizabilities along orthogonal axes correspond to the original spheroid shape. This is clearly an incomplete way of going from the scalar wave sphere to the electromagnetic non-sphere case but as we shall see, it provides the insight which we are looking for.

In application we let all deviations from sphericity be infinitesimal so that $m_1 = m + \Delta m$ and $m_2 = m$ where $\Delta m \rightarrow 0$.

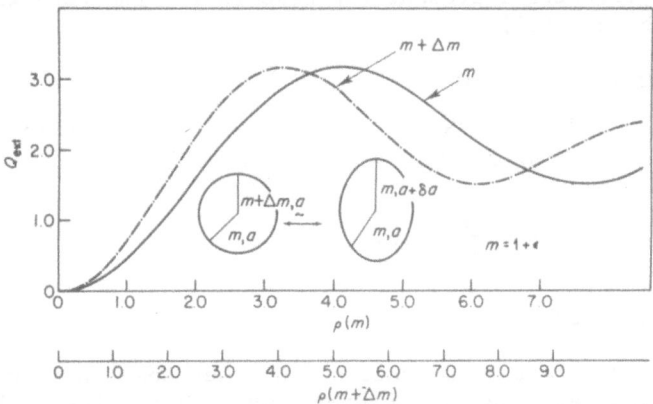

Fig. 10. Extinction efficiency curves for a birefringent sphere calculated from the ray approximation (Eq. 2.13). The curve for index of refraction m + Δm is used to simulate the extinction by a slightly elongated spheroid.

The relationship between extinction and linear polarization is readily seen by comparing the solid and dotted curves in Fig. 10. The dotted curve, being for a slightly larger index of refraction value than the solid curve, advances from small to large values of the inverse wavelength more rapidly. These solid and dotted curves bear a substantial (but not literally accurate) resemblance to the Q^H and Q^E curves respectively shown in Fig. 3 where it was noted (without explanation) that the polarization reaches a maximum at about $\frac{1}{2}$ the peak in the extinction. This is now seen to be related to the polarizability difference between the two orthogonal orientations of the cylinder and the "effectively" *different optical* phase shifts.

An explicit representation of the complex extinction efficiencies of the two spheres with slightly different indices is indicated in Fig. 8. The advance along the spiral is faster for $m + \Delta m$ than it is for m. Each pair of points represents a given wavelength. Since the projection on the Q axis gives the extinction we see that this reaches its first maximum at $\rho = 4$. Following the same arguments as given for the discussion of Fig. 6, the polarization is maximum at about $\rho = 2$ which is also where the crossover occurs in the predicted circular polarization produced by the difference in the projected pairs of points on the imaginary P axis.

We conclude that a unifying physical explanation of the qualitative interrelationships between the wavelength dependences of extinction, linear polarization and circular polarization by *dielectric* particles may be derived on the basis of the phase of the scattered radiation through the single parameter ρ.

The above discussion of the birefringent sphere analog to electromagnetic scattering could have been carried through analytically and, for completeness I present the results for the case where $\Delta m \to 0$ (see Greenberg, 1977).

Extinction coefficient $\kappa = Q$

Polarization coefficient $p = \Delta Q$ (2.14)

Phase cofficient $\Delta \epsilon = \frac{1}{2}\Delta P$

where $\Delta Q = \frac{dQ}{d\rho}\rho\left(\frac{\Delta m}{m-1}\right)$ and $\Delta P = \frac{dP}{d\rho}\left(\frac{\Delta m}{m-1}\right)$.

3. INTERSTELLAR DUST MODEL

This section consists first of an outline and then some annotated figures of observational results and some simple theoretical calculations which show how one can be led to a reasonably self consistent model of the dust grains and how they evolve. Even though (maybe because) the discussion is skimpy at least the structure of the theory is displayed.

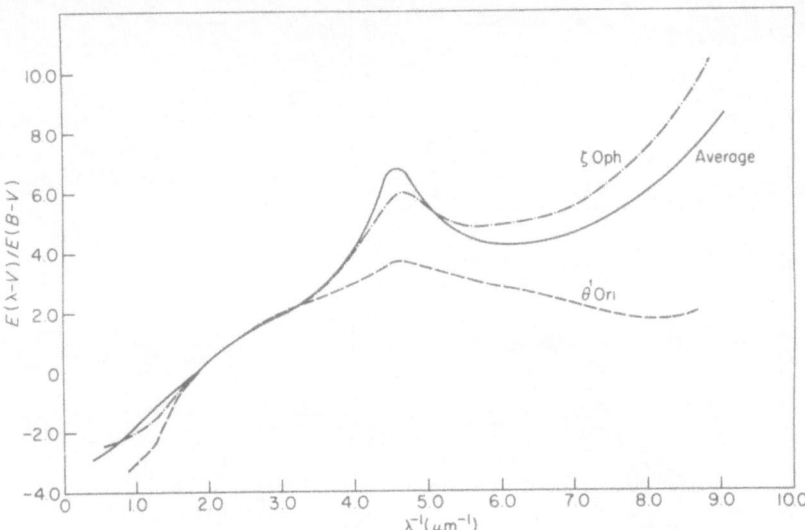

Fig. 11.   Observed (normalized) wavelength dependence of extinction from the infrared to the far ultraviolet.   The very exceptional star θ'Orionis is shown for comparison. (From Bless and Savage, 1972).

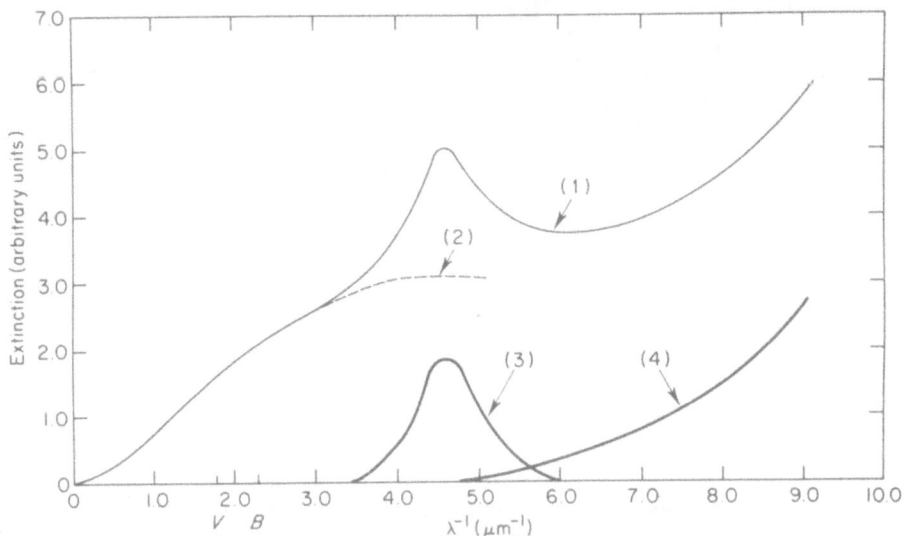

Fig. 12.   Schematic extinction curve.   The mean extinction curve (1) is arbitrarily divided into portions to which the major contributions are made by classical sized particles (2) and by very small particles (3) and (4).

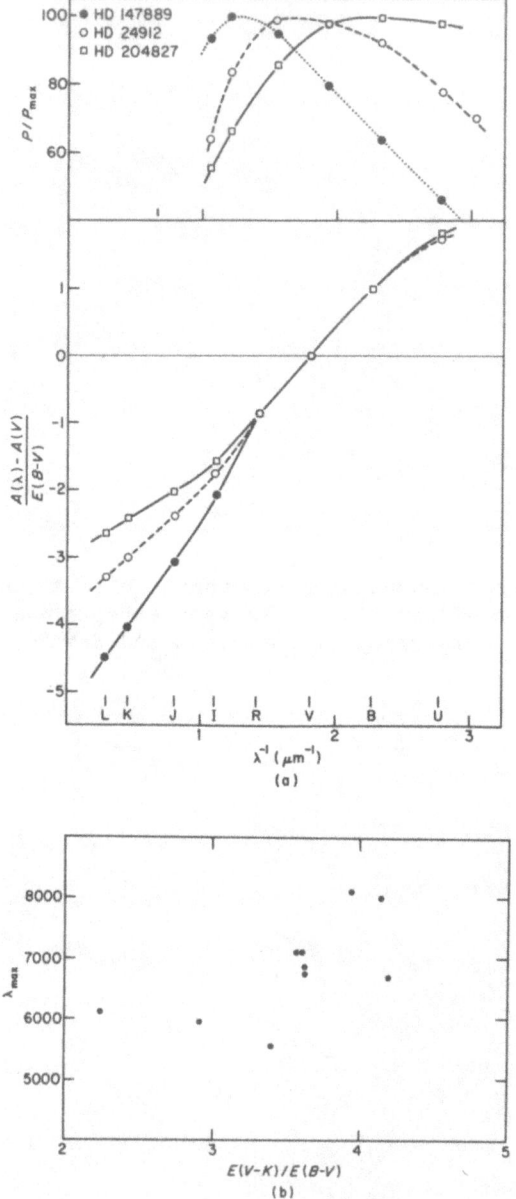

Fig. 13. a) Observations of the correlation of variations in the wavelength dependence of extinction and polarization (Serkowski et al., 1968). b) The value of the maximum of the polarization versus the total to selective extinction as given approximately by $E(V - K)/E(B - V)$. Observation in ρ Oph. (From Carrasco, Strom and Strom, 1973).

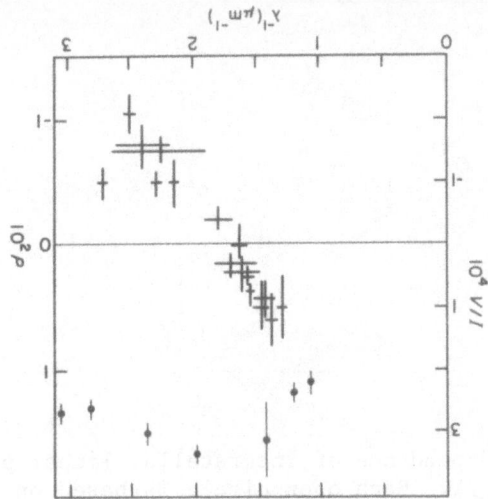

Fig. 14. Observations of linear and circular polarization for the star σ Scorpii (From Martin, 1974).

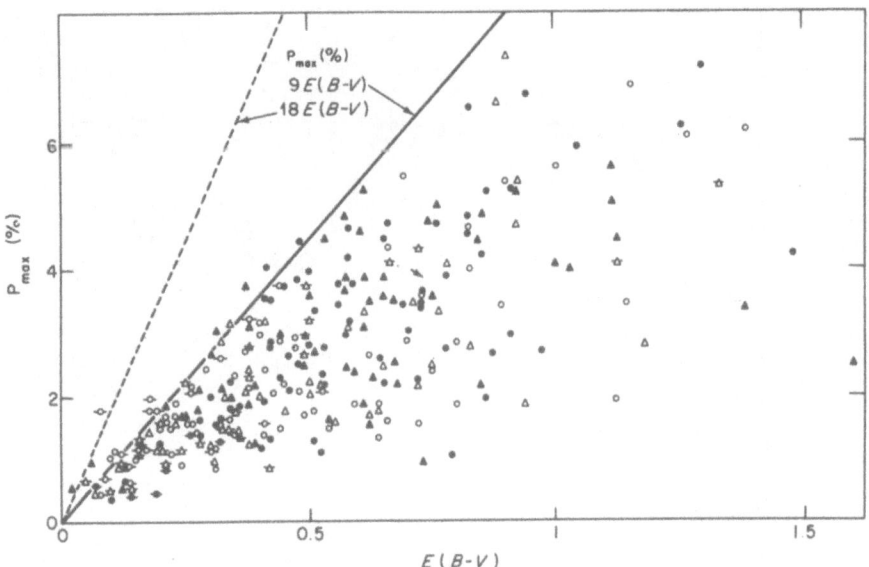

Fig. 15. Ratio of maximum polarization to color excess E(B-V). (From Serkowski, Mathewson and Ford, 1975). It is to be noted that P_{max} occurs at different wavelengths for different stars and therefore an inferred value of the ratio of polarization (at the visual wavelength) to the extinction at the same wavelength must take this into account. (See Fig. 16.)

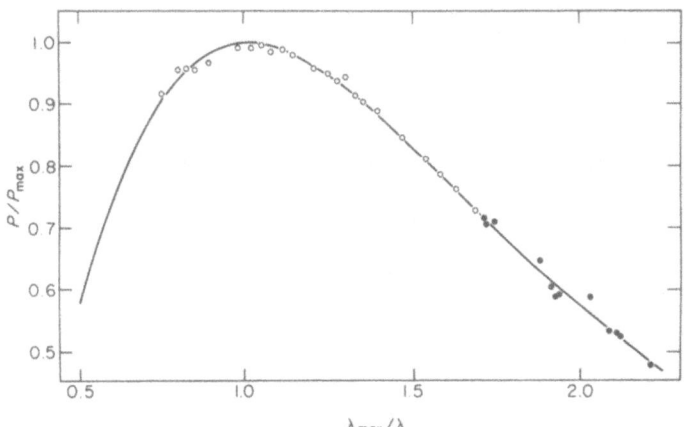

Fig.16. "Uniform" wavelength dependence of interstellar linear polarization: P/P_{max} versus λ_{max}/λ. Each open circle is based on
20 stars while each dot represents the observations of an individual star with a particular filter. (From Serkowski, Matthewson
and Ford, 1975).

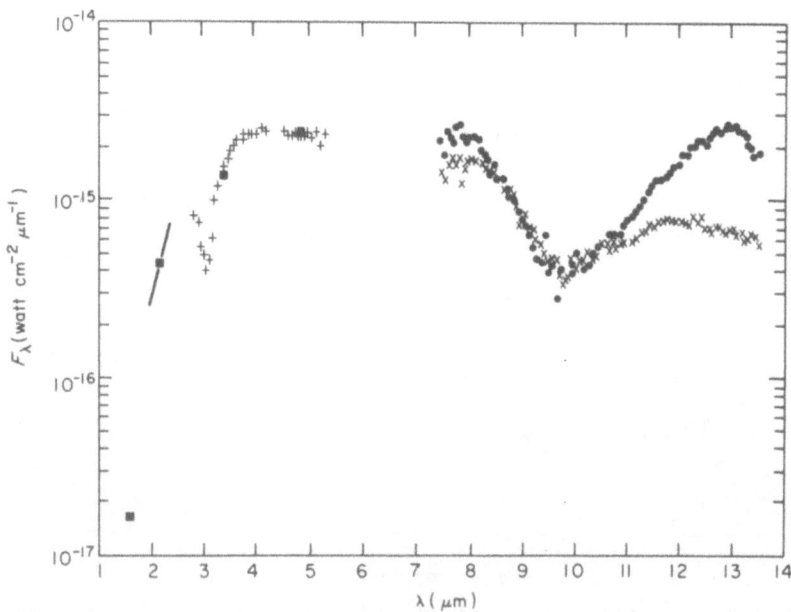

Fig.17. Spectrum of the Becklin-Neuegebauer star with wings embedded in the Kleinmann-Low infrared nebulae in Orion. The 3.07
μm band is identified as H_2O ice and the 9.7 μm feature is identified with a vaguely defined silicate. (From Gillet and Forrest,
1973).

3.1 Outline

Observation	Inference

Observation | Inference

a) $\Delta m(\lambda) \rightarrow \bar{a}_{\Delta m}$

Figs. 11,12.

For a given type of material and index of refraction the wavelength dependence of *extinction* leads to a characteristic size.

b) $P_L \rightarrow$ nonspherical particles

$P_L(\lambda) \rightarrow a_{P_L}$

(L ≡ linear)
Figs. 13a,13b.

For a given type of material and index of refraction the wavelength dependence of linear *polarization* leads to a characteristic size. If this size is different from that predicted by $\Delta m(\lambda)$ this is evidence that each is produced by different particles.

c) $\lambda_{P_{max}} \simeq \frac{1}{2} \lambda_{\Delta m_{sat}}$

$\bar{a}_{\Delta m} = \bar{a}_{P_L}$

$\lambda_{P_{max}} \simeq \lambda_{P_C} = 0$

(C ≡ circular)
Fig. 14.

This occurs if the particles producing extinction and polarization are the same and have an approximately constant index of refraction in the range $1 < \lambda^{-1} < 3$ μm^{-1} (i.e. are dielectric).

d) $P_L/\Delta m < 0.06$

Fig. 15.

The degree of alignment of the particles producing the polarization must be large enough to produce the observed maximum polarization relative to extinction. (The possibility that 0.06 may be an overestimate is under investigation).

e) P/P_{max} vs λ_{max}/λ is "uniform"

Fig. 16.
(see 3.2(b) below)

Essential constancy of the particle index of refraction from the near infrared to the near ultraviolet; i.e. dielectric material. Eliminates metal particles (graphite, iron, magnetite) contributing significantly to polarization.

f) variation of $\lambda_{P_{max}}$ correlated with variation of

$R = \dfrac{A(V)}{E(B-V)}$

(see 3.2(c) below)
Figs. 13a, 13b

Polarization and extinction produced by *same* particles which vary in size from region to region. This implies varying accretion of oxygen, carbon and nitrogen – the abundant condensible elements.

g) $A(V)/N_H$ = "const." =   The correlation of dust and gas shows that

$= 0.7 \times 10^{-21} \ cm^2$

they are generally well mixed. The *value* of the correlation constant shows how much *area* the dust grain must have relative to the hydrogen density. Combining this area with size from (a) leads to *volume* of *dust*.

h) Cosmic Abundance
   Table 1
   (see 3.2(c) below)

Limits the choice of dust models if the amount of an element is insufficient to supply the volume implied from (g); for example eliminates pure silicate grains or pure iron grains but allows for "ices" with significant excess which is required for accretion.

i) Infrared spectra
   9.7 μm and 3.1 μm
   interstellar ab-
   sorption or cir-
   cumstellar emis-
   sion.

The 9.7 μm absorption or emission implies a general silicate type (Si-O or perhaps MgO stretch). The 3.1 μm absorption probably indicates the OH stretch in *some* solid H_2O *but* could be produced by more complex molecules. Both the strength *and* the identification of these bands are not obviously interpreted (see 3.2 following and last section on irradiated dust mantle "ices").

Fig. 17.

Table 1

Selected Abundances of the Elements

Elements	Relative Number of Atoms
H	1
He	1.2×10^{-1}
C	3.70×10^{-4}
N	1.17×10^{-4}
O	6.76×10^{-4}
Ne	6.3×10^{-4}
Na	1.3×10^{-6}
Mg	0.34×10^{-4}
Si	0.32×10^{-4}
S	2.8×10^{-5}
Ca	1.6×10^{-6}
Fe	0.26×10^{-4}

3.2 Illustrative Calculations and Discussion

a. Sizes. Given the chemical composition and the visual index of
refraction of the grain material, the most defined property is
size.

Grain materials may be classified as dielectrics or metals.
Dielectrics have small values of m" in the visual but substantial
values of m" in the ultraviolet. Metals(graphite, iron, magnetite) have substantial (and wavelength dependent) values of m" in
the visual.

The observed extinction saturates at about $\lambda^{-1} \simeq 5$ μm^{-1} (ignoring for the moment the 4.6 μm^{-1} extra "hump") before *again* rising beyond $\lambda^{-1} \simeq 6$ μm^{-1}.

Since we have shown that once the extinction saturates for a
given particle size it can no longer rise we must assume that
there are two basically different particle sizes in interstellar
space. We call these the classical particles and the very small
particles. First consider the classical particles and then the
bare ones. We note that the particles producing the 4.6 μm^{-1} hump
must be quite small ($\lesssim 0.02$ μm) but they may constitute only a
fraction of all the bare particles.

For dielectric spheres (see Fig. 1) the saturation occurs at
around $\rho = (4\pi a/\lambda)(m'-1) \simeq 4$. Inserting the observed saturation value of $\lambda^{-1} = 5$ μm^{-1} into the equation for ρ and using $m'_{ices} = 1.33$
(ices) and $m'_{sil} = 1.66$ (silicates) we find respectively $a_{ice}^{(class)} \simeq$
0.19 μm and $a_{sil}^{(class)} \simeq 0.095$ μm as representative classical (dielectric) particle sizes. For metals we use the saturation value
of $x = 2\pi a/\lambda = 1.5$ to correspond to $\lambda^{-1} \simeq 5$ μm^{-1} to give $a_{metal}^{(class)}$
$\simeq 0.05$ μm. More exact calculations and use of size distributions
leads to very similar results.

In the far ultraviolet all particles, whether dielectric or
metallic have large m" and therefore act similarly. Since the far
ultraviolet part of the extinction is still rising with upward curvature at $\lambda^{-1} \simeq 10$ μm^{-1} we see from Figs. 2a and 2b that this must
correspond to $2\pi a/\lambda \simeq 0.5 - 0.75$ leading to particles with characteristic radii $a_b \simeq 0.01$ μm; i.e. about 1/20 as large as the classical particles.

b. The uniform polarization law. A comparison of the theoretical
polarization by aligned cylinders with the observed polarization
can be similarly made and produces comparable size values. A significant difference however exists between the dielectric and metallic particle polarization as one changes their sizes. First we
see that the "shape" of the visual polarization curve for dielectric (m = const.) particles is a function of a/λ alone (see Fig.3),
and therefore, as a is increased or decreased, the shape of P/P_{max}
vs λ_{max}/λ is a constant. On the other hand the introduction of a
substantial change in m across the visual band (as is the case for
metals) destroys this invariance.

c. Depletion of elements. Suppose all the interstellar dust is in
the form of spheres of a fixed characteristic radius $\bar{a}$. For mass
estimates this is acceptable. The visual extinction per unit
length is

$$\Delta m(V) = A(V) = n_d \pi \bar{a}^2 Q(V).$$ (3.1)

where n_d = number of dust particles per unit volume and $Q(V)$ is the
extinction efficiency factor which is $\simeq 1$.
    The mass density of grains is

$$\rho_d = n_d \frac{4}{3} \pi \bar{a}^3 s$$

$$= \frac{4}{3} A(V) s \bar{a}/Q(V),$$ (3.2)

where s = grain material specific density.
    Using $A(V) = 0.7 \times 10^{-21} N_H$ cm^2 in Eq.(3.2) we get

$$\frac{\rho_d}{n_H} = \frac{4}{3} \frac{\bar{a}}{Q(V)} (0.7 \times 10^{-21}) s.$$ (3.3)

    But ρ_d is simply the number density of molecules of the grain
material times the molecular weight $n_M M m_H$ so that

$$n_M/n_H = \frac{4}{3} \frac{\bar{a} s}{Q(V)} \frac{(0.7 \times 10^{-21})}{M m_H}$$ (3.4)

where m_H = mass of hydrogen and M = molecular weight of the grain
material.
    Equation (3.4) gives the *required* number of molecules in the
dust relative to hydrogen and it must be less than that given from
cosmic abundance in order to be acceptable for a given grain model.
    If we had grains made of pure H_2O of radius a = 0.2 μm, Equa-
tion (3.4) would give us $n_o/n_H = 5 \times 10^{-4} <$ C.A..
    On the other hand, 0.1 μm radius grains of olivine, $MgSiO_3$, as
a representative silicate give $n_{Si}/n_H = 6 \times$ C.A. which means that
more Si is required than is available. Similarly pure iron grains
of radius 0.05 μm give $n_{Fe}/n_H \simeq 6 \times$ C.A. which is again unreason-
ably large.

d. A simple unified analytical model for extinction, linear polarization and circular polarization. The ray approximation described
in Section 2 provides a remarkably simple basis for understanding
the interrelationships between the three wavelength dependencies:
extinction, linear polarization, circular polarization. In the
next section a detailed model calculation is compared with the set

of observations of all three for one star.

e. Average grain model. The basic form of the classical sized grains suggested by the observations is a core-mantle structure with the core consisting of a silicate material and the mantle of a generalized material consisting of complex combinations O, C, N and H. The more refractory cores are probably ejecta from cool evolved stars. The mantles are accreted by the cores in interstellar clouds and subsequently modified by the interstellar ultraviolet radiation to become a heterogeneous mixture of complex molecules which is substantially more refractory than the ices consisting of water, methane and ammonia. Using a cylindrical model with size distribution for mantles a_m on a single core size a_c

$$n(a_m) = \exp\{-5(\frac{a_m - a_c}{a_i})^3\} \qquad (3.5)$$

and indices of refraction "characteristic" of silicates and ices one arrives at mean classical and bare particle dimensions

$$\left.\begin{array}{l} a_i = 0.2 \ \mu m \\[2em] a_c = 0.05 \ \mu m \end{array}\right\} \rightarrow \bar{a}_m = 0.12 \ \mu m \qquad (3.6)$$

bare $a_b = 0.005 \ \mu m$ (assumed to be silicate, see later). These sizes produce $A(V)/E(B-V) = 3$.

The mean space density of such particles is (see Greenberg and Hong, 1974; Hong, 1975):

$$n_{c-m} = 9.5e^{-1} \times 10^{-13} n_H$$

$$n_b = 4.6 \times 10^3 en_{c-m} \qquad (3.7)$$

where e is the elongation of the classical particles.

It is interesting to note that the amount of silicate material in the cores is somewhat greater (by about a factor of two) than that which can be produced by M stars unless it is concentrated to the galactic plane relative to the M star distribution.

Using the above model for the dust and applying it to the observation of ζ Oph we find that the total number of accountable atoms of O, C and N along the line of sight as either atom + ions (Copernicus results) or in dust is only about 50% of those predicted by cosmic abundance. Since the most abundant molecule, CO, supplies at most the order of an additional 20% to C, 10% to O and nothing to N, there is implied a substantial gas fraction in undetected molecules forms - probably mostly quite complex. This sea

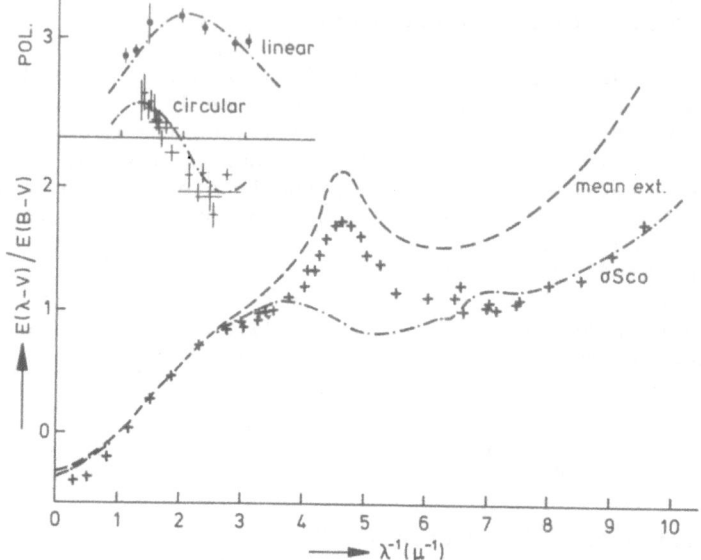

Fig. 18. Extinction, linear polarization, and circular polarization calculated from a bimodal grain model and compared with the
observations of these quantities for the star σ Sco. (From Hong,
1975).

of condensible gaseous material is actually required to produce
the growth of grains as we see them in dense clouds relative to
the average.

A star for which extinction, linear and circular polarization
are all observed is σ Sco. The model representing this is given
by (Hong, 1975; Greenberg, 1977):

$$
\left.
\begin{aligned}
a_i &= 0.26 \ \mu m \\
a_c &= 0.05 \ \mu m \\
a_b &= 0.005 \ \mu m
\end{aligned}
\right\} \rightarrow \bar{a}_m = 0.15 \ \mu m \qquad (3.8)
$$

The comparisons between observations and theory for σ Sco are
shown to be rather good in Fig. 18 where we have had to use slightly larger than average classical particles (see section f, next).

f. Evolution of grains. Clouds with densities $\geq 10^3$ cm^{-3} whether
dark or H II regions (not too close to the exciting star) have observationally implied larger than average dimensions for the classical sized particles. On the other hand, the extinction (wavelength dependence in the far ultraviolet) by the very small particles, appears to imply that their sizes (but not numbers) are relatively constant. The differences in the relative extinction con-

tributions by the classical and bare particles in the far ultraviolet seems to be easily accounted for by a combination of varying sizes of the larger particles with relative separation of the
bare particles due to the radiation pressure differential between
the two in the presence of strongly ultraviolet emitting stars.

For example, θ' Orionis has a wavelength dependence of extinction which bends over at a longer wavelength than average, thus
implying larger than average classical sized particles. At the
same time the far ultraviolet extinction is substantially reduced
relative to the visual extinction. If the classical sized particles are only 1.5 times larger than average in front of θ'Orionis
we can already account for a factor of ~ 2 in this ratio relative
to the average. Suppose, in addition, the bare particles have
been pushed outward from the star so that they fill a sphere of
radius $R + \Delta R$ while the core-mantle particles occupy a sphere of
radius R. (In addition to this separation we may expect that *both*
classical sized and bare particles are excluded from a region
close to the star but this small hole can be neglected in making
the relative comparisons). Because of the relative redistribution
of the bare particles and core-mantle particles, the line of sight
number density ratio is changed from $\dfrac{n_b}{n_{cm}} \dfrac{R}{R}$ to $\dfrac{n'_b}{n_{cm}} \dfrac{(R+\Delta R)-\Delta R}{R}$,
where $\dfrac{n'_b}{n_b} = \dfrac{R^3}{(R+\Delta R)^3-(\Delta R)^3}$. Thus a shift of only 20% in R ($\Delta R/R =$
0.20) produces an "effective" line-of-sight reduction in the num-
ber of bare particles by a factor of 1.7 (Greenberg and Hong, 1974
a). This factor combined with the factor due to larger classical
particles can account for a reduction of the "apparent" UV extinction (as commonly normalized) by as much as 1/3.

In the ρ Oph Cloud the visual extinction and polarization can
be explained by grains whose size implies very close to total depletion of the available condensible atoms O, C and N. It is notable that the heavy atom depletion observed in σ Sco is greater
than average consistent with the fact that the grain mantles have
to be thicker than average to produce the observed wavelength dependences of extinction and linear and circular polarization. At
T_d = 10 K the sticking coefficient for all atoms and molecules,
including even CO, which has a high vapor pressure, should be very
close to unity. Photodesorption on the classical particles is a
minor effect because energy is readily absorbed in the bulk of the
grain. The time scale for accretion of condensible species from a
cloud (*not* the grain growth time) may be shown to be $\tau_{ac} < 2.6$ x
$10^9/n_H$ yrs. By comparison the Jeans time (free fall) is $\tau_{ff} =$
4 x $10^7/(n_H)^{1/2}$ yrs which is already longer than τ_{ac} at a density
$n_H = 10^4$ cm^{-3} (see Greenberg, 1977). The maximum mean mantle dimension consistent with cosmic abundance is $\bar{a}_m = 0.21$ μm.

On the bare particles desorption by various mechanisms is
growth limiting. When a photon is absorbed the absorption occurs
either at the surface or within the grain. If the specific heat
of the grain material is low (low concentration of imperfections)
each photon induces a temperature spike to T = 30-100 K which, in

most regions, occurs much more frequently than the atom or molecule collision thus preventing sticking. If there is a high concentration of impurities the specific heat is raised and the temperature spikes are damped. However, in this case, it may be
shown that very small particles with a < 0.01 μm the number of
surface imperfections equals the number of volume of imperfections
so that ultraviolet produced photons are likely to be absorbed on
the surface and shake off any existing physisorbed species(Greenberg, 1977).

We are thus led to the following ranges and limitations for
interstellar particles

classical sized

no mantles: $\bar{a}_m = a_c \simeq 0.05$ μm: R <<3

maximum mantles: $\bar{a}_m \simeq 0.21$ μm: R>3

bare

$a_b \simeq 0.005$ μm (if silicates)

$a_b \simeq 0.02$ μm (if graphite)

(not discussed in detail here).

For $a_b \simeq 0.005$ μm (silicates):

$$\frac{n_b}{n_{c-m}} \left\{ \begin{array}{l} \simeq 4000 \text{ e: average} \\ \simeq 3000 \text{ e: near hot star} \end{array} \right.$$

e = elongation of classical
    particles

with the proviso that $\int n_{c-m} dx dy dz \sim \int n_b dx dy dz$ for the total cloud,
i.e. the *total* number of particles *relative* to each other is constant except, of course, *very* near hot stars where both core and
bares may evaporate.

The cores and bare particles (of whatever type) are resistant
to disruption except at temperatures > 1000 K. The mantle material of modified ices is substantially more refractory than simple
ices. It is relatively resistant to sputtering because of larger
sized molecules and has an evaporation temperature in the 300-400
K range rather than 100 K.

4. OPTICAL PROPERTIES OF GRAIN MATERIALS

We are predominately interested here in the infrared properties of
grains. However, it is useful to summarize briefly the expected
visible and ultraviolet properties. In both of these and the infrared there are substantial uncertainties which are not yet resolved either observationally or theoretically (e.g. in the laboratory).

The indices of refraction in the visible for dielectric materials are almost real and constant. However even a small value
of m" in the visible may contribute substantially in the determin-

ation of the grain temperature. Because "pure" ices and earth or lunar type silicates are often used as representing mantle and core materials some of the consequences must be treated as preliminary. Thus the names of the substances considered below are to be taken as generic rather than literal. Ice means accreted mantle material, silicate stands for core or bare particle material, graphite stands for a possible source of the 2200 Å hump in the extinction. I will note some effects of using more realistic grain materials.

4.1 Outline of Optical Properties: Indices of Refraction

Material	Used Here	Comment

a. The "visible" range, $1 < \lambda^{-1} < 3\ \mu m^{-1}$.

ice	m from Bertie et al., 1969, for H_2O ice, with and without m" = 0.02 *or* m = 1.33 with or without m" = 0.02	m for complex molecule ice mixture is probably more absorbing in the visual than H_2O, at least for $\lambda^{-1} > 2.5$ μm^{-1} because large molecules have a wider distribution of absorptivities as a result of more types of chemical bonding.
silicate	m from Huffman and Stapp, 1971 (Olivine) *or* m = 1.66	Probably should add a small but significant m" to represent highly amorphous mixture. This appears to be required in order to get a good representation of the IR spectrum from cool stars (Bedijn, 1977).
graphite	$m^2 = 4 - 10i\lambda$ *or* m from Taft and Philipp, 1965	?

b. The ultraviolet, $\lambda^{-1} > 3\ \mu m^{-1}$.

ice	m from Greenberg, 1968	Probably should have a smooth total ultraviolet absorption from the multitude of components.
silicate	m from Huffman and Stapp, 1973	Probably *less* absorption in near ultraviolet than that measured by H&S because of their including internal

scattering in derived m".

graphite $m^2 = 4 - 10i\lambda$

c. Near infrared, 1 μm < λ < 25 μm.

ice see Fig.21	m from Bertie et al. Note strong absorption at 3.1 μm and substantial absorption from 10 μm to 20 μm broadly *peaked* at ∿ 12 μm *or* m = 1.33 - 0.1i	m due to complex mixture would be similar but 3.1 μm band certainly differs from OH stretching mode in pure H_2O and is possibly even due to a different molecule (see last section).
silicate see Fig.19	m from Steyer and Huffman, 1974, (Olivine) strong absorption around λ ≃ 10 μm (not 9.7 μm) *or* m from Launer, 1955 (see Greenberg, 1973) with *medium* absorption at λ ≃ 10 μm *or* Lorentzian shape absorption at λ ≃ 10 μm	In amorphous or generalized silicate 9.7 μm band due to Si—O and/or Mg—O stretch. Basaltic glass has absorption at 9.5 μm (Pollack et al., 1973).
graphite	$m^2 = 4 - 10i\lambda$	see comment on far IR.

d. Medium to far infrared, λ > 25 μm.

ice see Table 2	m from Bertie et al., 1969 (Note m" ∿ $\lambda^{-1\cdot6}$ in region 100 < λ < 200 μm *or* Lorentzian ice (last absorption at λ ≃ 45 μm, m" ∿ λ^{-1} in *far* IR) extrapolated from Bertie et al. ice λ > 25 μm.	Complex molecular composition must have different absorption. Perhaps larger toward submillimeter region because of high internal damping. At sufficiently long wavelength, as determined by the *last* absorption, we must have m" ∿ λ^{-1}.
silicate	m from Pollack et al., 1973, ε_2 ∿ λ^{-1} for λ > 100 μm (extrapolated from last oscillator)	Perhaps qualitatively reasonable.
graphite	$m^2 = 4 - 10i\lambda$, ε_2 ∿ λ, λ ≥ 10 μm	ε_2 ∿ λ is characteristic of any metal at far infrared wavelength. This should be substantially temperature dependent in the sense that the coefficient of λ in m^2

graphite   is smaller at higher temper-
                                            atures (see Greenberg, 1968).

Table 2

Complex Indices of Refraction for H_2O ice and silicate (extrapolated), $\lambda > 25$ μm.

Wavelength (μm)	Silicate		H_2O	
	m'	m''	m'	m''
25.0	2.31	0.59	1.33	0.0223
33.3	2.23	.17	1.12	0.139
40.0	2.15	.097	.09	.450
41.7	2.14	.0891	.00	.557
43.5	.13	.0794	.10	1.250
43.7	.13	.0794	.22	1.310
44.4	.12	.0759	.65	1.140
45.5	.12	.0750	.75	0.864
47.6	.12	.0654	.81	.640
50.0	.11	.0610	.76	.497
52.6	.10	.0562	.77	.470
55.6	.10	.0501	.72	.320
58.8	.09	.0447	.65	.452
62.5	.08	.0398	.76	.523
71.4	.08	.0335	.92	.352
83.3	.07	.0275	.91	.173
100	.07	.0217	.84	.122
125	.07	.0170	.81	.107
167	.06	.0123	.82	.0834
250	.06	.0079	.82	.0358
333	2.05	.0054	.79	.0239

5. FAR INFRARED ABSORPTION ($\lambda \geq 25$ μm) BY SMALL GRAINS

I shall divide the following discussion between spherical and nonspherical particles with a major portion devoted to the spherical
case which generally provides the effects of major astrophysical
interest. The Rayleigh approximation (Eq. (2.4)) is clearly appli-
cable since, for a ≤ 0.2 μm and $\lambda \geq 25$ μm, the value of $x = 2\pi a/\lambda$
< 0.05.

5.1 Spheres - Homogeneous and Core-Mantle

In the previous section the indices of refraction in the far infrared were shown generally to depend on λ in a complicated way which only at very long wavelengths may have a simple representation.

Table 2 contains, in addition to the complex indices of refraction of basaltic silicate and H_2O ice, the absorption (extinction) and scattering efficiencies of concentric spherical particles with 0.05 µm silicate core and mantles of thickness 0, 0.05, 0.15, 0.25 µm. Thus for mantle thickness zero we have the absorption by the silicate alone, and for mantle thickness even as small as 0.05 µm (close to the interstellar mean of 0.07 µm) the absorption is dominated by the ice. This is not surprising since its volume is 7 times larger. This means that the core-mantle particles act, in the infrared, like pure mantle particles.

The absorption formula for homogeneous spheres (Eq.(2.4)) reduces to

$$C_{abs} = \frac{18\pi}{\lambda} V \left\{ \frac{\varepsilon_2}{(\varepsilon_1+2)^2 + \varepsilon_2^2} \right\} \qquad (5.1)$$

which takes on relatively simple forms in the long wavelength limit both for simple dielectrics and metals.

For dielectrics $\varepsilon_2 \sim \lambda^{-1}$ and $\varepsilon_1 \sim$ const while for metals $\varepsilon_2 \sim \lambda$, $\varepsilon_1 \sim$ const so that one gets the same *asymptotic relation* for both dielectrics and metals

$$C_{abs} \sim \lambda^{-2}; \quad \lambda \to \infty \qquad (5.2)$$

However, it is important to note that this is only the asymptotic limit and, that generally speaking, the absorption by dielectrics is greater than that by metals. Thus, at $\lambda = 333$ µm the .05 µm radius silicate or ice absorption efficiencies are 0.71 x 10^{-5} and 0.35 x 10^{-4} while the graphite absorption efficiency is 0.34 x 10^{-5} = 1/10 that of ice.

While Eq.(4.2) applies to the millimeter region of the spectrum it must be used with caution towards shorter wavelengths – even those as long as 300 µm!

5.2 Spheroids

It is well known that the long wavelength absorptivity (emissivity) of elongated particles is greater per unit volume than that for spheres. This fact has been, on occasion, assumed to imply unjustifiably remarkable properties of needle shaped metallic (conducting) particles. I have shown elsewhere that the really interesting properties for elongated (prolate) particles occur with the simultaneous restrictions $L|m^2-1| \lesssim 1$ and $m'x \ll 1$ (modified Rayleigh criterion taking into account the finite time for a light signal to traverse the particle). Combining these two restrictions leads to the fact that the maximum emissivity occurs for a particle of thickness $2b = 0.006$ µm at the rather modest elongation of $e = 6 =$

a/b and that the wavelength at maximum emission is only λ_{max} = 2.3 μm. The details of this calculation are a little beyond the scope of this lecture (Greenberg, in preparation).

The differences in absorption (emissivity) for *aligned* non-spherical particles leading to the possibility for observing polarized radiation have long been recognized as a diagnostic of the interstellar or circumstellar medium.

6. GRAIN TEMPERATURES

In the majority of situations, the temperatures of interstellar grains are determined exclusively by the balance between the rate at which electromagnetic radiation is absorbed from the ambient field and emitted by the particle. This is expressed by the equation

$$_0\int^\infty \epsilon(\lambda)R(\lambda)d\lambda = \,_0\int^\infty \epsilon(\lambda)B(\lambda,T_d)d\lambda, \qquad (6.1)$$

where $\epsilon(\lambda)$ is the emissivity of the particle at the wavelength λ, $R(\lambda)$ is the radiant energy distribution of the environment and $B(\lambda,T_d)$ is the Planck distribution at the temperature of the grain. In the case of spheres, the emissivities are the absorption efficiencies, $\epsilon(\lambda) = Q_{abs}(\lambda)$. For particles of other shapes one should use for $\epsilon(\lambda)$ the average of the absorption efficiencies over particle orientation and polarization. A black body has ϵ = 1 for all wavelengths. For spherical particles of arbitrary size and optical properties, Eq.(6.1) must be solved numerically using Mie Theory for the values of $Q_{abs}(\lambda)$.

6.1 Temperatures in the General Interstellar Medium

Even with a complete knowledge of the grain optical properties there remains the uncertainty of $R(\lambda)$ in various interstellar regions. Classically, calculations have been performed using as a starting point for the average interstellar region the Eddington field $R(\lambda) = WB(10.000K)$, where W is a dilution factor $W = 10^{-14}$. This leads to the well known result that dielectric interstellar grains have characteristic temperatures in the 10 K range. A few illustrative simple calculations are given to show how this comes about.

For a black body (ϵ = 1) and $R(\lambda) = WB(T_R)$, Eq.(6.1) reduces to

$$WT_R^4 = T_d^4, \qquad (6.2)$$

so that for T_R = 10.000 K and $W = 10^{-14}$, T_d = 3.16 K = T_{BB}.

It is clear from the absorption properties of interstellar sized small grains that ϵ for the short wavelengths is larger than for the wavelengths in the far infrared or even millimeter so that we must have $T_d > T_{BB}$.

A way of seeing quickly how much affect this has is to re-place Eq.(6.1) by

$$\bar{\varepsilon}_{abs}(a,T_R)u_R = \bar{\varepsilon}_{em}(a,T_d)u_{T_d}, \qquad (6.3)$$

where the energy density in the dust grain is $u_{T_d} = 4.73 \times 10^{-2}T_d^4$ ev cm^{-3}, and where $\bar{\varepsilon}_{abs}(a,T_R)$ and $\bar{\varepsilon}_{em}(a,T_d)$ are mean absorption efficiencies over the wavelengths where the radiation is absorbed from the medium and emitted by the dust. Suppose we use as a mo-del of the dust one whose size is such that $2\pi a/\lambda \gtrsim 1$ for $\lambda < 2000$ Å (i.e. a > 0.03 μm) and no absorption from longer wavelength in the radiation field (*pure* H_2O ice). Then $\varepsilon_{abs} \approx 1$ and $u_R = \int_0^{2000\text{Å}} u_\lambda d\lambda = 0.0316$ ev cm^{-3}. Then

$$T_d = (\frac{0.032}{0.047})^{1/4} (\bar{\varepsilon}_{em})^{-1/4} \approx (\bar{\varepsilon}_{em})^{-1/4} \qquad (6.4)$$

(absorption only in the ultraviolet)

If we further assume that the value of $\bar{\varepsilon}_{em}$ is approximately that corresponding to the wavelength at the maximum in the black body emission curve (actually the wavelength should be taken a bit larger), then using $\lambda_m T_d = 0.29$ cm deg we have $\bar{\varepsilon}_{em} \approx \varepsilon(a,0.29/T_d)$. At a *guess*, $T_d = 10$ K so that for, say, a = 0.1 μm from Table 3, $\bar{\varepsilon}_{em} \approx 0.8 \times 10^{-4}$ ($\lambda = 290$ μm) which when put into Eq.(6.3) gives $T_d \approx 9.6$ K, consistent with our guess.

This kind of circular argument may not be the best way to get an accurate result but it is useful for obtaining first approxima-tions and, in any case, is useful for demonstrating how much warm-er small particles become than black bodies. A detailed Mie Theo-ry calculation for 0.1 μm ice grains leads to $T_d = 14$ K, where, however, a value of m" = 0.05 in the visual (rather than m" = 0) is used so that there is as much absorption in the visual as in the ultraviolet. If m" had been chosen to be zero the temperature would have again turned out to be ≈ 10 K.

6.2 Temperatures in Clouds

In general, the interior of a dust cloud may be receiving radia-tion not only from outside the cloud (the general interstellar ra-diation field) but also from a source within the cloud. If there are no interior sources of energy, the grain temperatures within the cloud decrease with increasing depth because the outside ul-ultraviolet radiation field is attenuated by the grains themselves and converted to far infrared which is poorly absorbed by the grains. This effect is the basis of a very simple radiative transfer calculation which is shown to produce a reduction from the 10 K range by only a few degrees, with a value of $T_d \sim 8$ K being likely at the center of a spherical cloud with 10 magnitudes of visual extinction (Greenberg, 1971).

If there is an energy source at the "center" of a cloud one must treat the problem again by radiative transfer methods but here the problem is more complicated because those grains near the source are sufficiently hot that their wavelengths are in such a range as to be absorbed again by grains further out. This problem has received some detailed attention but for grain models which are probably oversimplified and which can lead to substantially different consequences than have been predicted. However, the theoretical methods have been carefully worked out (see for example Leung, 1976) and the inclusion in the calculations of more realistic bimodal grain models should be a reasonably straightforward numerical task.

6.3 Dust Near Hot Stars

It is relatively simple to show the consequences of using a bimodal grain model to predict the infrared radiation near a source with a high temperature. Again, as in section 6.1 I will not try to derive exact results but rather just show the kind of thing one may readily demonstrate, namely, a strongly bimodal temperature.

Let the temperature of the star be such that most of its radiation is at wavelengths such that not only is $x_{c-m} \gtrsim 1$ but also $x_b \gtrsim 1$. Since $a_b \lesssim 0.01$ the implied value of λ at the distribution maximum is $\lambda \lesssim 0.06$ μm in the H II region which implies $T_{star} \gtrsim 50,000$K. Actually to achieve a mean absorption efficiency = 1 one may even use a lower temperature because over a substantial range the value of Q_{abs} is *greater* than one (see Figs. 1b, 2, 3).

Clearly then $Q_{abs}(c-m) = 1$ and therefore the *relative* temperatures of bare and core-mantle particles in a high temperature field are given by

$$\frac{T_b}{T_{c-m}} = \left\{ \frac{Q_{c-m}(\lambda_{T_{c-m}})}{Q_b(\lambda_{T_b})} \right\}^{1/4} , \qquad (6.4)$$

where $Q_{c-m}(\lambda_{T_{c-m}})$, the emission (absorption) efficiency of the core-mantle particles at the peak of the black body emission at T_{c-m}, is taken as a reasonable approximation to the mean core-mantle emission efficiency at T_{c-m}. The same approximation is made for the bare particles.

It is now easy to show that near a hot star the bare particle temperatures tend to be about twice as high as the core-mantle temperatures. We can show this by using pairs of values for λ_{T_b} and $\lambda_{T_{c-m}}$ such that $\lambda_{T_b} = \frac{1}{2}\lambda_{T_{c-m}}$ and showing that the derived

Table 3

Approximate Temperature Ratios for Bare (a_b = 0.005 μm) and Core-Mantle (a_m = 0.12 μm) Particles.

λ_{T_b}	$\lambda_{T_{c-m}}$	Q_b	Q_{c-m}	T_b/T_{c-m}	$T_{c-m} = 0.29/\lambda_{c-m}$
(μm)	(μm)			(eq.6.4)	°K
100	200	0.86×10^{-5}	0.29×10^{-3}	2.4	14.5
50	100	0.47×10^{-4}	0.12×10^{-2}	2.3	29
33	66	0.18×10^{-3}	0.49×10^{-2}	2.3	44
25	50	0.77×10^{-3}	0.42×10^{-2}	1.5	58
20[†]	40[†]	0.21×10^{-2}	0.20×10^{-1}	1.75	73
10	20	0.82×10^{-2}	0.81×10^{-2}	1[††]	145
10	12.5	0.82×10^{-2}	0.15×10^{-2}	1.2	230

[†] Resonances at both silicate and ice.
[††] Assumption not satisfied.

temperature ratios are consistent with this assumption. This is shown in Table 3 for several such combinations. The ones for both temperatures ≳ 100 K are applicable to H II regions.

Table 3 may be taken as indicative of the fact that very small bare particles are substantially hotter by a factor between 1.5 and 2 than core-mantle particles when both are in a primarily ultraviolet field. On the other hand, in normal interstellar radiation fields there is no strong dependence of temperature on particle size so long as the temperature is definable (Greenberg, 1968). The next section takes up this point.

6.4 Temperature Fluctuations

When the volume of a particle is extremely small its heat content may rise substantially with the absorption of a single photon. Should this happen the grain temperature is not definable because it is an average over large fluctuations. For a grain whose size is ≲ 0.01 μm this situation may occur when the photon density is relatively small as in tenuous and medium density clouds in the interstellar medium. When the photon density is higher the fluctuations are reduced *relative* to the time averaged energy input so that in an H II region one does not expect temperature fluctuations even for very small grains. An important requirement for the grain to undergo temperature fluctuations is, in addition to its smallness, a very small low temperature specific heat as is achievable for some crystals. However, if the bare grains are amorphous or have at least a substantial amount of impurities – chemical disorder etc. – the specific heat is much higher than given by the usual Debye formula and the fluctuations are damped. Surface contribution to the specific heat can also be quite large for very small particles (Dupuis et al., 1960). If the bare particles have temperature fluctuations they may spend more time at temperatures lower than average so that their *effective* temperature in terms of radiant energy distribution is lower than their mean temperature (see Greenberg, 1977; Hong and Greenberg, 1976; Purcell, 1976).

7. INFRARED ABSORPTION SPECTRA

The spectral absorption by dust at 9.7 μm and 3.07 μm is well established but there remain serious problems in their interpretation. Although these bands are considered likely to be caused by *some* silicate and perhaps H_2O ice respectively, it has been difficult to obtain theoretical band shapes to match the observations (see for example Steyer et al., 1974). I will try to show how some of these problems arise. In particular, I will show through a series of illustrative examples why there would be uncertainties in the derived amounts of silicate or ice even if we knew positively that the dust materials involved were really a known silicate and pure H_2O ice with well defined optical properties. There may be some ways out of these difficulties but they are not likely to be simple or direct.

7.1 Relative Proportion of "Ice" and Silicate from Relative Absorptions at 9.7 μm and 3.07 μm

Let us take as our example the Becklin–Neugebauer object and note that the depths at 3.07 μm and 9.7 μm relative to a background continuum as calculated by Bedijn (1977) are equivalent to about 1 and 2 (or 2.5) magnitudes of absorption respectively. Then the optical depth for either material treated as *separate* spherical particles is

$$\tau_{ice} = C_{abs}^{ice} = \frac{18\pi}{\lambda_{ice}} V \left\{ \frac{\varepsilon_2}{(\varepsilon_1+2)^2 + \varepsilon_2^2} \right\}_{ice}$$

$$\tau_{sil} = C_{abs}^{sil} = \frac{18\pi}{\lambda_{sil}} V \left\{ \frac{\varepsilon_2}{(\varepsilon_1+2)^2 + \varepsilon_2^2} \right\}_{sil}$$

(7.1)

At the absorption maximum for H_2O ice from Bertie et al., 1969, we have $\varepsilon = 1.33 - 2.24i$. For silicate the complex dielectric constant obtained from Launer (see Greenberg, 1973) is $\varepsilon = 2.4 - 2.38i$ while that from Huffman is $\varepsilon = -2 - 0.25i$. Setting

$\tau_{ice} = \tau_{sil}$ we obtain for these two choices

$$\left(\frac{V_{ice}}{V_{sil}}\right)_{Launer} = 0.11$$

(7.2)

$$\left(\frac{V_{ice}}{V_{sil}}\right)_{Huffman} = 4$$

where $\tau_{sil} = 2$ is used. In other words, with this kind of calculation we can not tell whether there is more or less ice than silicate in the dust.

There are three assumptions which are obvious or possible
sources of error in this interpretation. (1) We have assumed *separate* ice and silicate particles which is certainly wrong if we
believe in core-mantle particles (how can solid ice grow without
some nucleation elements such as silicate cores), (2) uncertainties in the absorption strengths at 3.07 μm (relative to the a-
mount of mantle material) and at 9.7 μm (relative to the amount of
core material), (3) background emission by hot dust grains (see
Bedijn, 1977; Gillet, 1973).

With regard to (1), the effect of the core on the mantle 3.07
μm absorption is not significant (and can be shown so by exact
calculation) because the core material is not "seen" as a result of
the strong absorption by the mantle. On the other hand, if the
mantle material has any absorptivity it may mask the core entirely.
This will be demonstrated in more detail in 7.2.

With regard to (2), a possible question is whether the 3.07
μm band is really due to H_2O ice. If so we have a fairly direct
measure of the amount of solid H_2O. *But* since the mantle material
undoubtedly consists of a mixture of complex as well as simple
molecules and radicals it is not clear whether the 3.07 μm absorption can be unambiguously attributed to H_2O and in any case the
amount of ice (if present) is not simply related to the total a-
mount of mantle accreted material.

Assumption (3) is too complex to discuss here.

7.2 Structure of the "9.7 μm" Band in Core-Mantle Particles

Is the 9.7 μm band due to pure silicate particles? If so we might
have some hope of identifying the material and its amount. If it
is due, in part or in entirety, to the cores of core-mantle grains
then the identification problem becomes much more complex. A number of examples of the degree of complexity introduced by the mantle have been calculated.

First let us consider some idealized cases. Let the silicate
absorption be of a Lorentzian shape given by

$$m' = m'_o + \frac{k_o \gamma (1-\alpha^2)}{(1-\alpha^2)^2 + \alpha^2 \gamma^2}$$

$$m'' = \frac{k_o \alpha \gamma^2}{(1-\alpha^2)^2 + \alpha^2 \gamma^2} \quad .$$

$$(7.3)$$

where $\alpha = \lambda/\lambda_o$, $\lambda_o = 9.5$ μm or 11.5 μm, $m'_o = 1.7$, $k_o = 0.7$, $\gamma = 0.2$. This gives a half-width $\Delta\lambda = 1.8$ μm and $m'' = 0.7$ at resonance. The choice of a resonant wavelength at 11.5 μm instead of 9.7 μm (as observed in the interstellar medium) was made to correspond roughly to olivine. We then let the mantle material be of either real pure H_2O ice or a fictitious ice with constant complex index $m = 1.33 - 0.05i$ or $m = 1.33 - 0.10i$.

Fig. 19 shows the Lorentzian silicate m' (with $\lambda_o = 9.5$ μm) and the m'' for real ice. We see in Fig. 20a,b that real ice mantles *increase* the absorption efficiency and therefore even *more* increase the absorption cross section because cores with mantles are, of course, larger. For comparison we see that fictitious ice mantles of 0.05 μm thickness *decrease* the absorption efficiency relative to that for the cores. It is seen that real ice mantles

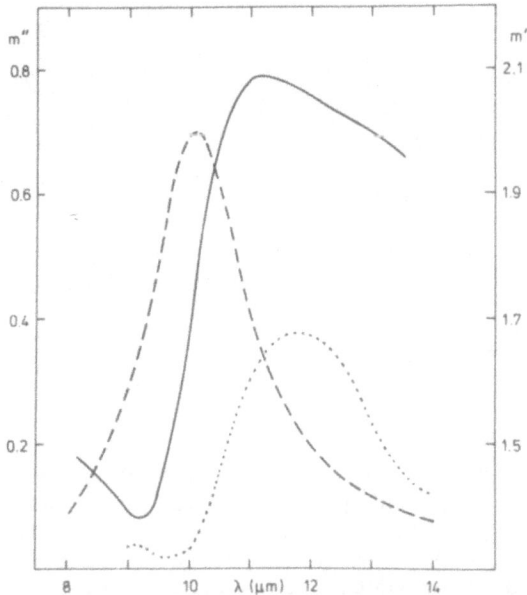

Fig. 19. Real(solid) and imaginary (dash) parts of the complex index of refraction m = m'-im", for silicate absorption with Lorentzian shape (Eq.(7.3)). The m" for ice is shown for comparison as the dotted curve.

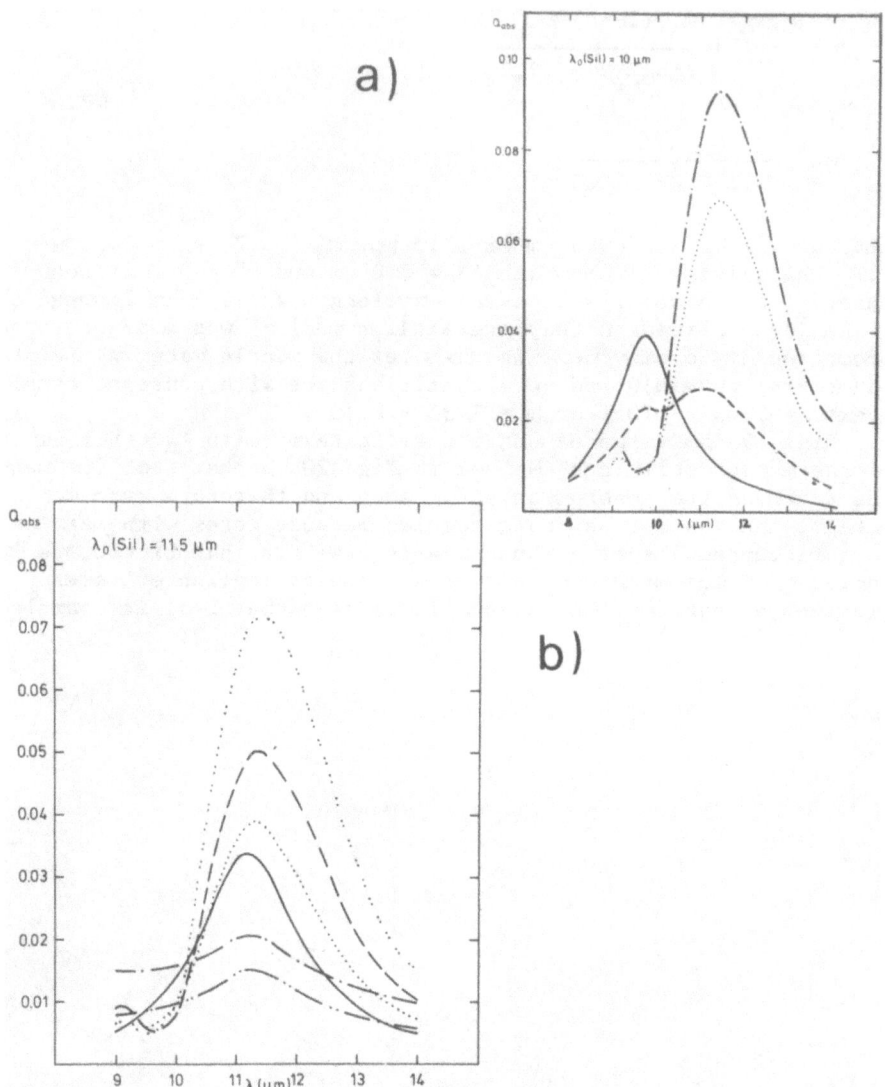

Fig. 20. a) Absorption efficiencies for spherical core-mantle
particles with core absorptivity given by a Lorentzian "silicate"
with λ_0 = 10.0 μm (see Fig. 19). Core radius a_c = 0.05 μm. Mantles of H_2O ice with radii: a_m = 0.05 μm (solid curve, core only);
a_m = 0.07 μm (dashed); a_m = 0.15 μm (dots); a_m = 0.20 μm (dash dot).
b) Same as 20a but λ_0 = 11.5 μm. Mantles of H_2O ice with radii:
a_m = 0.05 μm (solid, core alone); a_m = 0.07 μm (dots); a_m = 0.10
μm (dashed); a_m = 0.15 μm (double dot). Also shown are results for
mantles of fictitious ice (a_m = 0.10 μm) with m' = 1.33, m" = 0.10
(dash dot), m" = 0.05 (dash double dot).

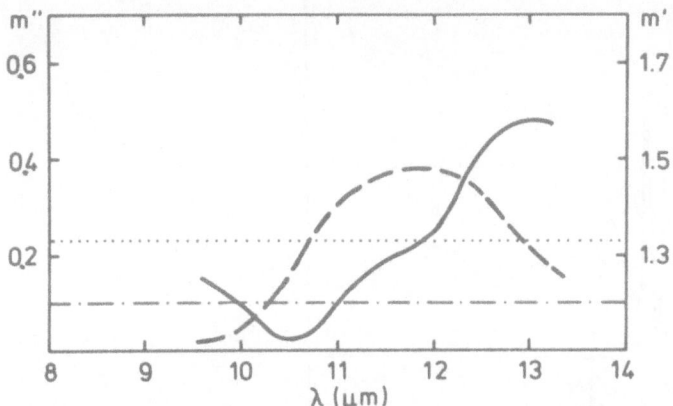

Fig. 21. Real (solid) and imaginary (dashed) parts of the indices of refraction for H_2O ice between 9 μm and 14 μm. Comparison with constant indices of a fictitious mantle material with m'=1.33, m" = 0.10.

do not drastically change the shape of the given absorption curve shown for the bare core particles. Fig. 21 shows the values of the complex indices of pure ice in the vicinity of 12 μm and we see a broad peak of m". This degree of absorption but not its shape would also be characteristic of the realistic icy mantles.

Fig. 22 shows how even a *resonant* type silicate absorption ($\varepsilon = -2 - i\varepsilon_2$ produces an *infinite* Q in the *Rayleigh approximation*) is overtaken by an absorbing mantle. In this case we use the Huffman measures for olivine for the core and for the various mantles we use pure H_2O ice. For mantles with $\bar{a}_m = 0.12$ μm we infer that the absorption by the core is already hidden by the mantle and provides little of the observed absorptivity. We note also that for this highly resonant silicate absorption the *initial* tendency with very small mantle thickness is to reduce the absorption but that this trend is reversed with additional mantle thickness.

It is interesting to compare the apparent ice to silicate ratio based on isolated particles and on core-mantle particles. If we use a mean mantle thickness $a_m = 0.12$ μm then, from Fig. 20 a) the value of Q at the "silicate" peak for the core-mantle grain is about 0.07 as compared with a Q ≈ 0.04 for the bare silicate particle. The greater Q (by a factor of two) in combination with the greater grain area (by a factor of $(0.12/0.05)^2 \approx 5.8$) means that silicate cores with mantles would absorb 11.6 times as much at the 10 μm band than the same number of bare silicate particles. In terms of the assumption of separate silicate particles the amount of silicate material implied by the band depth may be, in reality, considerably less than one has calculated – for a particular silicate absorption strength. The ratio of ice to silicate then would be actually greater than calculated on a separate particle basis.

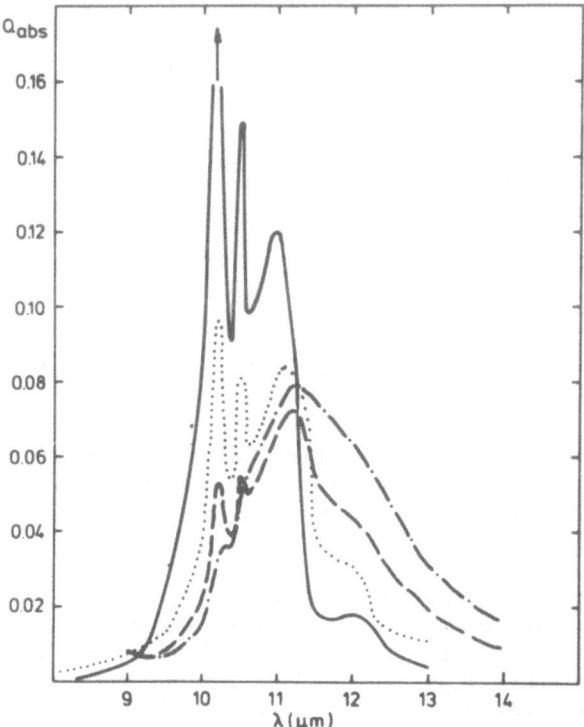

Fig. 22. Absorption efficiencies for spherical core-mantle particles with strongly resonant core (olivine) absorptivity and mantle
of H_2O ice. Core radius, $a_c = 0.05$ μm. Mantle radii: $a_m = 0.05$
μm (solid, core only); $a_m = 0.07$ μm (dots); $a_m = 0.10$ μm (dash);
$a_m = 0.15$ μm (dash dot).

7.3 $A(V)/\tau_{9.7}$

The ratio of visual extinction to the strength of the 9.7 μm absorption is used to give the visual extinction to an unseen object
if one has the 9.7 μm absorption. In terms of core-mantle particles this ratio is given by

$$\left\{ \frac{A(V)}{1.08\tau_{9.7}} \right\}_{\text{core-mantle}} = \frac{\bar{Q}_V \pi \bar{a}_m^2}{Q_{9.7}\pi \bar{a}_m^2} = \frac{\bar{Q}_V(a_m)}{Q_{9.7}(a_c,a_m)} \qquad (7.4)$$

Using the values of Q_{abs} from either Fig. 20 (a), (b) or Fig.
22 we find that $Q_{9.7}(a_c,a_m)$ is bracketed between about 0.06 and
0.07 for $a_m \simeq 0.12$ μm (depending *largely* on mantle rather than core
absorptivity).

We may use $\bar{Q}_V \simeq 1.5$, for $\bar{a}_m \simeq 0.1$ μm - 0.15 μm, to obtain from

Eq. (6.8)

$$\left\{ \frac{A(V)}{1.08\tau_{9.7}} \right\}_{core-mantle} \approx 25, \qquad (7.5)$$

The somewhat unexpected consequence of this result is that the value of $A(V)/\tau_{9.7}$ is rather insensitive to the mantle thickness if we are dealing with interstellar type grains. The ratio $A(2.2 \ \mu m)/\tau_{9.7}$ is not as insensitive to mantle thickness because $Q_{2.2}$ is more sensitive to a_m than is Q_V. One thing which we have not included in our calculated example is the possibility that some or all of the bare particles are silicate and may also contribute to the 9.7 μm strength while making no contribution to the visual extinction. If all the bare particles are silicates they are roughly equal in mass to the cores so that, for the same $A(V)$, one could have $\tau_{9.7}$ increased by a factor of two.

Suppose the silicate absorption is due to core-size particles alone? This is the kind of situation one might expect around late type stars if they are producing the silicate particles or in *very* hostile high temperature region where the mantles are blown off. In this case, the extinction to absorption ratio of the star by these grains is given by

$$\frac{A(V)}{1.08\tau_{9.7}} = \frac{Q_V(a_c)}{Q_{9.7}(a_c)} \qquad (7.6)$$

For the core particles alone $Q_{9.7}$ is probably (see Fig. 20 and 22) in the 0.03 - 0.10 range (depending on the type of silicate absorption) but $Q_V(a_c) \approx 0.1$, i.e., is substantially less than one if $a_c \approx 0.05 \ \mu m$. In fact the extinction curve produced by such small particles is completely different from a normal extinction curve, being strongly concave upwards in the visual region rather than bending over downward in the visual. Thus, although the implied visual extinction is only $A(V)/\tau_{9.7} \approx 1-3$, the ultraviolet extinction could still be quite high - certainly higher than inferred from the visual extinction.

7.4 Some Examples

Let us look again at the 9.7 μm and 3.07 μm absorption in the B.N. source spectrum assuming that they are produced by interstellar type particles rather than separate ice and silicate particles. First of all using $A(V)/(1.08\tau_{9.7}) = 23$ (from Eq.(7.4)) we find for $\tau_{9.7} \approx 2.5$ that the total visual extinction to the source is about 58 magnitudes. The volume of 0.05 μm core silicates is then, as obtained from *this* (*not* directly from the 9.7 μm absorption strength), found to be (for spheres)

$$V_{sil} = n_d \frac{4}{3}\pi a_c^3 = \frac{A(V)}{Q_V \pi a_m^2} \frac{4}{3}\pi a_c^3$$

$$= \frac{A(V)}{Q_V} \frac{4}{3} (\frac{a_c}{a_m})^2 a_c \qquad\qquad (7.7)$$

$$= 4.5 \times 10^{-5} \text{ cm}^3$$

The H_2O ice volume (from Eq.(7.1) and $\tau_{ice} \simeq 1$) is $V_{ice} = 3.7 \times 10^{-5}$ cm^3, from which we get $V_{ice}/V_{sil} = 0.8$ which is neither of the values obtained in Eq.(7.2).

I am not prepared to go more deeply into the physics of this at this time but it is certainly clear, I think, that there is a strong case to be made for doing the absorption problem properly keeping the basic dust model in mind.

Towards the galactic center, the 9.7 μm absorption appears to be about 3 magnitudes (any heating by this dust would fill in the absorption and tend to make it *appear* smaller). Using again the standard interstellar model (Eq.(7.5)) we find that the extinction to the galactic center is about 69 magnitudes which is a factor of ∿ 4 larger than one would estimate from the distance and an *average* extinction per unit length of 1-2 mag/kpc. The implication of this larger than average value (if true) is that the abundance of heavy elements towards the galactic center is ∿ 4 times more than in the neighborhood of the sun if the hydrogen density is the same (Greenberg and Hong, 1974).

7.5 Some Remarks on Infrared Dust Spectra

The need for more laboratory studies of the optical properties of possible interstellar dust materials is obvious. Most of the effort to date has been on investigations of the silicate absorption. With the indication of the relative importance of the mantle optical properties we have started a program of measuring the infrared absorption by a variety of possible low temperature mantle materials in the 2.5 μm to 25 μm range. We anticipate, in the near future that we will also be able to measure the absorption as far as 1 mm so that we will span the entire infrared range of interest either in determining the broad band emissivity of low temperature grains in the far infrared or the specific absorption properties in the spectrally interesting range 1-25 μm.

An example of the kinds of absorption spectra we obtain is shown in Fig. 23. These spectra also show some of the effects to be expected to result from exposure of the interstellar grains to ultraviolet irradiation of their mantles and subsequent chemical and optical modification. The photochemical processing has been performed in the laboratory by irradiation of a mixture of methane

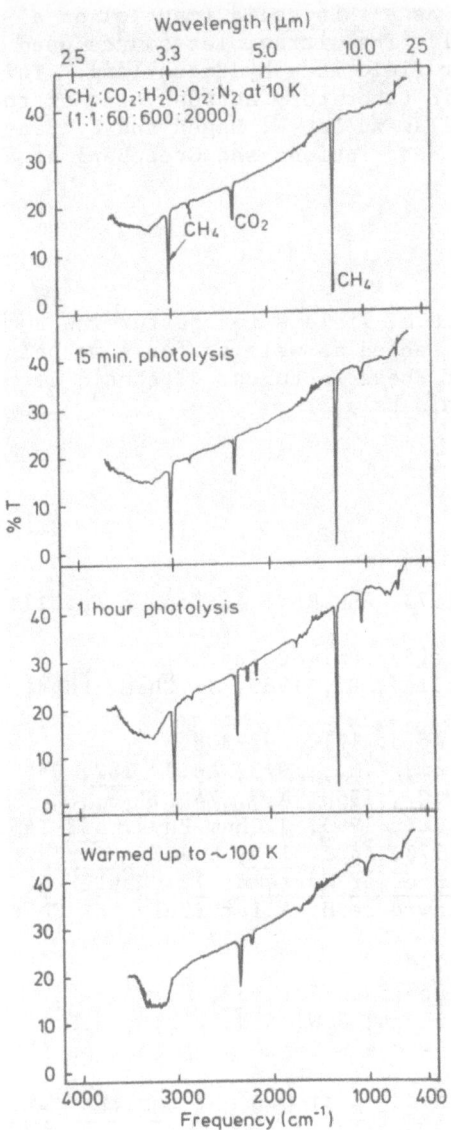

Fig.23. Infrared absorption spectra of a mixture of CH4 and air deposited as a thin film (∿ 0.5 μm) on a cold (10 K) finger of a cryostat and subsequently irradiated by ultraviolet radiation from a hydrogen lamp. Uppermost spectrum is before irradiation. Second spectrum is result after 15 minutes of photolysis (note new peaks appearing). Third spectrum is after 1 hr photolysis. Last spectrum is for what remains after irradiation stopped and sample warmed slowly to ∿ 100 K (note disappearance of CH4).

and air which has been deposited as a thin solid (mantle) on a
cold window (10 K) in the crystal. The ultraviolet source used
here to simulate the interstellar field is a hydrogen lamp. This
work is a sample of the program in Laboratory Astrophysics at the
University of Leiden with L.J. Allamandola, W. Hagen and F. Baas
participating. For some earlier descriptions see Greenberg et al.,
1972 and Greenberg, 1976.

ACKNOWLEDGEMENTS

        I am deeply indebted to A.G.G.M. Tielens for performing some
of the numerical calculations presented as well as for some helpful suggestions in the writing of these lectures. I should also
like to thank Dr. S.S. Hong for his help.

REFERENCES

* = Review

*Aannestad, P.A., Purcell, E.M., 1973, Ann.Rev. Astron. & Ap. 11,
309.
Bedijn, P., 1977, Ph.D. Thesis, Leiden University.
Bertie, J.E., Labbé, H.A. and Whalley, E., 1969, J. Chem. Phys.
50, 4501.
Bless, R.C., Savage, B.D., 1972, Ap.J. 171, 293-308.
Carrasco, L., Strom, S.E. and Strom, K.M., 1973, Ap.J, 182, 95.
Coyne, G.V. and Wickramasinghe, N.C., 1969, A.J. 74, 1179.
Dupuis, M., Maxo, R. and Onsager, L., 1960, J.Chem.Phys. 33, 1452.
Gillet, F.C. and Forrest, W.J., 1973, Ap.J. 179, L 483.
*Greenberg, J.M., 1968, Stars and Stellar Systems, 7, Chapter 6,
eds. Barbara M. Middlehurst and Lawrence H. Aller (Univ. of Chica-
go Press), p. 221.
Greenberg, J.M., 1971, A.&A. 12, 240.
Greenberg, J.M., 1972, J. Coll. Interface Sci. 39, No. 3.
Greenberg, J.M., Yencha, A.J., Corbett, J.W. and Frisch, H.L.,
1972, Mémoires de la Société Royale des Science de Liège, 6e série,
tome III, pp. 425-436.
Greenberg, J.M. and Hong, S.S., 1974a, H II Regions and the Galac-
tic Center, Eslab. Symp. no. 8, Frascati, ed. A.F. Moorwood, ESRO
SP-105, p. 153.
Greenberg, J.M. and Hong, S.S., 1974b, H II Regions and the Galac-
tic Center, Eslab. Symp. no. 8, Frascatie, ed. A.F. Moorwood, ESRO
SP-105, p. 221.
Greenberg, J.M., 1976, Ap.&Sp. Sc. 39, 9.
Greenberg, J.M., 1978, Paper presented at Liège Astrophysical Sym-
posium on Small Molecules, June 1977.
*Greenberg, J.M., 1978, Cosmic Dust, Chapter 4, ed. A.J.M. McDon-
nell (Wiley, London) p. 187.

Hong, S.S., 1975, Unified Model of Interstellar Grains, Ph.D. Thesis, State Univ. of New York at Albany.
Hong, S.S. and Greenberg, J.M., 1976, Far Infrared Astronomy, ed. M. Rowan-Robinson (Pergamon Press), p. 299.
*Huffman, D.R., 1977, Advances in Physics 26 H2, 129.
Huffman, D.R. and Stapp, J.L., 1971, Nature Phys. Sc. 229, 45.
Huffman, D.R. and Stapp, J.L., 1973, Interstellar Dust and Related Topics, IAU Symposium no. 52, eds. J.M. Greenberg and H.C. van de Hulst (Reidel, Dordrecht), p. 297.
Hulst van de, H.C. 1957, Scattering by Small Particles (New York: J. Wiley&Sons, Inc., Chapman&Hall, Ltd.).
Launer, P.J., 1952, Am. Mineralogist 37, 764.
Leung, C.N., 1975, Ap.J. 199, 340.
Martin, P., 1974, Ap.J. 187, 461.
Pollack, J.B., Toon, C.B. and Khare, B.N., 1973, Icarus 19, 372.
Purcell, E.M., 1976, Ap.J. 206, 685.
Serkowski, K., Mathewson, D.S. and Ford, V.L., 1975, Ap.J. 196, 261.
Steyer, T.R., 1974, Ph.D. Thesis, University of Arizona.
Steyer, T.R., Day, K.L. and Huffman, D.R., 1974, Appl. Opt. 13, 1586.
Taft, E.A. and Philipp, H.R., 1965, Phys. Rev. 138; A 197.
Werner, M.W. and Salpeter, E.E., 1969, M.N. 145, 249.

PHYSICS OF MOLECULAR CLOUDS FROM MILLIMETER WAVE LINE OBSERVATIONS

P.M. Solomon

Astrophysics Program, SUNY, Stony Brook, N.Y. 11794

1. INTRODUCTION (MOLECULAR HYDROGEN)

The advent of millimeter wave molecular line astronomy has made it possible over the past seven years for astronomers to study the physics and chemistry of dense interstellar clouds. The dominant gaseous constituent of these clouds is molecular hydrogen, H_2, and hence the name molecular clouds. In fact all interstellar clouds with hydrogen density greater than 100 cm^{-3} are composed primarily of molecular and not atomic hydrogen. The dominance of molecules in dense clouds is due to the self shielding of H_2 from the photodissociating starlight at the wavelengths of the ultraviolet Lyman bands. At a density of 100 cm^{-3} the production rate of H_2 on grain surfaces is sufficient to build up a layer which is optically thick to the photodissociation occurring in the lines of the Lyman band. Inside the clouds, the photodissociation, which is limited to the line wings, is reduced by several orders of magnitude and the production rate dominates with almost all hydrogen converted to H_2. Since H_2 does not have a true continuum photodissociation at $\lambda >$ 912 Å, the self shielding of the bound-bound-free process of the Lyman bands is critical to the distinction between atomic and molecular clouds.

The cross section for one of the photodissociating lines (the upper electronic state decays into the continuum of the ground electronic state) is about 10^{-14} cm^2. Therefore the core of the line becomes opaque when a column of H_2 is of the order $N_{H_2}\ell \simeq 10^{14}$ cm^{-2}, where ℓ is the mean free path of dissociating photons. Assuming that H_2 forms every time two H atoms strike a grain, and with about 1% by mass in the form of grains the formation rate of H_2, β, is:

$$\beta = \tfrac{1}{2}N_H N_{grain} \sigma_{grain} V_H = 3 \times 10^{-17} N N_H, \qquad (1.1)$$

97

G. Setti and G. G. Fazio (eds.), Infrared Astronomy, 97-114.

where $N = N_H + 2N_{H_2}$. The photodissociation rate from the background UV radiation field is $\gamma = 10^{-11}$ sec^{-1}. Setting formation and photodissociation equal and $N = N_H$ we have:

$$10^{-11}N_{H_2} = 3 \times 10^{-17}N_H^{\ 2}, \tag{1.2}$$

which gives

$$N_{H_2} = 3 \times 10^{-6}N_H^{\ 2} \tag{1.3}$$

The condition for becoming optically thick in the center of the line is

$$3 \times 10^{-6}N_H^{\ 2}\ell \gg 10^{14} \tag{1.4}$$

or

$$N_H^{\ 2}\ell \gg 3 \times 10^{19} \tag{1.5}$$

If the mean free path in the outer layer of the cloud is small enough compared to the cloud size, the absorption inside the cloud will be reduced by a factor of $3 \times 10^5/N_H$ and molecules will dominate. It can be shown by treatment of the line transfer problem that in typical clouds if $N < 10$ cm^{-3} this never occurs and atoms dominate ($\ell \sim r_{cloud}$), if $10 < N < 100$ cm^{-3} there will be a mixture with the balance depending on the exact geometry and radiation fields and if $N > 100$ cm^{-3} ($\ell < 10^{-4}r_{cloud}$) molecules will dominate. This latter group comprising all dense clouds and star formation regions is the source of most interstellar molecular line radiation and infrared emission.

2. SUMMARY OF GIANT MOLECULAR CLOUD PROPERTIES

Although molecular clouds have a wide range of sizes, it has become apparent that a large fraction of interstellar molecules are in what I refer to as Giant Molecular Clouds (GMC) or Giant Molecular Cloud Complexes. Nearby examples of GMC's are the molecular clouds connected with the Orion Association extending over 10° in the sky and the Perseus clouds. Fig.1 is a contour map of column densities in the SgrB2 molecular cloud located about 200 parsecs from the galactic center. Typical physical conditions in these clouds have been deduced by many observers from millimeter wave observations of rotational line emission in several molecules, most importantly ^{12}CO and ^{13}CO. These clouds are the most massive objects in the galaxy containing from $10^4 - 10^7$ M$_\odot$ with linear extent from 10 - 100 pc. Within this giant cloud there may be several core regions of high density containing infrared sources, compact H II regions, maser sources and regions of peak CO temperature. All of these are associated with recently formed stars. Table 1 gives typical properties of GMC's. In general any lumin-

Table 1

Basic Physical Parameters of Giant Molecular Clouds. Dominant
Constituent is Molecular Hydrogen.

Dust/Gas = 0.01 by mass

Visual Optical Depth
$\tau_V = 1.08\ A_V \sim 20-100$
$n(H_2)\ \ \ = 500-2,000\ cm^{-3}$
$\bar{n}(H_2)\ \ \ = 700\ cm^{-3}$
$D = 2R\ \ \ = 10\ pc - 80\ pc$

Mass $\ \ \ = 10^4-10^7\ M_\odot$ typically $10^5\ M_\odot$
Mass $\ \ \ = 0.16\ R^3$ (pc) $n(H_2)\ M_\odot$
Kinetic Temperature of Gas $T_K = 10-20\ K$
Distance $\ \ 400\ pc - 15,000\ pc$
All of the luminosity will emerge in the far IR or Sub-millimeter.

Core Regions

$\ \ \ n(H_2) = 10^4-10^6\ cm^{-3}$
$\ \ \ R < 1\ pc$
$\ \ \ T_K\ \ \ = 40-100\ K$
$\ \ \ Mass\ \ 10^2-10^3\ M_\odot$

Core Regions Contain

$\ \ \ $IR Sources
$\ \ \ $OH Masers
$\ \ \ H_2$O Masers
$\ \ \ $Young Stars
$\ \ \ $Compact H II Regions

---

ous object within these clouds will be observable only in the IR
since all of the radiation will be degraded to IR by the dust.
The core regions may be strong 10 μm, 20 μm or 100 μm sources but
the GMC as a whole will radiate predominately in the 300 μm region
due to its low temperature.

Radio continuum and recombination line observations of dense
H II regions throughout the galaxy provide a tracer of young massive stars. Much of the earliest work in molecular line observations was directed toward these well known sources and virtually
all strong radio H II regions are observed to be associated with
molecular clouds. By contrast only a small fraction of molecular
clouds are associated with observed H II regions. In those cases
where there is a close association the molecular cloud as traced
by CO observation is found to be one or two orders of magnitude
larger and more massive than the giant dense H II regions. In
terms of size or mass the H II region is just a perturbation of

the GMC, although it may be the most energetic region within the
cloud.

Most of the interstellar medium in the galaxy, interior to the
sun, is in the form of molecular clouds. The total mass in the
clouds has been estimated (Scoville and Solomon, 1975) to be about
3×10^9 M_O. Our recent extensive survey of ^{12}CO and ^{13}CO emission
in the galaxy confirms this estimate. For our present purposes
these estimates lead to an interesting result regarding the relationship of star formation to molecular clouds. For the densities
and sizes in Table 1, it is easily shown that the clouds are gravitationally bound, the observed internal velocity dispersions falling in the range of a few km/s. The gravitational collapse time
for a cloud is then

$$t_c = \frac{3 \times 10^6}{(n_{H_2}/10^3)^{1/2}} \quad yrs. \tag{2.1}$$

Under free fall most of these clouds would collapse in a few
million years. The average star formation rate in the galaxy is a
few solar masses per year; consequently only one part in a thousand
of the mass in dense clouds collapses into stars at the free fall
rate. The clouds either have a lifetime of about 10^9 years or
break up after only a very small fraction of the total mass has
been converted to stars.

3. MILLIMETER OBSERVATIONS: INTENSITIES AND TEMPERATURES

The importance of millimeter waves to the observation of interstellar molecules results from the energy level spacing of molecules
containing two or more heavy atoms. For any molecule with two or
three heavy atoms of the astronomically abundant elements O, C, N,
S, or Si the fundamental rotational transition occurs at millimeter
wavelengths. In the case of diatomic molecules with reduced mass μ
the rotational constant is

$$B = h/8\pi^2 \mu r^2 \tag{3.1}$$

The transition frequencies are proportional to the rotational quantum number J and are

$$\nu = 2BJ \tag{3.2}$$

The fundamental rotational frequency and wavelength for a few
of the more abundant molecules found in dense clouds are listed in
Table 2. For heavier molecules the ground state transition occurs
at longer wavelengths but low lying excited states will be observable at millimeter wavelengths. Thus all molecules with two or
more heavy atoms have millimeter wave transitions. The strength

Table 2

Rotational Transition: Wavelengths and Spontaneous Decay Rates For Some Frequently Observed Interstellar Molecules

Molecule	B(MHz)	λ(J+1-J) mm	λ(J+1-J) mm	A(J+1-J) S^{-1}
CO	57635	2.6(1-0)	1.3(2-1)	6×10^{-8}(1-0)
HCN	44316	3.4†(1-0)	1.7(2-1)	2×10^{-5}(1-0)
CS	24495	3.1(2-1)	2.0(3-2)	6×10^{-5}(3-2)
H$_2$CO	(B) $\left\{ \begin{matrix} 38834 \\ 34004 \end{matrix} \right.$ (C)	2.0(2_{11}-1_{10})		8×10^{-5}
		2.1(2_{12}-1_{11})		7×10^{-5}

† HCN has three hyperfine components

of the transition depends on the magnitude of the permanent electric dipole moment; symmetric molecules, such as O_2 and N_2, having no dipole moment are therefore unobservable by this technique. As can be seen from Table 2 the wavelength region 2 mm < λ < 6 mm is extremely useful and has proved to be by far the richest part of the radio spectrum.

To date forty-one molecules have been identified in interstellar clouds. All except H_2 and CH$^+$ have observed radio lines. Table 3 lists all molecules identified, grouped according to the number of atoms. All of the molecules excepting H_2 are minor constituents in the sense that only a small fraction of an available element is tied up in any specific molecule. The hydrogen molecule, having no dipole moment is unobservable except in the ultraviolet which does not penetrate molecular clouds and through very weak quadropole lines, excited only at high temperatures, in the infrared. It can not be observed directly by any technique over most of the volume or mass in molecular clouds.

The most abundant molecule observed at millimeter wavelengths is carbon monoxide, CO, which has estimated concentration relative to H_2, CO/H_2 $\sim$ 1-3 x 10^{-5}. H_2O may have a comparable abundance but estimates are extremely uncertain since only excited states have been observed. Potentially abundant species not observed are CH$_4$ and CO$_2$. All other molecules listed in Table 2 have an abundance relative to H_2 << 10^{-7}, with typical values H_2CO/H_2 $\sim$ 3 x 10^{-9}, CS/H_2 $\sim$ 10^{-8}-10^{-9}, HCN/H_2 $\sim$ 10^{-7}-10^{-8}. Accurate abundance determination for many molecules is complicated by optical depth effects.

For the purpose of studying the physical conditions in molecular clouds the most useful spectral lines are those of ^{12}CO and ^{13}CO. The relatively high abundance of CO coupled with the long lifetime of the transition combines to produce widely observable

Table 3

Molecules Identified In Molecular Clouds

				Number of Atoms			
2	3	4	5	6	7	8	9
H_2	HCN	H_2CO	CH_2NH	CH_3OH	CH_3CHO	$CHOOCH_3$	CH_3CH_2OH
CH	H_2O	HNCO	HC_2CN	NH_2CHO	CH_2CHCN		$(CH_3)_2O$
CH^+	H_2S	H_2CS	CHOOH	CH_3CN	CH_3NH_2		CH_3CH_2CN
OH	OCS	NH_3	NH_2CN		HC_2CH_3		$HC_2C_2C_2CN$
CN	,HCO	C_3N			HC_2C_2CN		
CO	SO_2						
CS	HCO^+						
SiO	HN_2^+						
SiS	C_2H						
NS	HNC						

and high intensity emission. For this reason CO observations have become the major technique for studying the astrophysics of inter-stellar clouds, acting as a tracer of H_2 density, temperature and kinematics in the clouds. The following sections discuss the formation of millimeter wave emission lines and the relationship between observations and the physical conditions in molecular clouds.

4. FORMATION OF MILLIMETER WAVE EMISSION LINES

The intensities of radio spectral lines are usually measured in terms of antenna temperature defined by the Rayleigh-Jeans law,

$$I(\nu) = \frac{2kT_A}{\lambda^2} \quad (\text{ergs cm}^{-2} \text{ sec}^{-1} \text{ Hz}^{-1} \text{ ster}^{-1}), \tag{4.1}$$

where $I(\nu)$ is the specific intensity of the incident radiation and T_A is the antenna temperature for a perfect antenna.

The brightness temperature, T_b, is defined by the Planck function

$$I(\nu) = \frac{2h\nu^3}{c^2 (\exp(\frac{h\nu}{kT_b}) - 1)} \tag{4.2}$$

At millimeter wavelengths the high temperature condition $h\nu/kT_b \ll 1$ is not always valid and the antenna temperature may be significantly less than the brightness temperature. Ignoring the cosmic background radiation we have

$$T_A = \frac{h\nu}{k} \frac{1}{\exp{(h\nu/kT_b)} - 1} \qquad (4.3)$$

The equivalent antenna temperature of the 2.7 K cosmic background radiation at $\lambda = 2.6$ mm is T_A (background) = 0.8 K.

An important parameter of the molecules in determining the emitted intensity is the excitation temperature T_{ij} governing the relative populations in the two states of the transition. The ratio of state populations is

$$\frac{N_j}{N_i} = \frac{g_j}{g_i} \exp{(\frac{-h\nu}{kT_{ij}})}. \qquad (4.4)$$

In general the excitation temperatures of the various energy levels will be different and knowledge of T_{ij} is essential for interpreting line intensities. Letting τ be the transition optical depth, the basic equation of radiative transfer for a homogeneous source which is larger than the antenna beam, filled with molecules in levels i and j is

$$T_A = \left\{ \frac{h\nu}{k} \frac{1}{\exp{(\frac{h\nu}{kT_{ij}})} - 1} - T_A(bg) \right\} (1 - \exp(-\tau)) \qquad (4.5)$$

Neglecting $T_A(bg)$ and with $h\nu/kT_{ij} \ll 1$, we have the familiar relation

$$T_A = T_{ij}(1 - e^{-\tau}) \qquad (4.6)$$

which gives

$$T_A = T_{ij}\tau, \qquad \text{for } \tau \ll 1$$
$$\qquad\qquad\qquad\qquad\qquad\qquad (4.7)$$
$$T_A = T_{ij}, \qquad \text{for } \tau > 1.$$

For small optical depths the intensity is therefore a measure of excitation temperature and optical depth. For a fixed population in the lower state i, the optical depth depends inversely on the excitation temperature and the intensity is a measure of the total number of molecules in state i. For large optical depths the intensity is only a measure of the excitation temperature T_{ij}.

The optical depth in a cloud of length L with total velocity dispersion Δv is given by

$$\tau = \alpha A_{ji} \, (1 - \exp(-\frac{h\nu}{kT_{ij}})) N_i L = \alpha A_{ji} (N_i - \frac{g_i}{g_j} N_j) L \, , \qquad (4.8)$$

where $\alpha = \dfrac{g_j}{g_i} \left(\dfrac{c^3}{8\pi\nu^3 \Delta v} \right)$.

As can be seen the optical depth depends on the Einstein A coefficient and on the total column density of the *difference* in population between the two states. Thus the optical depth has a very complex dependence on the total number of molecules and on the physical conditions in the cloud, such as temperature and density. While optical depth is a useful concept for radiative transfer it does not have, in general, a simple relationship to the true physical properties of the cloud such as abundance or density.

As can be seen from the equations above, the emission line intensity T_A provides a direct measure of the excitation temperature T_{ij} if the cloud size is greater than the antenna beam and if $\tau \gg 1$. The kinetic temperature can then be directly measured if $T_{ij} = T_{kinetic}$. The relevant question is then: Under what conditions is the population of the observed levels governed by the gas kinetic temperature?

Both collisions and radiation lead to changes in rotational energy. Continuum radiation at these frequencies is negligible, but the radiation field at the line frequency may itself be important. The simplest case to consider is pure collisional excitation ignoring the effect of trapped line photons. For simplicity we will derive the condition for $T_{ij} = T_{kinetic}$ in a two level system. Let N_0, N_1 be the number density of molecules in J = 0 and 1, respectively. The collision rate between levels determined by the cross section and hydrogen density, is given by $C_{10} = N_{H_2}\sigma v$ and $C_{01} = C_{10}(g_1/g_0) \exp(-h\nu/kT_{kin})$. Equating the total rate into J = 0 with the rate out of J = 0, we have $N_0 C_{01} = N_1(C_{10} + A_{10})$. From the definition of excitation temperature

$$\frac{N_1}{N_0} = \frac{g_1}{g_0} \exp(-\frac{h\nu}{kT_{01}}) \qquad (4.9)$$

Equating the above expressions for N_1/N_0 we have

$$(\frac{k}{h\nu}) T_{01} = \frac{(\frac{k}{h\nu}) T_{kin}}{1 + (\frac{k}{h\nu}) T_{kin} \ln(1 + \frac{A}{C})} \, . \qquad (4.10)$$

The condition for $T_{01} \rightarrow T_{kin}$ becomes

$$C \gg AkT_{kin}/h\nu \qquad (4.11)$$

The two level treatment provides a reasonable estimate for the thermalization condition. However in the multi-level problem collisions to higher levels may even result in inversions in J = 1-0 as a result of cascades down the energy level ladder. This inversion is damped out by stimulated emission preventing the operation of a maser.

An important factor in millimeter line emission is the effect of the line radiation field on the population of levels and consequently on emitted intensity. Obviously, in the limit of infinite optical depth, black body radiation at the gas temperature is emitted from a cloud even with an extremely low collision rate C << A. In this case the radiation field is coupled to the rotational energy levels. Each photon produced by a collision interacts with other molecules before escaping the cloud resulting in a net decrease in the effective spontaneous decay rate. This process which is frequently referred to as "line trapping" must be taken into account in the interpretation of molecular lines. Several authors have treated this in the multi-level problem. A very useful formulation of this problem can be obtained for a cloud in which the total velocity spread is very large compared to the thermal Doppler width and with radial velocity a function of position in the cloud. This regime applies generally to either collapse, expansion or rotation.

In the velocity gradient approximation, the optical depth becomes a local property. For a cloud of dimension r with gradient dv/dr and thermal Doppler width Δv_{th} << (dv/dr)r, the line frequency is then a function of position in the cloud with

$$\nu = \nu_0 (1 - \frac{dv}{dr}\frac{r}{c}). \qquad (4.12)$$

The effect of line trapping in this situation can be described by the escape probability, a function of the line optical depth dependent on the geometry of the model. For spherical geometry the escape probability is $p = \frac{1}{\tau}(1-e^{-\tau})$; for plane parallel motion it can be shown to be $p = \frac{1}{3\tau}(1-e^{3\tau})$. The optical depth depends inversely on the velocity gradient and is

$$\tau = \alpha A_{ji}(1-\exp(-\frac{h\nu}{kT_{ij}}))N_i, \qquad (4.13)$$

with
$$\alpha = \frac{g_j}{g_i}\frac{c^3}{8\pi\nu^3 dv/dr}.$$

The net loss of population from level j to i by spontaneous emission is then pA_{ji} which for a spherical geometry at large τ becomes A_{ji}/τ. In other words the previous treatment of excitation for two levels remains the same except that A is replaced by A/τ. The most interesting result of this effect occurs for the case $A \ll C$ but $C\tau > A$. The excitation temperature then becomes

$$(\frac{k}{h\nu})T_{01} = \frac{(\frac{k}{h\nu})T_{kin}(\frac{C\tau}{A})}{\frac{C\tau}{A} + \frac{k}{h\nu}T_{kin}}$$ (4.14)

Since the optical depth is large and each layer in the clouds emits radiation to the outside due to the velocity gradient, the emitted intensity at each velocity will be $B_\nu(T_{01})$. The condition for thermalization becomes

$$C\tau \gg A\frac{kT_{kin}}{h\nu}$$ (4.15)

The complete multi-level problem can be solved as a function of T_{kin}, N_{H_2}, molecular abundance, and the collision cross sections. Sample results are presented in Figure 2. The curves of growth show that the line intensity is a slowly varying function of the product of the hydrogen density and the molecular density. For a fixed hydrogen density the line grows linearly with abundance when $C\tau \ll A$ gradually flattening out and approaching the black body radiation at the kinetic temperature at approximately the thermalization condition above. Over much of the range of interest the intensity varies as the .3 to .5 power of the abundance or , with fixed abundance, the .3 to .5 power of the hydrogen density. This explains why $^{13}CO/^{12}CO$ isotopic intensities have only a moderate observed range in their line centers.

As a consequence of line trapping the intensity in the optically thick regime does not depend on the transition strength but rather on $C\tau/A$. Since τ is itself a function of A, the line intensity for a particular molecule depends only on the collision rate and molecular abundance. Thus even in the optically thick regime line intensity is a function of abundance and therefore a measure of abundance for intensities below black body radiation at the kinetic temperature ($T_{ij} < T_{kinetic}$).

5. DETERMINATION OF H_2 DENSITY AND MASS

The hydrogen density and consequently the mass is the most important parameter to be determined from observations. Several methods for deducing N_{H_2} from millimeter observations of CO, ^{13}CO, CS and H_2CO appear in the literature, each with its own set of assumptions. For the GMC's CO and ^{13}CO are the only molecules observed over the full extent of the clouds. The mass determination of GMC's and total mass of interstellar matter depend critically on the interpretation of this data. A summary of methods and assumptions used to determine density is presented below.

5.1 Optically Thin ^{13}CO

The total column density of ^{13}CO can readily be determined assuming that the line is optically thin and with a partition function given by the kinetic temperature as measured by ^{12}CO intensities, $\Sigma N_i = KT^{12}/hB$. The optical depth of ^{13}CO,

$$\tau(^{13}CO) = \ln \frac{1}{1 - T_A^{13}/T_A^{12}}, \tag{5.1}$$

can be determined from observations of both lines. The column density obtained from the integrated optical depth can be expressed in terms of the ^{12}CO and ^{13}CO intensities with T_{01} determined from ^{12}CO,

$$N(^{13}CO) = 2.8 \times 10^{14} \int \frac{\tau(v)(T_{01} + 1)}{1 - \exp(-5.5/T_{01})}. \tag{5.2}$$

This method underestimates the column density in the observed states by assuming an optically thin line at the apparent optical depth, and overestimates the ratio of total ^{13}CO to J=0 ^{13}CO through the use of a thermal partition function at ^{12}CO temperature. These errors are related and partially cancel each other for moderate optical depths yielding a column density which is a better estimate than the assumptions imply. The next step requires knowledge of $N(H_2)/N(^{13}CO)$. Observations of local molecular clouds visible on the sky survey allow a comparison between visual optical depth A_v obtained from star counting and ^{13}CO column density. Ultraviolet observations of H_2 and H in thin clouds supply the gas to dust ratio in the form $N(H_2)/A_v$. CO observations of nearby clouds can only be calibrated for $A_v < 5$; in thicker clouds A_v is indeterminate. ^{13}CO and extinction measurements yield $N(^{13}CO)/N(H_2) \sim 1.5 \times 10^{-6}$[†]. This corresponds to about 10% of all available C in CO and $^{12}CO/^{13}CO$ isotope ratio of 40. (Analysis of $^{13}CO/H_2$ ratios by other techniques indicate somewhat lower ^{13}CO abundances). The hydrogen column densities obtained from ^{13}CO can then be translated into cloud mass and densities if an entire region is mapped. (See Fig. 1). The main uncertainty is the $^{13}CO/H_2$ abundance ratio.

5.2 Use of Curve Growth From Radiative Transfer to Determine Hydrogen Density

Observations of ^{12}CO and ^{13}CO may also be interpreted using the

† Several authors used a slightly different method for calculating $N(^{13}CO)$ than the one described here.

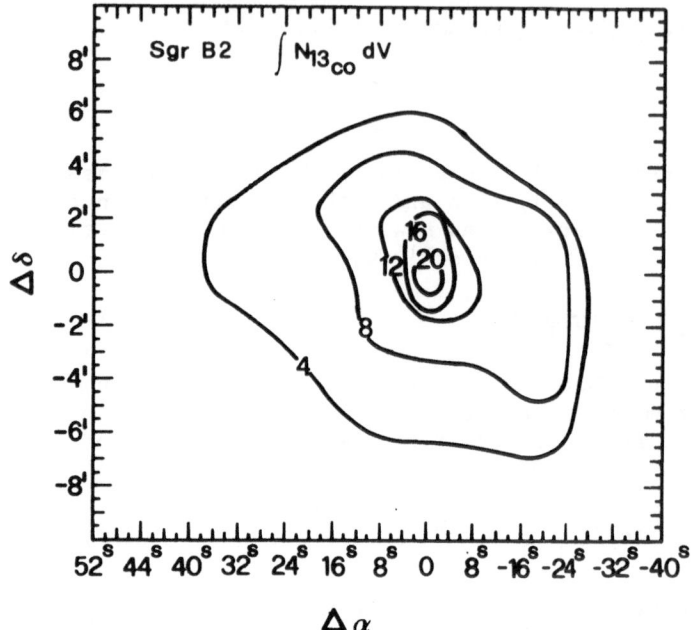

Fig. 1.   Column density of ^{13}CO in the Sgr B2 Molecular Cloud.
Contour units are 3×10^{16} cm^{-2}. 1 arc minute equals 3 pc.   The
total mass of Sgr B2 is 3×10^{6} M$_\odot$ if 10% of C is in CO and 10^{7} M$_\odot$
if 3% of C is in CO.   (Drawing from N.Z. Scoville, P.M. Solomon
and A.A. Penzias, 1975).

curves of growth in Fig. 2.   The ^{13}CO intensity is a function of
kinetic temperature, the hydrogen density N_{H_2} and the ^{13}CO abundance per unit velocity $\varepsilon/(dv/dr)$ where $\varepsilon = ^{13}CO/H_2$.   The kinetic
temperature again comes from ^{12}CO observations; therefore for a
given ε and velocity gradient the hydrogen density can be determined directly from the ^{13}CO intensity even where ^{13}CO is optically thick.   The parameter (dv/dr) can be estimated from the line
width and cloud size, $\Delta v/\Delta r$; even in the absence of a true gradient the line trapping effects will be much better accounted for by
this approximation than by ignoring them.   As can be seen from
Fig. 3 the optical depth of ^{13}CO in a cold cloud is $\tau > 1$ even for
$T(^{13}CO)/T(^{12}CO) > 1/8$.

5.3   Excitation Ladder Observations To Determine Density

Another method for determining density is to measure the relative
intensities of 2 or more transitions from the same molecule and
obtain the ratio $T_A(2-1)/T_A(1-0)$, or $T(3-2)/T(2-1)$, etc.   With two
intensities from the same molecule both the abundance and hydrogen
density can be simultaneously determined providing that neither of

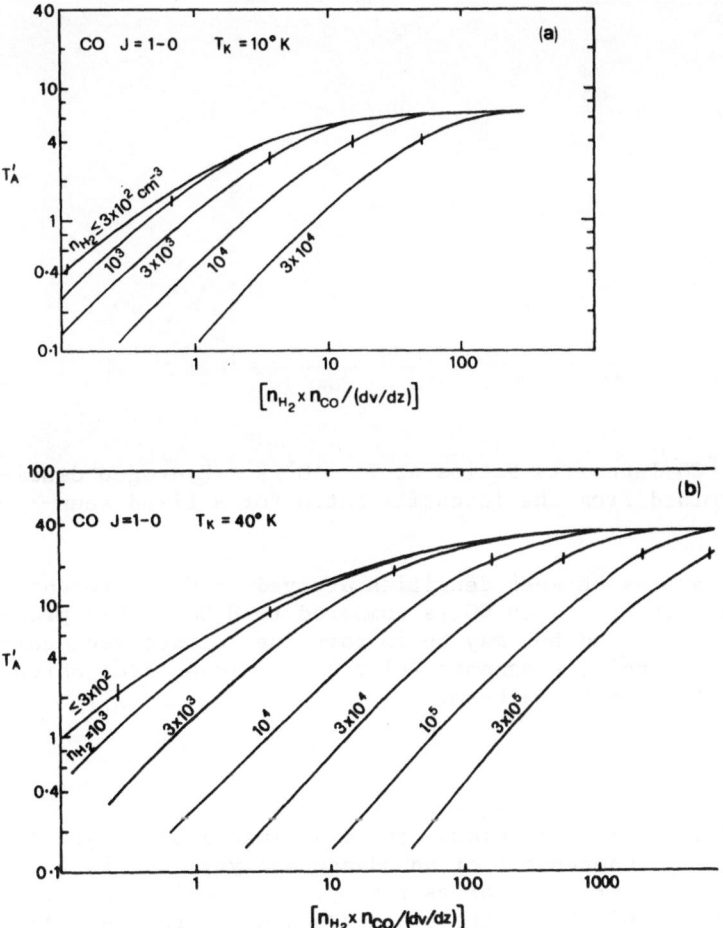

Fig. 2.   a) Curve of Growth for CO emission for T = 10° K from Radiation Transfer Calculations.   b) Curve of Growth for CO emission
at T = 40 K.

the levels is thermalized. The kinetic temperature can as usual
be obtained from ^{12}CO. This technique has been applied to CS,
^{13}CO, HC_3N and other molecules. Fig. 4 shows theoretical curves
for ^{13}CO. Very accurate intensity calibrations are required for
this method to yield significant results. This technique can also
be applied to H_2CO which has 2 cm and 6 cm doublet transitions that
are refrigerated (below 2.7 K) by collisions. By comparing observed rotational emission with the doublet absorption hydrogen densities may be deduced using the results of radiative transfer and
excitation calculations. These methods tend to indicate that cloud
core densities are in the range of $10^4 - 10^6$ cm^{-3}. There are sig-

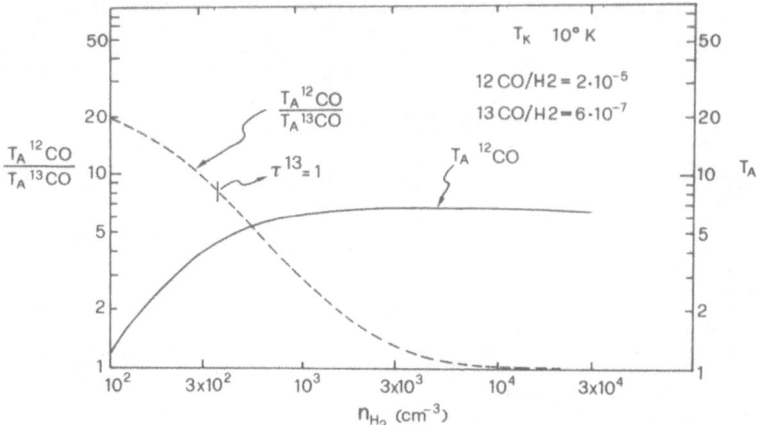

Fig. 3. $^{13}CO/^{12}CO$ Intensity Ratios at T = 10° K.   Hydrogen density can be determined from the intensity ratio for a fixed abundance.

nificant descrepancies between densities arrived at by different
observations particularly when CO is compared to H_2CO. This problem has not been resolved but may be in part due to incorrect calibrations or high density fragments (clumps) in cloud cores which
fill only part of the antenna beam.

5.4 Virial Theorem

The mass of a Giant Molecular Cloud can be estimated by a method
which is completely independent of abundance analysis and is frequently used in other branches of astronomy. Unlike low density
clouds in pressure equilibrium these objects are in gravitational
equilibrium. The observed line width Δv provides a lower limit to
the total energy of the system. Application of the virial theorem
yields a mass $M = \beta R(\Delta v)^2/G$, where β is a constant of order 0.5
depending on the geometry and model of the cloud equilibrium. Typical velocity dispersions are in the range 3-6 km/s, which gives
masses in the range $10^4 - 10^6 M_\odot$ when combined with observed dimensions.
     Order of magnitude estimates are the best that can be obtained
with the virial theorem since the processes maintaining equilibrium
are not understood. To the extent that magnetic fields contribute
to the internal pressure the mass estimated from the velocity dispersion is a lower limit.

6. HEATING AND COOLING OF MOLECULAR CLOUDS

The primary source of heating for molecular clouds is stellar ra-

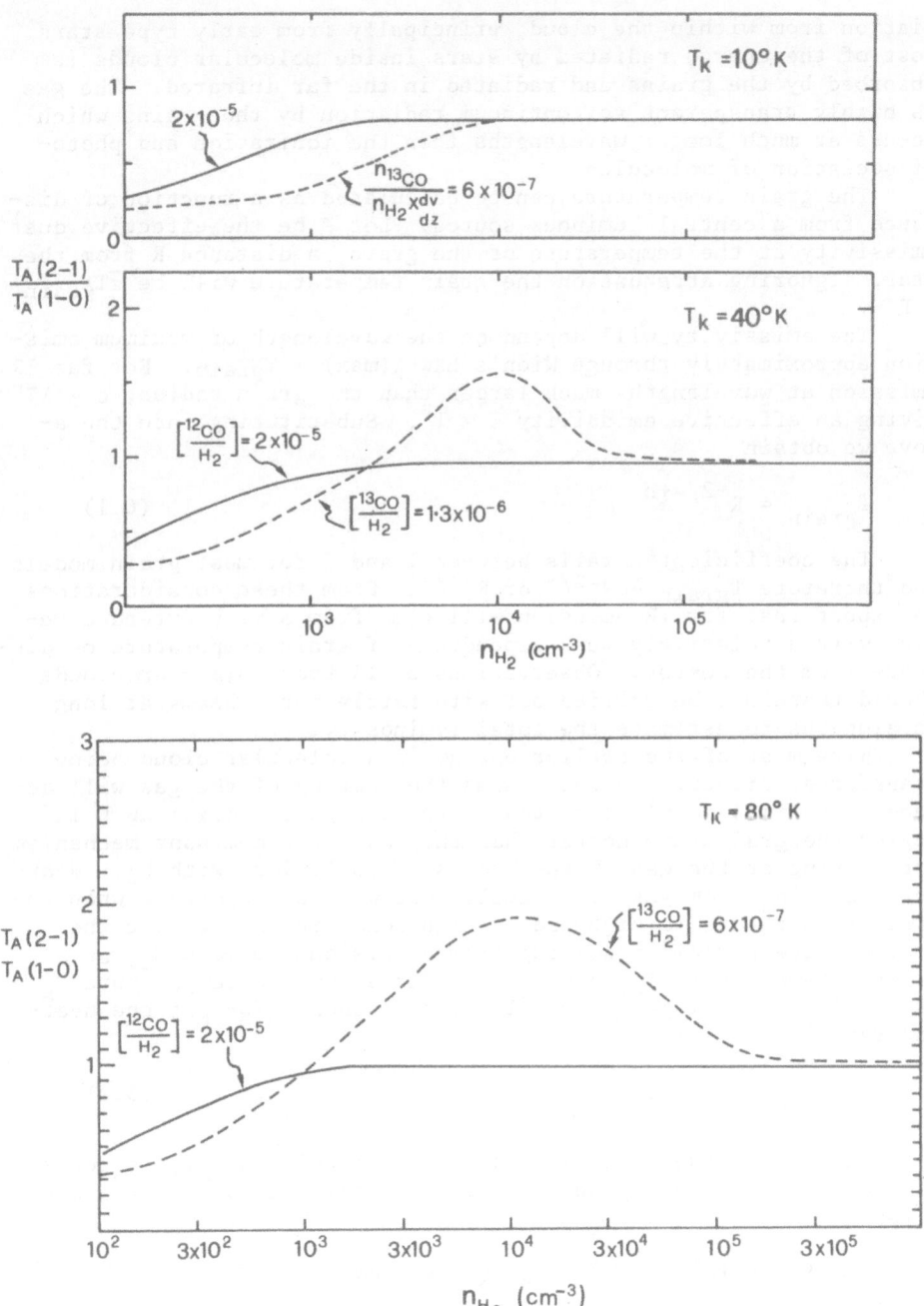

Fig. 4. Ratio of J = 2-1 line compared with J = 1-0 as a function of density and temperature.

diation from within the cloud, principally from early type stars.
Most of the energy radiated by stars inside molecular clouds is
absorbed by the grains and radiated in the far infrared. The gas
is highly transparent to continuum radiation by the grains which
occurs at much longer wavelengths than the ionization and photodissociation of molecules.

The grain temperature can be calculated as a function of dis-
tance from a central luminous source. Let $\bar{\varepsilon}$ be the effective dust
emissivity at the temperature of the grain, a distance R from the
star. Ignoring attenuation the grain temperature will be $\bar{\varepsilon}T_{grain}^4$
$\propto R^{-2}$.

The emissivity will depend on the wavelength of maximum emis-
sion approximately through Wien's Law $\lambda(max) \propto T_{grain}^{-1}$. For far IR
emission at wavelengths much larger than the grain radius, $\varepsilon \propto \lambda^{-n}$
giving an effective emissivity $\bar{\varepsilon} \propto T^n$. Substituting into the a-
bove we obtain

$$T_{grain} \propto R^{-2/4+n}$$ (6.1)

The coefficient n falls between 1 and 2 for most grain models
and therefore $T_{grain} \propto R^{-2/5}$ or $R^{-2/6}$. From these considerations
we expect that far IR emission will come from a very extended region with a relatively weak dependence of grain temperature on distance from the center. Observations of IR from molecular clouds
should therefore be carried out with fairly large beams at long
wavelengths to estimate the total luminosity.

With most of the stellar energy in a molecular cloud being
transferred directly to the grains the heating of the gas will depend on thermal coupling of the grains and gas. Unlike an H II
region the grains are hotter than the gas. The dominant mechanism
for heating of the gas by the grains is collisions with H_2. Heating resulting from gas-grain collisions will be important when the
grain temperature is high and at high densities. Let Γ be the
heating rate per cm^3 resulting from collisions between H_2, at a
kinetic temperature T_K, and grains with a surface temperature T_g.
The maximum energy gain per collision is then $k(T_g-T_K)$; the heating rate is

$$\Gamma = (n_{H_2} N_g \sigma_g vk)(T_g-T_K).$$ (6.2)

Expressing the collision rate per molecule, $N_g\sigma_g v$, in terms
of the hydrogen density and assuming all collisions are effective
gives

$$\Gamma = 1.4 \times 10^{-33}N_{H_2}^2 T^{1/2}(T_g-T_K) \text{ ergs cm}^{-3} \text{ s}^{-1}.$$ (6.3)

The dominant gas cooling mechanism is line radiation from rotational transitions in abundant molecules. This is just the sum
of all photons emitted from all levels and may be calculated from

the radiative transfer approximation discussed previously. The most effective cooling occurs from levels of abundant molecules with energy spacing equal to a few kT. For our present purpose we will adopt an analytic fit to the cooling (Scoville and Solomon, 1974) which can be approximated in two regimes corresponding to cooling transitions being optically thin and thick respectively. The major cooling is from the most abundant molecule CO. H_2 is important only at much higher temperatures, $T \gg 100$ K. Let Λ be the energy loss per cm^3:

$$\Lambda \sim 6.10^{-29} N_{CO} N_{H_2} T_K^2, \qquad \text{for } N_{CO} N_{H_2} < 10^2;$$

$$\Lambda \sim 2 \times 10^{-27} T_K^3, \qquad \text{for } N_{CO} N_{H_2} \gtrsim 10^3. \tag{6.4}$$

In the limit of high optical depth in the cooling lines which corresponds approximately to $N_{CO} N_{H_2} > 10^3$, the cooling rate saturates and becomes constant for a fixed T_K. Equating the heating and cooling rates in the case $N_{CO} N_{H_2} < 10^2$ gives

$$\frac{T_g - T_K}{T_K} \sim 4 \times 10^4 \frac{N_{CO}}{N_{H_2}} T_K^{1/2} \tag{6.5}$$

Using $N_{CO}/N_{H_2} = 2 \times 10^{-5}$ the cooling is extremely effective and $T_g \sim T_K (1 + T_K^{1/2})$, or $T_g \sim T_K^{3/2}$ for high T_g. Thus, with grain temperatures as high as 30K, the gas will be heated to 8K. Grain-gas coupling is therefore not effective in this regime.

At high density the heating becomes more efficient relative to the cooling and

$$\frac{T_g - T_K}{T_K} \sim 1.6 \frac{10^4}{N_{H_2}} \frac{T_K^{3/2}}{30} \tag{6.6}$$

Close thermal coupling with the gas temperature approaching the grain temperature will therefore occur for $N_{H_2} > 4 \times 10^4$ cm^{-3} at $T_g \sim 30$K and $N_{H_2} > 10^5$ cm^{-3} at $T_g = 100$K. Any additional radiative cooling in the gas from H_2O or ^{13}CO will of course lower T_K and raise the threshold density for $T_K \rightarrow T_g$.

Far infra-red observations of the cores of molecular clouds where $N_{H_2} \sim 10^5$ should therefore show a close correspondence between IR emission and CO emission. However, the densities over most of the volume in a Giant Molecular Cloud, $N_{H_2} \sim 10^3$ cm^{-3}, are too small for effective grain-gas coupling. For $10^3 < N_{H_2} < 10^5$ cm^{-3} the gas temperature as indicated by the CO brightness temperature will be lower than the far infra-red color temperature (grain temperature); there should still be a general correlation between the two types of maps.

At very long wavelengths corresponding to $h\nu \ll KT_g$ the far infrared observations indicate primarily total optical depth or column density since the temperature enters only linearly. In this case the IR data should correlate very well with ^{13}CO integrated intensities which`are also a measure of column density. Observations in molecular lines from less abundant molecules with strong dipole moments such as CS, HCN or H_2CO emission are indications of high density. Correspondence between these maps and IR sources at $\lambda \sim 100$ μm indicate that regions of high density are also regions of high temperature, confirming the general model developed here of a central heating source in a high density core.

REFERENCES

Scoville, N.Z. and Solomon, P.M., 1974, Astrophys. J. <u>187</u>, L 67.
Scoville, N.Z. and Solomon, P.M., 1975, Astrophys. J. <u>199</u>, L 105.
Scoville, N.Z., Solomon, P.M., and Penzias, A.A., 1975, Astrophys.J. <u>201</u>, 353.

THEORETICAL ASPECTS OF THE INFRARED EMISSION FROM HII REGIONS

Nino Panagia

Laboratorio di Radioastronomia, CNR
Bologna, Italy

1. INTRODUCTION

Treatment of the subject of IR emission from HII regions will ne-
cessitate considerable discussion of the parameters and properties
of dust grains. Therefore I feel that I should first justify why
and how dust problems are strictly related to the IR astronomy and
also why they are interesting.
    There are several good reasons to believe that dust grains are
responsible for the IR emission observed in and near HII regions:
1) The IR spectrum consists essentially of a true continuum with an
overall shape typical of thermal emission at low temperature; 2)
The IR emission is greatly in excess relative to the radiation of
the ionized gas (either f-f and b-f continuous emission or permitted and forbidden lines of any kind); 3) Strong emission from either the neutral or the molecular gas cannot explain these observations because they are rather "inefficient" emitters and would
emit mostly line radiation, which is not characteristic of the bulk
of the IR; in addition, molecules need an efficient shielding by
dust absorption (say $A_v \gtrsim 5^m$) in order to survive radiative dissociation and, thus, dust emission would clearly dominate; 4) In
several cases there is clear evidence of heavy and patchy absorption which is associated very closely with HII regions. Therefore,
the argument may be constructed as follows: There is no simple explanation of the IR emission as due to radiation from the gas. On
the other hand there is plenty of dust associated with HII regions
and dust grains are expected to emit in the IR. Thus it is most
natural to accept that the dust is actually doing so, i.e. that the
IR emission is due to dust grains.
    The next question to which we will direct our attention is why
it is interesting to study dust grains. First of all, the grain
properties such as size, amount, composition etc. can provide val-

115

G. Setti and G. G. Fazio (eds.), Infrared Astronomy, 115-136.

uable information on the abundance of the heavy elements and, therefore, on the physical conditions and the evolution of our Galaxy
and the galaxies in general. In particular, from the knowledge of
the dust properties in HII regions and the immediate vicinities,
one can gain insights as to the interaction of dust and gas in both
hot and cold environments, and on the properties and the dynamical
evolution of interstellar clouds and thus clarify and eventually
solve the problem of star formation. Therefore, it is undoubtedly
worthwhile to make IR observations, to interpret them and to derive
as much information as possible on the physical conditions and properties of both dust and gas.

     Let us now turn our attention to the theoretical aspects of
the IR emission from HII regions. In these lectures, rather than
even attempting to review all possible topics, I will devote my
attention to those problems which either need clarification or
whose solution may add to our knowledge concerning HII regions and,
more generally, diffuse matter in space.

2. ENERGY BALANCE OF DUST GRAINS

The first and most important aspect to consider is the thermal
equilibrium of dust grains because the grain temperature determines
the observed features of the IR emission. In fact, once the temperature distribution is computed, the derivation of fluxes and
brightness distributions is only a matter of straightforward numerical calculations.

     The simplest form of the thermal balance equation can be writ-
ten as Heating = Cooling. It expresses the concept that the grain
temperature is determined by the statistical equilibrium of the processes of the energy gains and the energy losses. For a simplified
but yet sufficient picture, we assume that a region be described as
shown in Fig. 1, i.e. as consisting of a cloud of gas and dust which
surrounds a central star. The gas is ionized up to a certain radius, R_H. Outside this radius, the gas is mostly neutral (only carbon, magnesium, silicon, iron and other trace elements may still be
once ionized) and farther away it becomes molecular. Here the HII
region is "ionization bounded" (as opposed to "density bounded") in
the sense that its boundary is determined by the condition that all
available Lyman continuum photons ($\lambda \lesssim 912$ Å) be absorbed within
the region (rather than the region being limited by the amount of
matter). The energy which flows into a grain in an HII region and
in the surrounding medium is provided by radiation. For conven-
ience we divide the spectrum of the input radiation into five
"bands", each characterized by a different radiative transfer problem discussed below.

     a) Lyman continuum radiation is the radiation emitted at $h\nu$
$\gtrsim 13.6$ eV and therefore, capable of ionizing the gas. Dust grains
compete with the neutral gas (mainly neutral H) in absorbing the
available radiation. However, the absorption by neutral H takes

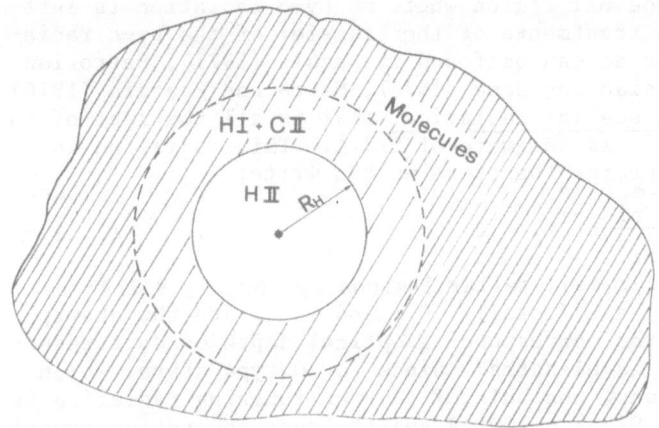

Fig. 1. A schematic representation of a cloud with an early type
star embedded in it.

place predominately near the boundary of the HII region because the
farther from the ionizing star the more neutral hydrogen becomes.
Therefore, a dust grain will receive and absorb the Ly-c radiation
of the central star almost unattenuated by gas absorption until
very near to the external boundary. There gas absorption takes
over and in a relatively short distance ($\Delta R/R_H \simeq 0.2$) the Lyman
continuum flux is reduced to zero. The contribution of the Ly-c
radiation to the heating of a grain can be written as

$$H_{Ly-c} = \pi a^2 Q_{abs}(Ly-c) \frac{L^*(Ly-c)}{4\pi r^2} g(r) e^{-\tau_r(Ly-c)} \quad \text{erg s}^{-1}/\text{grain}, \quad (2.1)$$

where πa^2 is the geometric cross section of a grain, $Q_{abs}(Ly-c)$ is
the average absorption efficiency of a grain in the Ly-c, $L^*(Ly-c)$
is the luminosity of the central star for $\lambda \leq 912$ Å, which is 0.67
to 0.20 of the total stellar luminosity for O type stars between
04 and 09 (Panagia, 1973), and $g(r)$ is a suitable function which
accounts for the attenuation by gas absorption. As mentioned above
it is approximately unity in the central parts and drops rapidly
to zero near the external boundary of the HII region. For example,
in a model with total optical depth $\tau(Ly-c) = 1$ it is $g = 0.999$,
0.941 and 0.562 for $r/R_H = 0.1$, 0.45, and 0.8 respectively. The
dust optical depth, $\tau_r(Ly-c)$, is given by

$$\tau_r(Ly-c) = n_d \pi a^2 Q(Ly-c) r, \quad (2.2)$$

n_d being the number density of grains. Note that in this section
a subscript "r" will denote the radial optical depths measured from
the center to any point in the cloud, whereas the optical depth
from the center to the edge of the HII region will be given without
any subscript.

Of course, outside the HII region where no Ly-c radiation is left
H_{Ly-c} = 0. Complete treatments of the transfer of the Ly-c radiation and useful formulae can be found in Mathis (1971), Petrosian
et al. (1972), Petrosian and Dana (1975), Natta and Panagia (1976).

   b) <u>Non-ionizing stellar radiation,</u> that is all the rest of the
stellar radiation emitted longward of 912 Å. This radiation is only absorbed by dust grains therefore we can write:

$$H_{ni} = \pi a^2 \ Q_{ni} \ \frac{L^*_{ni}}{4\pi r^2} \ e^{-\tau}r,ni \ , \qquad\qquad (2.3)$$

where L^*_{ni} is the integrated stellar luminosity for $\lambda \geq 912$ Å; of
course L^*_{ni} + L*(Ly-c) = L*(total). Q_{ni} and $\tau_{r,ni}$ are the absorption efficiency and the corresponding optical depth of dust respectively averaged over wavelengths. Since for O type stars, which
ionize most HII regions, the bulk of the non ionizing radiation is
emitted in the range 912 Å – 2000 Å and the dust absorption roughly
behaves like λ^{-1}, one can reasonably assume $Q_{ni} \simeq 0.5 \ Q(Ly-c)$ and
correspondingly, $\tau_{r,ni} \simeq \tau_r (Ly-c)/2$.

   Besides stellar radiation, there is the radiation emitted by
the ionized gas; it is conveniently divided into two additional
bands.

   c) <u>Lyman-alpha line radiation,</u> as well as any other resonance
line. This emission corresponds to the permitted transition from
the <u>first</u> excited level to the ground level of an ion. Consequently, it cannot be degraded into any secondary radiation because,
once a photon is absorbed making an electron jump to the excited
level, the photon is subsequently re-emitted by spontaneous decay
with only a small frequency shift due to the thermal motion of the
ion. Therefore, for these photons one speaks more properly of
<u>scattering</u> rather than absorption. On the other hand, the optical
depth for resonance line radiation is in general very high because
of the nature of the involved transition. In the case of the hydrogen Ly-α (λ = 1215.67 Å) the optical depth for absorption by a
neutral H-atom within an HII region is of the order of $\tau_H \simeq 3 \times 10^4$
so that a Ly-α photon undergoes approximately 10^5 scatterings before reaching the boundary of the HII region, where it arrives with
a frequency shift of about 4 Doppler widths. However, the path
length between most of these scatterings is very short (about R_H/τ_H
$\sim 10^{-4} \ R_H$) and thus a Ly-α photon may reach the boundary after
having travelled a path of about 10 times the radius of the HII
region. This implies that the absorption by dust is about 10 times
more efficient than it is for non-resonant radiation with a similar
frequency. Thus, a radial optical depth of about 0.1 may be enough
to reduce the Ly-α intensity by a factor of 2 (Panagia and Ranieri,
1973). Furthermore, a Ly-α photon which escapes from an HII region
may be scattered by the surrounding HI gas where it has a much
shorter mean free path because of the highly increased density of
the neutral hydrogen. After a short path travelled within the HI
region, the photon may be scattered back into the HII region, which
because of the frequency shift, will now be crossed without any

further scattering after travelling a path of about $4/3$ R_H. So the
Ly-α photon reaches the HII-HI boundary at the opposite side and
may again be scattered back, and so on. The total path travelled
by a Ly-α photon may become so long that even with a very small radial absorption optical depth of dust it may nevertheless be absorbed. However, if dust is strongly depleted within an HII region
but is present with normal abundance in the neutral envelope, a
Ly-α photon will eventually be absorbed more easily in the neutral
region. This may happen because in the neutral region a Ly-α photon must pass through a depth of neutral hydrogen of about 4×10^{19}
H-atoms cm^{-2} before being scattered back into the HII region. With
normal dust abundance and properties, this thickness corresponds to
an optical depth of $\delta\tau_{abs} \sim 0.02$. Therefore, when the optical
depth for dust absorption within an HII region is lower than 0.02,
the Ly-α radiation will be preferentially absorbed in a narrow layer at the very edge of the HII region, where hydrogen is mostly
neutral and dust abundance normal. Actually the optical depth must
be higher than about 0.2 because even for $\tau(Ly-\alpha) \simeq 0.15$ about 10
percent of the Ly-α radiation would be absorbed in a narrow neutral
layer with $\delta_r/R_H \sim 10^{-2}$. As a consequence, one would expect a
strong and sharp increase of the IR brightness just at the edge of
the HII region which has never been observed in any source. There
are other evidences which indicate that the optical depth of dust
in dense HII regions is of the order of unity (cf. chapter 3).
Therefore, one can usually assume that Ly-α radiation is entirely
absorbed in the ionized region. The corresponding heating will be
nearly uniform because of the characteristic long random path of
these photons. Thus, we can write

$$H(Ly-\alpha) \cong \frac{n_e^2 \; \beta_{2p} h\nu(Ly-\alpha)}{n_d} , \tag{2.4}$$

where β_{2p} represents the effective recombination rate to the 2p
level. β_{2p} may vary from 1 to 2/3 the total recombination rate to
all the excited levels, β_2, depending on the electron density (e.g.
see Gerola and Panagia, 1968). In a ionization bounded HII region,
the total number of recombinations must balance the total number of
ionizations due to gas absorption of Ly-c photons so that

$$\frac{4\pi}{3} n_e^2 \beta_2 R_H^3 = N_L^* e^{-\tau(Ly-c)} , \tag{2.5}$$

where N_L^* denotes the flux of Ly-c photons emitted by the central
star and the term $\exp(-\tau(Ly-c))$ approximately accounts for dust absorption. Therefore, eq. (2.4) can also be written as

$$H(Ly-\alpha) = \gamma N_L^* h(Ly-\alpha) e^{-\tau(Ly-c)} / (\frac{4\pi}{3} R_H^3 n_d) \tag{2.6}$$

γ being defined as $\gamma = \beta_{2p}/\beta_2$. Typically it is $N_L^* h\nu(Ly-\alpha)/L^* = 0.3$
to 0.1 for stars between spectral types O4 and O9 (Panagia, 1973).

d) <u>Other nebular radiation</u>: The energy emitted by the ionized
gas is entirely gained from absorption of Ly-c photons which have
an average energy of $\sim$ 20 eV in the case of an O type star. About
10eV are lost by emission of Ly-α photons, and thus the rest of
the nebular emission contains as much energy as the Ly-α radiation,
i.e. $\sim$ 10 eV per ionization. Most of this energy ($\sim$80 %) is emitted in the optical and ultraviolet in the form of permitted lines
($\sim$16%), forbidden lines ($\sim$37%) and continuous emission ($\sim$27%).
Only the remaining 20% is emitted in the IR and radio domains.
Since the emitters (ions) and the absorbers (grains) are intimate-
ly mixed, the resultant heating within the HII region is rather
uniform: in fact, the heating due to this diffuse radiation is only two times higher at the center of the region than near the boundary. Also, since most of the radiation is emitted in the optical
range, an effective wavelength for this "band" is around 5000 Å.
Thus, $Q(neb)/Q(Ly-c) \approx 0.25 - 0.5$ and correspondingly $\tau_{neb}/\tau(Ly-c)$
$\approx 0.25 - 0.5$. Within the HII region the heating can approximately
be written as

$$H_{neb} \approx \frac{L^*(Ly-c) - N_L^* \gamma h\nu(Ly-\alpha)}{\frac{4\pi}{3} R_H^3 n_d} e^{-\tau(Ly-c)}(1-\varepsilon), \quad r \leq R_H \qquad (2.7)$$

where ε is the fraction of the nebular radiation which can escape
from the HII region and is of the order of $(1+\tau_{neb})^{-1}$ (cf. Osterbrock, 1974; Panagia, 1974). In the surrounding neutral and molecular cloud the heating corresponds simply to what escapes the HII
region, further attenuated by the intervening dust, thus

$$H_{neb} = \pi a^2 Q_{neb} \{L^*(Ly-c) - N_L^* \gamma h\nu(Ly-\alpha)\} \varepsilon \frac{e^{-\tau(Ly-c)-\Delta\tau_{r,neb}}}{4\pi r^2}, \quad r > R_H,$$
$$\qquad (2.8)$$

with

$$\Delta\tau_{r,neb} = \pi a^2 Q_{neb}(r-R_H). \qquad (2.9)$$

e) <u>Infrared radiation emitted by the dust itself</u> may also
provide some heating. Since the absorption efficiency in the IR
is much lower than in the UV and in the optical range, this additional contribution will be important only when the other contributions become small, that is, where the optical depth from the
central star becomes large. In order to estimate the IR heating
let us assume that the radiation moves only radially and in the
outward direction. Moreover, for sufficiently high optical depth,
$\tau(Ly-c) \geq 1$ (Panagia, 1974), it is $L_{IR} \approx L^*$ and, since the resulting IR spectrum peaks at some wavelength between 20 and 100 μm, the
effective absorption efficiency will be $Q_{IR} \approx 10^{-2}$, say within a
factor of 2. Thus, the heating due to IR reradiation can be estimated as

$$H_{IR} \cong \frac{L^*}{4\pi r^2}\pi a^2 Q_{IR}e^{-\tau}r,IR \tag{2.10}$$

Comparing this to the other heating contributions we find that H_{IR} represents about one half of the total heating when $\tau_r(Ly-c) \simeq 6$ and therefore only in rather thick clouds does it become important for the thermal balance of dust grains. For the same reason, in a thick cloud the dust temperature will not decrease as fast as it would if this "secondary" heating were neglected. Rather, the heating of a grain, being dominated by the absorption of IR radiation for which $\tau_{IR} \cong 0$, will vary approximately like r^{-2} (cf. eq. (2.10)) but will be lower by a factor of $Q_{IR}/Q_{UV} = 10^{-2}$ than it would be in the case of heating by unattenuated ($\tau_{UV} << 1$) stellar radiation.

The cooling processes one may consider are: radiative losses, photoelectric effect, and grain-atom collisions. This last process which may work both for cooling and heating a grain depending on the difference in temperature between grains and atoms (Spitzer, 1968), proves to be never important for the thermal equilibrium of dust, even if it may be of primary importance for the heating of molecules in dense clouds (cf. Solomon, this volume). Therefore, here we consider only the first two.

a) <u>Radiative losses</u>. These are due to the thermal radiation from grains which can be written as

$$\Lambda_{rad} = 4\pi a^2 \int_0^\infty Q_{abs}(\nu)\pi B(\nu,T_d)d\nu, \tag{2.11}$$

where T_d represents the grain temperature and $B(\nu,T_d)$ is the Planck function. In general this integral must be evaluated numerically because $Q_{abs}(\nu)$ does not usually display a nice analytical behavior. However, there are cases in which simple approximations can be used. If we assume Q to behave like $Q = Q_0\lambda^{-m}$, the integral in eq. (2.11) is readily evaluated to be

$$Q_{abs}(\nu)\pi B(\nu,T_d)d\nu = \frac{2\pi h}{c^{2+m}}\left(\frac{kT}{h}\right)^{4+m} Q_0\zeta(4+m)\Gamma(4+m)$$

$$= 2\pi\left(\frac{k}{hc}\right)^{3+m}kcQ_0\zeta(4+m)\Gamma(4+m)T_d^{4+m}$$

$$= A_m Q_0 T_d^{4+m} \tag{2.12}$$

where $\zeta(4+m)$ and $\Gamma(4+m)$ are the Riemann zeta function and the gamma function of argument $4+m$, respectively. Some values of the coefficient A_m are presented in Table 1. These power law behaviors are not only convenient for easy calculations but also may provide reasonable approximations for the real quantities. In fact, $Q = 2\pi a/\lambda$ is an approximate upper limit to the IR absorption efficiency for any type of grain (Purcell, 1969), $Q \propto \lambda^{-2}$ is expected for con-

Table 1

Some Values of the Coefficient A_m in eq. (2.12)

m	0	1	2	3
A_m	5.670×10^{-5}	1.510×10^{-4}	5.148×10^{-4}	2.128×10^{-3}

ductors and for most materials it behaves like λ^{-2} or λ^{-3} in the far IR ($\lambda \gtrsim 50$ μm), and thus eq. (2.12) may be used for $T_d \lesssim 50$ K.

b) Photoelectric effect. An energetic photon is absorbed by a grain. If its energy exceeds some threshold energy (the work function W ≈ 5 eV plus the electric potential U if the grain is positively charged), there is a finite chance that the event results in the ejection of an electron. The yield for this occurrence becomes significantly high only for photon energies of the order of 8-10 eV and is approximately constant for energies higher than ∿ 15 eV. The value of the yield for energetic photons depends strongly on the physical properties and the size and the shape of a grain. For planar surfaces it is about 3% for graphite and may be ∿ 30% for some metals (Feuerbacher and Fitton, 1973); for small grains it should be systematically higher, possibly of the order of unity (Jura, 1976). The importance of this process for the cooling of a grain can be estimated in an HII region assuming that a grain is positively charged (Feuerbacher et al., 1973) so that W+U ≈ 14 eV and only Ly-c photons may produce photoejection with a yield Y = 1/3. Then, for every three Ly-c photons absorbed by a grain one photoelectron would be ejected with an average energy of E(photoelectron) ≈ hν(Ly-c)-W-U ≈ 6 eV. Because the process is in statistical equilibrium, each ionization must be balanced by a recombination such that thermal electrons from the ionized gas recombine onto the grain surface at a rate equal to the ejection rate. However, each recombining electron carries on the average only an energy E(thermal) ≅ 2 eV, thus the energy balance will be negative resulting in a cooling of the grain. The corresponding rate can be written as

$$\Lambda_{ph} \cong \pi a^2 Q(Ly-c) Y \frac{N_L^*}{4\pi r^2} e^{-\tau} r^{(Ly-c)} [E(photoel)-E(thermal)] \quad (2.13)$$

Comparing this expression with eq. (2.1) we see that this energy loss is only a fraction $Y\{E(photoel.)-E(thermal)\}/E(Ly-c) \cong 0.07$ of the total energy absorbed by a grain in the Ly-c and even a smaller fraction when compared with the total heating of a grain (∿3%). Thus it can usually be neglected. However, it is worth noting that, while this process is negligible for the equilibrium of a grain, it may be important for the thermal equilibrium of the gas. This fact was pointed out by Watson (1972), who studied the contribution of the photoelectric effect to the heating of hydrogen in neutral clouds and found that it may become the dominant source of heating in some cases. Here, we note that for HII regions the photoelectric effect on grains has an unexpectedly high efficiency for heating the gas. In fact, in an HII region the gas absorbs about $N_L^* \exp(-\tau(Ly-c))$ Ly-c photons per second and the dust the remaining $N_L^*(1-\exp(-\tau(Ly-c)))$. Each photon absorbed by the gas releases an energy $\Delta E_g = h\nu(Ly-c)-13.6$ eV ≈ 6 eV which directly goes into heating of the gas so that the direct input rate of energy is $N_L^* \exp(-\tau(Ly-c))\Delta E_g$. This is to be compared with the energy that dust grains are giving to the gas through the photoelectric effect. The input rate is given by (number of Ly-c photons absorbed by dust grains) x (net energy output of photoelectrons) $\approx N_L^* \{1 - \{\exp(-\tau(Ly-c))\}\}\{E(\text{photoel.})-E(\text{thermal})\}Y$. In a typical case it is $\tau(Ly-c) = 1$, $\Delta E_g \approx 6$ eV, $\Delta E(\text{photoel.}) \approx 4$ eV, which implies that the gas heating through the photoelectric effect on grains may amount to 20-30% of the direct heating through gas ionization. As a consequence, the gas temperature would be higher by about 5-15% relative to what one might estimate neglecting this effect. It is interesting to note that this additional heating would most easily explain the relatively high temperatures found in many HII regions, which otherwise would be hard to account for.

As mentioned at the beginning of this section, the equilibrium temperature of a grain is determined by equating heating to cooling. To get the feeling of what to expect in the case of HII regions we evaluate the temperature at three positions of a nebula (r/R_H = 0.1, 0.8 and 2) for two types of grains, one with high infrared emissivity and big size, i.e. $Q_1 = 2\pi a_1/\lambda$, $a_1 = 10^{-5}$ cm (cold dust), and another with low emissivity and small size, i.e. $Q_2 = 7 \times 10^{-4} a_2/\lambda^2$, $a_2 = 2 \times 10^{-6}$ cm (warm dust). We adopt $L^* = 1.5 \times 10^{39}$ erg s^{-1}, $N_L^* = 2.3 \times 10^{49}$ s^{-1} (i.e. an O5.0 type star), $n_e = 3 \times 10^3$ cm^{-3}, $(n_{d,1}\pi a_1^2 + n_{d,2}\pi a_2^2) = 10^{-18}$ cm^{-1}; furthermore, for both grains Q(Ly-c) = 1, $Q_{ni} = 0.5$ and $Q_{neb} = 0.25$. The derived equilibrium temperatures (Natta and Panagia, 1976) are shown in Table 2. We see that these are just the temperatures needed to explain the observations, because they imply strong emission occur-

Table 2

Grain Temperature at Three Positions in a Model HII Region; the parameters are specified in the text.

r/R_H	T(cold dust)	T(warm dust)
0.1	182 K	372 K
0.8	73 K	173 K
2.0	39 K	96 K

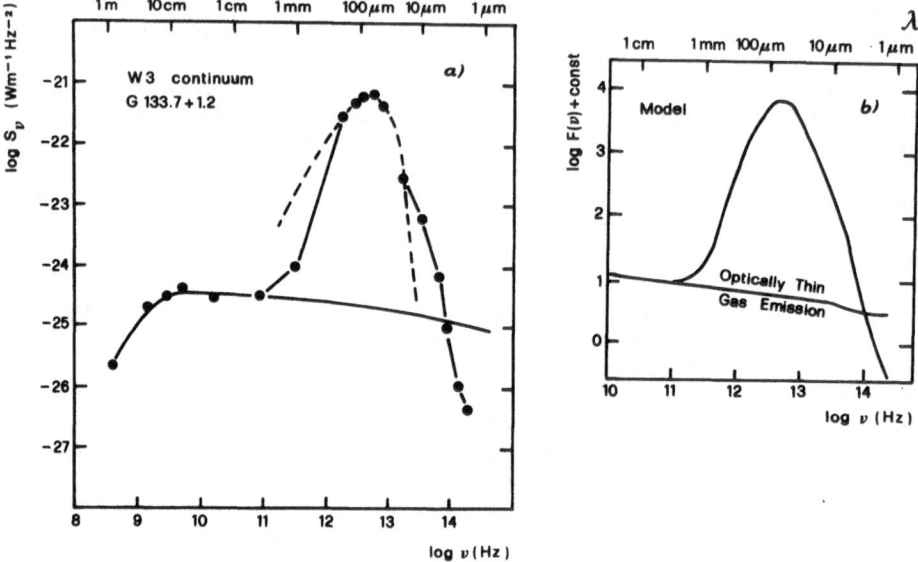

Fig. 2. a) The observed spectrum of the region G 133.7+1.2 (adapted from Wynn-Williams and Becklin, 1974). The dashed curve rep--
resents a black-body spectrum. The optically thin extrapolation
of the f-f radio spectrum is also shown. b) The computed spectrum
for a model HII region (from Natta and Panagia, 1976). The model
parameters are specified in the text.

ring between 5 and 200 μm. Moreover, the temperature is rather
insensitive to changes in most parameters because it appears with
a high power in eq. (2.12) and thus the emitted spectrum does not
depend critically on the adopted dust parameters. Therefore, we
can be sure that the IR emission in HII regions is correctly attributed to the dust grains.

3. THEORETICAL MODELS AND INTERPRETATION OF EXPERIMENTAL DATA

Several papers have been recently published on model calculations
of the IR emission from HII regions and molecular clouds (among
others, for clouds with moderate optical depths: Natta and Panagia, 1976; Bollea and Cavaliere, 1976; Aannestad, 1976; for very
thick clouds: Leung, 1975, 1976; Scoville and Kwan, 1976). Such
models were able to produce spectra and brightness distributions
which may closely represent the observations of actual sources.
For example, in Fig. 2a the observed spectrum of the region G 133.7
+1.2 is shown. At long wavelengths (λ > 100 μm) it is steeper than
a black-body spectrum, implying that the grain absorptivity, and
thus the emissivity, decreases rapidly with λ . Conversely, the

observed spectrum is shallower than a black-body curve at short
wavelengths, which implies a wide range of grain temperatures.
For comparison in Fig. 2b the spectrum derived from a purely theoretical model is shown. Note that it was not meant to fit any particular set of data and was computed adopting somewhat idealized
grain properties (Natta and Panagia, 1976). Yet the resemblance of
the two spectra is astonishing in that they agree almost to minute
details. This remarkable facility with which the observations were
reproduced indicates that, although model calculations are needed
in order to understand the basic facts, the properties of a source
often cannot be uniquely determined by fitting theoretical models
to the observational data concerning either the spectrum or the
brightness distribution. (See Natta and Panagia, 1976; Schmid-
Burgk and Scholz, 1976). Discouragingly enough, it has also been
demonstrated that the IR fluxes on the high frequency side of the
spectrum cannot be used for a reliable estimate of the absolute
amount of the dust in HII regions, firstly because grains at different temperatures contribute to the observed flux. Therefore,
the color temperature, which can readily be derived from observational data, is not a good estimate of an average physical temperature but rather it would lead to systematic overestimates of the
grain emissivity and thus underestimates of the dust amount (Panagia, 1975; Natta and Panagia, 1976). Secondly, there are convincing evidences that several dust components coexist in the same
volume (Becklin et al., 1976; Natta and Panagia, 1976; Panagia,
1977; Zeilik, 1977). They have very different properties from one
another and, therefore, significantly different temperatures and
emissivities, making any direct interpretation of flux data practically impossible as far as the absolute amounts of the various
components are concerned. Instead, the identification of dust components and the determination of the behavior of their emissivities
may rather safely be done because they involve only relative comparisons. Such analyses have permitted one to ascertain the presence of silicates from the observations of the 10 μm band (Aitken
and Jones, 1973; Gillet et al., 1975; Persson et al., 1976) and of
water ice, associated with dense clouds, from the observations of
the 3.1 μm absorption band (Gillet and Forrest, 1973; Soifer et al.
1976). However, even in these cases, which apparently require only relative fluxes and little interpretation, there is some possibility for ambiguity and/or high uncertainty concerning the quantitative results, as shown by Kwan and Scoville (1976) for the case
of the interpretation of the 10 μm band observed in very optically
thick sources. On the other hand, integrated IR luminosities are
rather insensitive to many parameters and depend strongly only on
the optical depth of dust and on stellar parameters which can easily be estimated from the radio and IR observations themselves (cf.
Panagia, 1974). Therefore, in the following discussion we shall
examine the available observations on IR luminosities in order to
derive information on the dust properties. We start by comparing
radio and middle infrared (MIR, $\lambda \leq 20$ μm) data relative to high

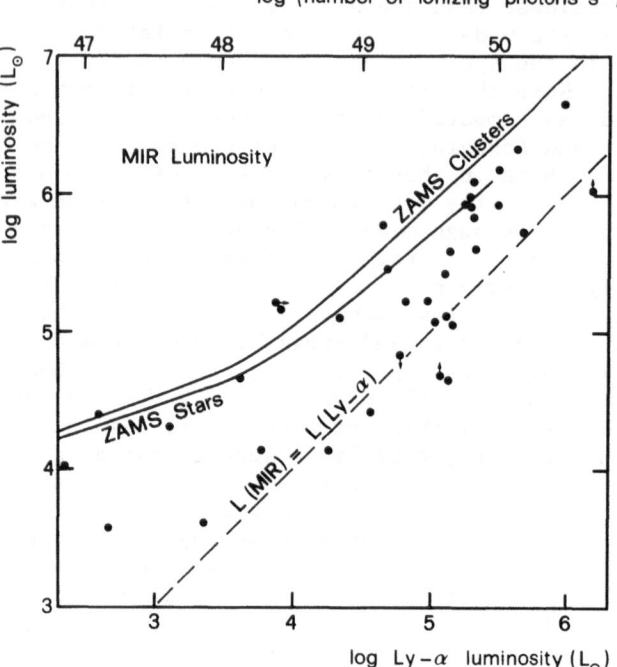

Fig. 3. The luminosity in the middle infrared as a function of the
Ly-α luminosity for 36 HII regions. For comparison, the curves
for ZAMS clusters, ZAMS stars and the line L = L(Ly-α) are also
shown.

angular resolution observations of HII regions. Since it has been
ascertained that the MIR radiation mostly comes from within an HII
region (e.g., Frogel and Persson, 1974), we are certain to deal with
emission by dust mixed with the ionized gas. In Fig. 3 the MIR luminosity is shown as a function of the Ly-α luminosity for 36 HII
regions (adapted from Panagia, 1977). We see that the energy emitted at $\lambda \lesssim 20$ μm is on the average a factor of two higher than the
energy that the Ly-α line could provide. Therefore, as much energy
as $\sim L(MIR)/2$, which also corresponds to about $L^*/5$ (cf. Fig. 3),
must have been absorbed by internal dust directly from the stellar
radiation field. Judging from model calculations, (Panagia, 1974,
and unpublished work) this implies an average UV optical depth of
internal dust of about 0.3. Furthermore, we must consider that the
estimated MIR luminosity of internal dust is in fact a lower limit
to its total emission because, lacking additional detailed information for most regions at $\lambda > 20$ μm, we cannot include the possible
emission by internal dust at longer wavelengths, which may indeed
be conspicuous (cf. Natta and Panagia, 1976). This fact strengthens
the conclusion that substantial amounts of dust are present within

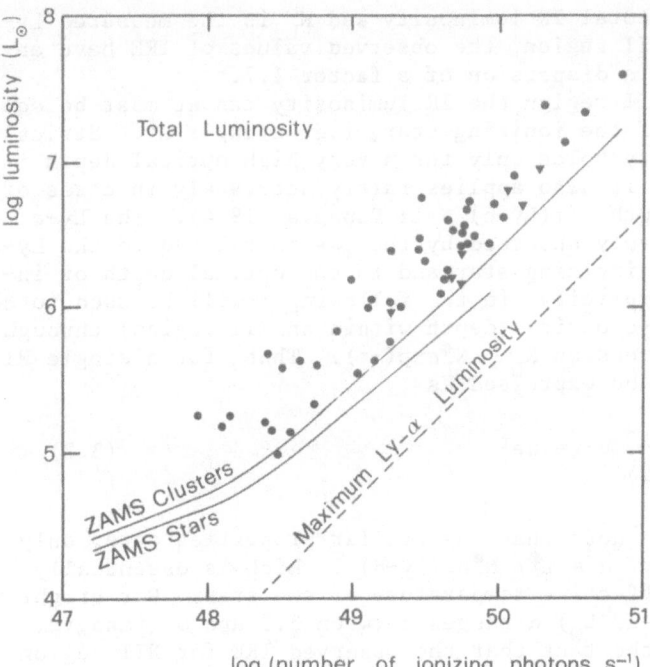

Fig. 4. The total IR luminosity is plotted against the Ly-c photon flux for 46 HII regions. 3σ upper limits for some undected regions are shown (triangles). For comparison, the curves of ZAMS clusters and ZAMS stars and the line $L_{IR} = L_{max}(Ly-\alpha)$ are also shown.

HII regions, but makes it difficult, if not impossible, to derive quantitative estimates from these data.

Then, we consider a similar graph (Fig. 4) in which the *total* IR luminosity is plotted as a function of the Ly-c photon flux for 46 HII regions. The IR data (radio data) are those obtained (compiled) by the IR group of the University College London (Emerson et al., 1973; Furniss et al., 1975). These data have the advantage of referring to a relatively large, fairly complete and, most importantly, homogeneous sample of HII regions. Additionally, the observations have been performed with a broad band photometer (40-350 μm); thus, most of the IR spectrum is included in the observational band and only small bolometric corrections may be needed. Looking at Fig. 4 we see that all sources not only greatly exceed the Ly-α luminosity but also they lie well above the curves defining the loci of ZAMS stars and ZAMS clusters (adapted from Panagia, 1973 and Natta and Panagia, 1977). If we define infrared excess as

$$IRE = L(IR)/L_{max}(Ly-\alpha) = L(IR)/\{N_L h\nu(Ly-\alpha)\} \qquad (3.1)$$

where L(IR) is the total IR luminosity and N_L is the measured Ly-c photon flux of an HII region, the observed values of IRE have an average of ∿14 with a dispersion of a factor 1.7.

For a single HII region the IR luminosity can at most be equal to the luminosity of the ionizing star, i.e. $L(IR) \lesssim L^*$. Strictly speaking the equality holds only for a very high optical depth in the cloud; however, it also applies fairly accurately in cases of moderate optical depths ($\tau(Ly-c) \gtrsim 1$: Panagia, 1974). The Ly-c photon flux effectively absorbed by the gas is related to the Ly-c flux emitted by the ionizing star and to the optical depth of internal dust (for simplicity, in the following τ will be used to denote the average Ly-c optical depth within an HII region) through the approximate expression $N_L \sim N_L^*\exp(-\tau)$. Thus, for a single HII region, the IRE may be expressed as

$$IRE = \frac{L^*}{N_L^* h\nu(Ly-\alpha)} e^\tau = \alpha e^\tau \tag{3.2}$$

It is interesting to note that the stellar properties enter only through the quantity, $\alpha = L^*/ N_L^* h\nu(Ly-\alpha)$, which is essentially determined by the effective temperature of the star. For bright O type stars ($L > 5 \times 10^4 L_\odot$) α ranges between 3.5 and 8 (Panagia, 1973). Therefore, the fact that the observed IRE for HII regions is about 14 implies an optical depth of internal dust of the order of unity. However, a quantitative and detailed interpretation of the observational data is not this easy. In fact, in Fig. 4 we see that many sources are brighter than $10^6 L_\odot$: thus, the presence of a rich OB cluster is needed in these sources in order to account for the high luminosity. In these cases, the interpretation of the observational data may not be as straightforward as for a single HII region because the geometric configuration of a source may not be well determined. The possible configurations will be intermediate between two extreme cases, namely: 1) all stars are concentrated near the center of a single big HII region, or 2) the stars are sparsely distributed within the cluster volume so that each forms its own Strömgren sphere which does not merge with any other (see Fig. 5). The first case is analogous to that of a single star with $L' = \Sigma_i L_i^*$ and $N_L' = \Sigma_i N_{L,i}^*$; thus for a young cluster the parameter α, which appears in eq. (3.2), will acquire a value of ∿ 6 (Natta and Panagia, 1977). The formulae given above can be directly used to estimate the dust optical depth from the measured IRE. Also, the derived optical depth can be correctly compared with the gas column density, $n_H R_H$, because they both refer to the same integration path i.e. the radius of the HII region. Thus, the dust-to-gas ratio can be estimated as

$$\frac{M_d}{M_g} = (\frac{4}{3} \rho a \frac{\tau}{Q})/\{n_e R_H m_H (X+Y)\} \tag{3.3}$$

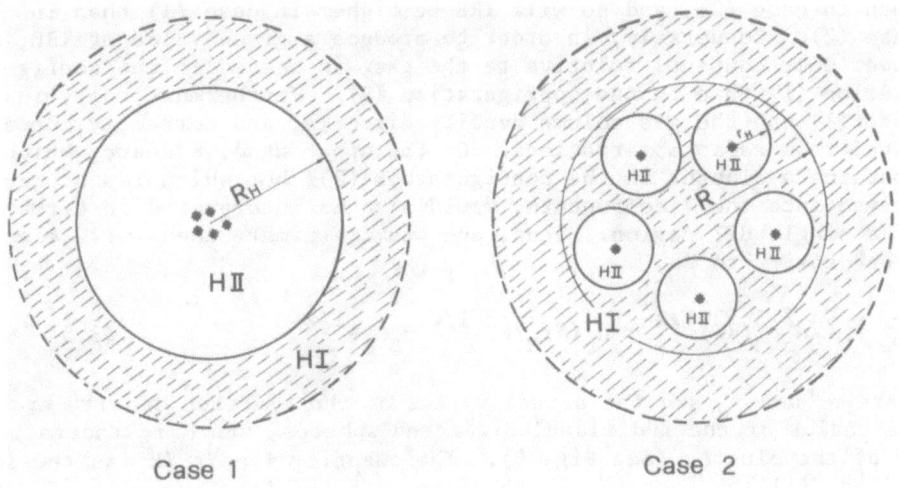

Case 1   Case 2

Fig. 5. A schematic picture of two extreme configurations of giant HII regions.

where ρ is the specific density of the grain material; m_H is the mass of the H atom; X and Y are the fractional abundances by number of hydrogen and helium in the gas (cf. Panagia, 1974). Instead, when each star of the cluster forms its own independent HII region the situation becomes much more complicated. Now, eq. (3.2) is no longer valid and it must be replaced by

$$\text{IRE(indep. regions)} \cong \frac{\Sigma_i L_i^*}{\Sigma_i N_{L,i}^* e^{-\tau_i}} \tag{3.4}$$

In this case by using eq. (3.2) as well to estimate τ , it follows that one obtains an *average value* of the optical depth of the dust contained within *each individual* Strömgren sphere. No more adequate analysis can be performed until the complex structure of the source is well elucidated. Therefore, while it is clear that the estimated τ may be compared to the gas column density of an *individual* Strömgren sphere rather than that evaluated for the entire source, it is also clear that a correct estimate of the corresponding column density is not possible if the real structure cannot be determined. Thus, no reliable estimate of the dust abundance can be obtained in these cases. To better realize the extent of the problem, let us consider the simplified case of a cluster made of N identical stars. In the configuration (1) the radius of the big HII region is approximately (not exactly because of dust absorption) $N^{1/3}$ times the radius of the individual Strömgren spheres of case (2). Therefore, the dust optical depth in case (1) will be higher

than in case (2), and so will IRE be higher in case (1) than in
case (2). Conversely, in order to produce a given value of IRE, a
lower dust content, relative to the gas, is needed in the configuration (1) than in the configuration (2). Furthermore, only in
case (1) can the gas column density directly, and correctly, be derived from radio observations. On the other hand, a source which
contains a cluster in the configuration (2), but which is not resolved into the N components, would also be interpreted in terms
of a single HII region. Thus, one would estimate the r.m.s. electron density to be

$$\bar{n}_e = (Nn_e^2 r_H^3/R^3)^{1/2} = n_e (Nr_H^3/R^3)^{1/2} = n_e \Phi^{1/2}, \qquad (3.5)$$

where n_e and r_H are the actual values of the electron density and
the radius of the individual Strömgren spheres, and R is the radius of the cluster (see Fig. 5). The quantity $\Phi = Nr_H^3/R^3$ is the so-
called "filling factor", which represents the fraction (<1) of the
cluster volume which is effectively occupied by the sources. The
gas column density would then be estimated to be

$$N_H = n_H R = \bar{n}_e R = n_e R (Nr_H^3/R^3)^{1/2} = n_e r_H (Nr_H/R)^{1/2}$$
$$= n_e r_H N^{1/3} \Phi^{1/6} \qquad (3.6)$$

A typical value of N for a cluster with L $\sim$ 3 x $10^6 L_\odot$ is $\sim$ 20.
Therefore, as long as $\Phi > N^{-2} \sim 1/400$, the value of the column density obtained by using eq. (3.5) is an overestimate. For example,
with N = 20 and Φ = 1/20, one obtains N_H = 1.65 $n_e r_H$. As a consequence, the derived dust-to-gas ratio, which is proportional to
$\tau/(n_e r_H)$ (cf. eq. (3.3)), would systematically be underestimated.
Unfortunately, it is just case (2) that most frequently occurs in
complex regions (Panagia et al. 1977b; Natta and Panagia, 1977).
Therefore, the data relative to bright and unresolved sources
(L $> 10^6 L_\odot$) cannot be used to derive absolute values of the opacity
and of the amount of the dust contained within HII regions.
    For sources with L $< 10^6 L_\odot$, instead, either one is dealing
with HII regions ionized by just one star or, if a cluster is present,both the total luminosity and the Ly-c flux are largely dominated by the brightest star (Natta and Panagia, 1977) and, thus the
approximation of a single region is fairly accurate. Therefore,
from the UCL sample we have selected the 16 regions for which L <
$10^6 L_\odot$. The absorption optical depth of dust averaged over frequencies in the Ly-c, has been derived from eq. (3.2) adopting α = 5.4
for all sources. If the dust properties (i.e. grain size, composition, abundance) were the same for all HII regions, at least in a
statistical sense, the derived optical depths,τ , would be linearly
related to the corresponding column densities, $n_H R_H$. Then, the

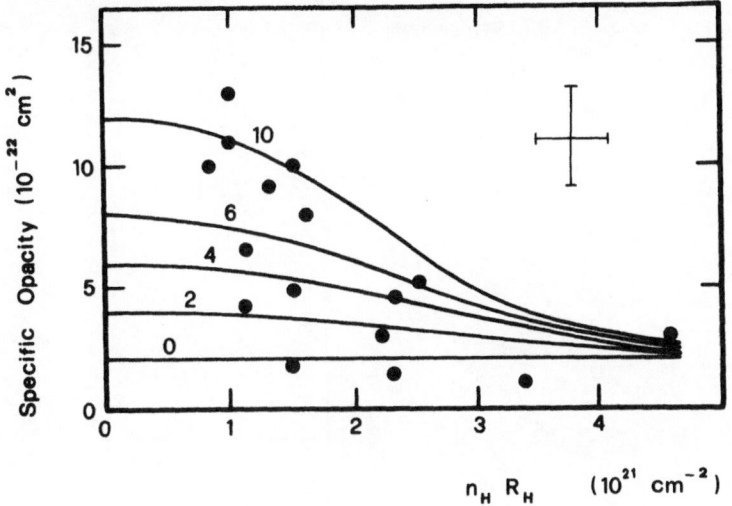

Fig. 6. The specific absorption opacity of dust, averaged in the Ly-c, is shown as a function of the gas column density for 16 HII regions with $L_{IR} < 10^6 L_\odot$. The typical error is shown in the upper right corner of the figure. The curves represent the computed relationships for models with two dust components which have different melting temperatures (see text). The labels on the curves denote the values of the specific opacity of the highly volatile component.

"specific opacity", defined as $K = \tau/n_H R_H$, would have to be essentially a constant, with some possible dispersion due to observational errors (about ± 30%) and intrinsic differences from source to source.

However, the observational data give a different picture, as shown in Fig. 6, where the derived values of K for the 16 selected HII regions are plotted against the corresponding column densities. By inspecting this figure one clearly notes that there is a systematic behavior, i.e. for low values of $n_H R_H$ there is a large dispersion of K-values (say, 2 - 12 x 10^{-22} cm^2 for $n_H R_H$ < 2 x 10^{21} cm^{-2}), whereas for higher $n_H R_H$ both the mean value and the dispersion of K are smaller (K ∿ $\bar{I}$ - 5 x 10^{-22} cm^2 for $n_H R_H$ ≳ 2 x 10^{21} cm^{-2}). To this we can add the fact that a similar behavior is also followed by all other HII regions for which it is possible to find observational data in the literature (Felli et al. 1977). Therefore, this result cannot be due to an observational bias and must be real. Qualitatively, it implies that in HII regions the smaller the radius (and therefore the higher the column density because it is approximately $n_H R_H \propto n_e^{1/3} \propto R_H^{-1/2}$), the smaller the dust opacity and, presumably, the dust amount relative to the gas. It also implies that the most compact, and therefore the youngest, HII re-

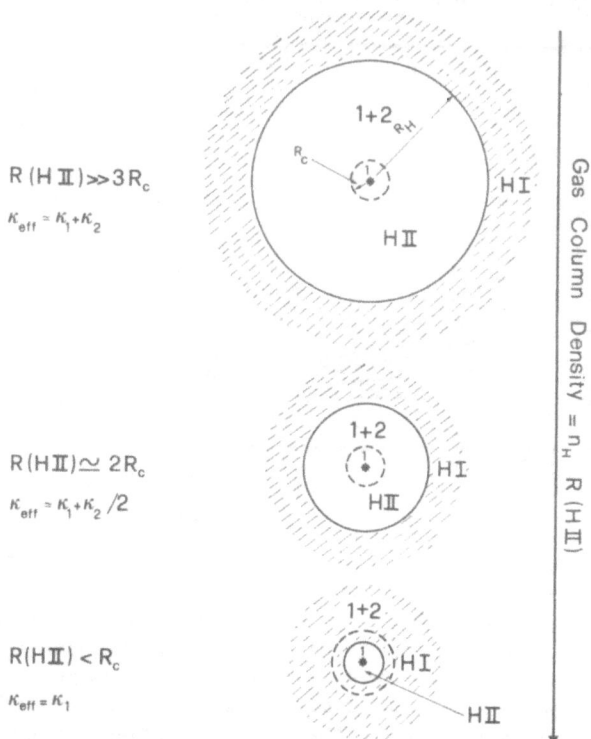

Fig. 7. Schematic representation of three configurations of an HII region which contains two dust components with different melting temperatures. A dashed circle indicates the region, of radius R_c, within which the component 2 cannot survive evaporation. The trend of the gas column density is also shown on the right hand side of the figure.

gions are the most dust depleted, and vice versa. This result can easily be explained in terms of a mixture of different types of grains, each with a different melting temperature: the lower the melting temperature of a grain the greater the distance from the exciting star within which the grain could not survive evaporation. A schematic representation depicting this mechanism is shown in Fig. 7, which illustrates the case of a cloud in which two types of dust grains are mixed together. One type (component 1) exists virtually at all distances from the central star while a second one (component 2) is only present outside a radius R_c, within which it is destroyed due to evaporation. When the radius of the HII region is much larger than R_c, the central zone devoid of component 2 has negligible importance, and the absorption effects correspond to the straight sum of the two opacities. On the other hand, for very small HII radii ($R_H \ll R_C$), only the resistant component 1 can exist within the HII region and thus, one observes

the absorption effects due only to this component. In intermediate
cases, the component 2 can survive and absorb only in the outer
parts of the HII region whereas the component 1 fully absorbs
throughout the region. Then, the absorption effects will correspond to an effective opacity which is $K_1 < K_{eff} < (K_1 + K_2)$. As
illustrated in the figure, most of the change of the effective opacity occurs for values of the HII radius from $\sim 4R_c$ to R_c; correspondingly, the column density varies by about a factor of 2. Several models of HII regions which contain two types of grains, one
with T_1(melting) = 10^3 K and the other with T_2(melting) in the
range 50 to 150 K, have been computed (Panagia et al., 1977a; Panagia and Natta, 1977). From these we have selected the models in
which the highly volatile component melts at $T_{m,2}$ = 10^2 K and as a
specific opacity K_2 ranging from 0 to $10^{-22} cm^2$. The resistant component was assumed to have $T_{m,1}$ = 10^3 K and K_1 = 2 x $10^{-22} cm^2$. The
derived curves are shown in Fig. 6. It is immediately clear that
they can account satisfactorily for all the observations.

A very detailed analysis is unwarranted because the observational errors are rather large and the statistics not very rich.
Nevertheless, it is possible to conclude that: 1) For the resistant
component, the possible values are $K_1 \sim (1-3)$ x $10^{-22} cm^2$; this
dispersion may not be real because it is comparable to that due to
observational errors; 2) The melting temperature of the highly volatile component must be $T_{m,2} < 10^2$ K, otherwise the turnover of
K_{eff} would occur at overly high values of $n_H R_H$; 3) the values of K_2
before any efficient evaporation (i.e. for low $n_H R_H$) *intrinsically*
vary from region to region in the range $(1-10)$ x 10^{-22} cm^2. This
result may be the straightforward consequence of different efficiencies for growth of mantles in various clouds, possibly because of
differences in age, density, temperature, etc.

It is interesting to compare these values of the specific absorption opacity to those measured in the diffuse interstellar medium (DIM). From the studies of the interstellar extinction by Jenkins and Savage (1974), Lillie and Witt (1976) and Mathis et al.
(1977), the specific absorption opacity around λ = 1000 Å is found
to be K_{abs} (DIM) $\sim (6-8)$ x 10^{-22} cm^2. By extrapolation to shorter
wavelengths, one may expect somewhat higher values of K_{abs} (DIM),
say of the order of 1 x 10^{-21} cm^2. The found value of K_{abs} (HII regions) $\sim (4-12)$ x 10^{-22} cm^2 then indicates that the properties (i.e.
size distribution, chemical composition, abundance, etc.) of grains
in HII regions may be comparable to those in the DIM but that some
additional and concurrent mechanisms of enrichment (condensation
and growth of icy mantles?) and depletion (evaporation, radiation
pressure, sputtering, etc.) must also be working in, and near to,
HII regions, thus modifying the detailed properties of the grains.

As we have previously noted, the data relative to giant HII regions $(L > 10^6 L_\odot)$ cannot be directly used for absolute determinations of the dust opacity because these regions may comprise several unresolved sources. On the other hand, in giant HII regions,

which necessarily are ionized by an OB cluster, the spectrum of
the ionizing radiation has approximately the same behavior in all
cases. Therefore, if any difference in the ionization of the elements is found in these regions, it can only be ascribed to effects
of selective dust absorption, i.e. absorption which becomes more
efficient as the frequency increases. A pioneer study in this direction has been performed by Churchwell et al. (1974). They noticed that the ratio of the ionized helium abundance to that of
ionized hydrogen, $y^+ = He^+/H^+$, decreases as the infrared excess of
an HII region increases. Since an increase of the latter is due to
dust absorption, the variation of the He ionization, being correlated to that of IRE, must also be due to the same cause, i.e. dust
absorption. Incidentally, we note that the argument can be reversed as follows. Since i) y^+ and IRE are related to each other,
and ii) the ionization of He and H can physically be determined only by phenomena which occur within the ionized region, then the
value of IRE must also be determined largely by what occurs solely
within an HII region. Therefore, even if a large part of the dust
emitting in the far IR may be outside the HII region, possibly in a
surrounding molecular cloud, the energetics of the dust, *both inside and outside the HII region,* is dominated by the supply provided by the ionizing star(s) and thus, possible additional energy
sources are not important.

By analyzing the then available data with the aid of a simplified model, Mezger et al. (1974) found that the absorption cross
section of dust should increase with frequency in such a way that
the ratio of the average value in the He-continuum (i.e. λ = 228 to
504 Å) to that in the H-continuum (λ = 504-912 Å) be $a = K_{He}/K_H =$
7 ± 3. However, this result is difficult to understand because
there is no known material whose opacity behaves in such a fashion.
This result cannot be justified even theoretically without unrealistic "ad hoc" assumptions. Therefore, a similar, but more refined,
analysis has recently been performed by Panagia and Smith (1977).
The IR data were taken from the measurements of the UCL-IR group;
the radio measurements were updated to the most recent determinations. Moreover, a more detailed theoretical model was employed
for the interpretation of the data. Such a model explicitly takes
into account the variation of the spectral distribution in the IR
as a function of the dust optical depth. Thus, the fraction of the
luminosity which is emitted within the observing band was not taken
to be a constant but rather was allowed to vary according to the model parameters. The results are: 1) If one accepts that the currently adopted model atmospheres and the temperature scale of O type
stars are 100% correct, then $a = K_{He}/K_H \simeq 4 \pm 1$; 2) If one adopts a
steeper spectrum for the stellar UV radiation, as suggested by several independent evidences (see the discussion in Panagia and Smith,
1977), then a value as low as $a \simeq 2$ is also possible. Since there
are definite evidences in favor of a rather steep UV spectrum,
whereas there is none to support the validity of the flat spectra

predicted theoretically, we argue that it is most likely that α $\simeq$ 2. This value corresponds to a frequency dependence of the absorption cross section $\sigma(\nu) \propto \nu$, which is that expected for small grains ($\lesssim$100 Å) with a constant index of refraction; 3) Absolute values of the specific opacities cannot be directly estimated because of the aforementioned difficulties. However, it has been shown that, by assuming giant HII regions to consist of a single big Strömgren sphere, the derived opacities are lower limits to the true values.

Using this assumption it is found that $K_{He} \gtrsim 10^{-21}$ cm^2 and K_H $\gtrsim 4 \times 10^{-22}$ cm^2. Moreover, from independent considerations on the most probable geometric configurations of giant HII regions, it has been estimated that the true values should be a factor of about 2 higher than the lower limits (Panagia and Smith, 1977; Natta and Panagia, 1977). Therefore, it should be $K_{He} \simeq 2 \times 10^{-21}$ cm^2 and $K_H \simeq 8 \times 10^{-22}$ cm^2. It is interesting to note that these absolute values, although rather uncertain because of the needed correction factor, compare favourably with those determined previously. In fact, being approximately 80% and 20% the fraction of Ly-c photons emitted in the H-continuum and the He-continuum, respectively, the specific opacity averaged over the whole Ly-c would turn out to be $\bar{K}(Ly-c) \simeq (0.8 \times 8 \times 10^{-22} + 0.2 \times 2 \times 10^{-21}) \simeq 10^{-21}$ cm^2, which agrees rather well with the value $K_{eff} \simeq (8\pm4) \times 10^{-22}$ cm^2 found for non-giant HII regions.

REFERENCES

Aannestad, P.A.,1976, in Far Infrared Astronomy, ed. M. Rowan-Robinson (Pergamon Press, Oxford, England), p. 257.

Aitken, D.K. and Jones, B., 1973, Astrophys.J., 184, 127.

Becklin, E.E., Beckwith, S., Gatley, I., Matthews, K., Neugebauer, G., Sarazin,C., and Werner, M.W., 1976, Astrophys. J., 207, 770.

Bollea, D., and Cavaliere, A., 1976, Astron. and Astrophys., 49, 313.

Churchwell, E., Mezger, P.G., and Huchtmeier, W., 1974, Astron. and Astrophys., 32, 283

Emerson, J.P., Jennings, R.E., and Moorwood, A.F.M., 1973, Astrophys.J., 184, 401.

Felli, M., Natta, A., and Panagia, N., 1977, in preparation.

Feuerbacher, B., and Fitton, B., 1972, J. Appl. Phys., 43, 1563.

Feurbacher, B., Willis, R.F., and Fitton, B., 1973, Astrophys. J., 181,101.

Frogel, J.A., and Persson, S.E., 1974, Astrophys. J., 192, 351.

Furniss, I., Jennings, R.E., and Moorwood, A.F.M., 1975, Astrophys. J., 202, 400.

Gerola, H., and Panagia, N., 1968, Astrophys. Space Sci., 2, 285.

Gillett, F.C., and Forrest, W.J., 1973, Astrophys.J., 179, 483.

Gillett, F.C., Forrest, W.J., Merrill, K.M., Capps, R.W., and Soifer, B.T., 1975, Astrophys. J., 200, 609.

Jenkins, E.B. and Savage, B.D., Astrophys. J., 187, 243.

Jura, M., 1976, Astrophys. J. 204, 12.

Kwan, J., and Scoville, N., 1976, Astrophys. J., 209, 102.

Leung, C.M., 1975, Astrophys. J., 199, 340.

Leung, C.M., 1976, Astrophys. J., 209, 75.

Lillie, C.F., and Witt, A.N., 1976, Astrophys. J., 208, 64.

Mathis, J.S., 1971, Astrophys. J., 167, 261.

Mathis, J.S., Rumpl, W., and Nordsieck, K.H., 1977, Astrophys. J., 217, 425.

Mezger, P.G., Smith, L.F., and Churchwell, E., 1974, Astron. and Astrophys., 32, 269.

Natta, A., and Panagia, N., 1976, Astron. and Astrophys., 50, 191.

Natta, A., and Panagia, N., 1977, in preparation.

Osterbrock, D.E., 1974, Astrophysics of Gaseous Nebulae, (W.H. Freeman and Co., Chicago), p. 241.

Panagia, N., 1973, Astron. J., 78, 929.

Panagia, N., 1974, Astrophys. J., 192, 221.

Panagia, N., 1975, Astron. and Astrophys., 42, 139.

Panagia, N., 1977, in Infrared and Submillimeter Astronomy, edited by G.G. Fazio (Reidel, Dordrecht, Holland), p. 43.

Panagia, N., Natta, A., and Felli, M., 1977a, Mem. Soc. Astr. Ital. 48, in press.

Panagia, N., Natta, A., and Preite-Martinez, A., 1977b, Astron. and Astrophys., in press.

Panagia, N., and Ranieri, M., 1973, Mem. Soc. Roy. Sci. Liège, 6e Ser., tome V, p. 275.

Panagia, N., and Smith, L.F., 1977, Astron. and Astrophys., in press.

Persson, S.E., Frogel, J.A., and Aaronson, M., 1976, Astrophys. J., 208, 753.

Petrosian, V., Silk, J., and Field, G.B., 1972, Astrophys. J., (Letters), 177, L69.

Petrosian, V., and Dana, R.A., 1975, Astrophys. J., 196, 733.

Purcell, E.M., 1969, Astrophys. J., 158, 433.

Schmid-Burgk, J., and Scholz, M., 1976, Astron. and Astrophys. 51, 209.

Scoville, N., and Kwan, J., Astrophys. J., 206, 718.

Soifer, B.T., Russell, R.W., and Merrill, K.M., 1976, Astrophys. J. 210, 334.

Spitzer, L., Jr., 1968, Diffuse Matter in Space, (John Wiley and Sons, New York), p. 142.

Watson, W.D., 1972, Astrophys. J., 176, 103.

Wynn-Williams, C.G., and Becklin, E.E., 1974, Publ. Astron. Soc. Pacific, 86, 5.

Zeilik, M., II, 1977, Astrophys. J., 213, 58.

STAR FORMATION AND RELATED TOPICS

Richard B. Larson

Yale University Observatory, Yale University
New Haven, Connecticut U.S.A.

1. THE EVOLUTION OF PROTOSTARS

Many processes and many types of observations are relevant to an
understanding of star formation. In these lectures I shall
attempt to sketch briefly some of the problems involved and some
results, mostly theoretical, of studies of star formation, con-
centrating on qualitative or "back-of-the-envelope" aspects. The
first lecture will consider the collapse of individual protostars,
as understood mainly from spherical collapse models; the second
will consider multidimensional collapse and the fragmentation
problem; and the third will consider some aspects of star formation on a galactic scale.

1.1 Initial Conditions for Protostars

The circumstances under which protostars form and begin to collapse
depend in detail on the prior evolution of collapsing clouds,
but one can estimate from the Jeans criterion that a minimum
density of the order of 10^4 - 10^5 cm^{-3} is required for a one-solar-
mass protostar to collapse (Larson, 1974). Such a collapsing
fragment will almost certainly have a significant angular momentum
and a nonspherical shape, and some coalescence of fragments may
even occur during star formation; however, many of the basic qualitative features of protostellar collapse are illustrated by the
simpler case of the collapse of a spherical, nonrotating cloud
fragment. Therefore we consider first the collapse of a spherical
condensation in an extended cloud of similar material.

G. Setti and G. G. Fazio (eds.), Infrared Astronomy, 137-158.
All Rights Reserved. Copyright © 1978 by D. Reidel Publishing Company, Dordrecht, Holland.

1.2 The Nature of Gravitational Collapse

The evolution of a collapsing cloud or protostar is governed by
the basic "runaway" nature of gravitational collapse: the density
in a collapsing cloud tends to approach infinity in a finite time
and, since the free-fall time varies as $\rho^{-1/2}$, the denser parts
of the cloud collapse first, leading to the runaway growth of
density fluctuations. Moreover, any anisotropy in the shape of
a collapsing cloud is also amplified if pressure forces are not
important, since gravity is strongest along the shortest axis of
a nonspherical object. Thus a collapsing protostar may be expected
to develop a very inhomogeneous and perhaps nonspherical structure,
with a "core" of very high density and an envelope of much lower
density. It will also tend to continually accrete more material
from the surrounding cloud, so that a protostar will in general
possess an extended accretion envelope of matter gravitationally
bound to the core.
     The best studied collapse problem is that of isothermal
collapse, which is relevant (approximately) to the early stages
of the collapse of a protostar. If the protostellar mass is
approximately equal to the Jeans mass, detailed calculations
show that the density distribution always becomes sharply peaked
at the center, regardless of the details of the initial and
boundary conditions. Even an initially uniform density distribution is soon altered by the propagation of rarefaction waves inward from the boundary in a time comparable to the collapse time.
The resulting density distribution is determined by the fact that
at any stage of the collapse the dense core of the cloud must
have a mass and radius satisfying $M/R \propto T$ = const. in order to
continue collapsing; thus the core must have a density $M/R^3 \propto R^{-2}$,
and since the density of the outer parts of the cloud hardly
changes during the runaway collapse of the core, a density distribution of the form $\rho \propto r^{-2}$ is built up. If the cloud mass substantially exceeds the Jeans mass, the central part of the cloud
can collapse more nearly uniformly, but detailed calculations
are then necessary to establish the form of the density distribution (eg. Bodenheimer and Sweigart, 1968).
     Isothermal collapse can continue only until the core of the
cloud becomes opaque at a density of $\approx 10^{10}$ - 10^{12} cm^{-3}, causing
the temperature and pressure to rise rapidly (Larson, 1974). The
core may then experience a brief period of hydrostatic equilibrium,
but this is soon destroyed when hydrogen molecules begin to
dissociate, reducing the ratio of specific heats γ below 4/3.
The collapse is not finally halted at the center until the hydrogen is largely ionized and essentially stellar conditions are
reached, with $\gamma > 4/3$. The condition $\gamma > 4/3$ for stability of
a hydrostatic stellar object is satisfied only if the radius is
$\lesssim 100$ R$_\odot$, so the stellar core of a protostar must be tiny compared
to the envelope, which may have a radius of ≈ 0.01 - 0.1 pc.
Since the initial isothermal collapse generates a mass distribution

$M \propto R$ in the outer envelope, most of the mass is still at large
radii when the core forms, and the core initially has only a very
small mass.

Thereafter the evolution of the protostar consists essentially
of the building up of the core by infall of material from the
envelope. Accretion phenomena thus become of dominant importance:
supersonic infall creates an accretion shock at the surface of the
core, and a density distribution of the form $\rho \propto r^{-3/2}$, which is
characteristic of a spherical accretion flow, is gradually established throughout the envelope. The timescale for the infall of
the remaining envelope matter onto the core is essentially the
free-fall time for the outermost layers of the envelope, and this
is still an appreciable fraction of the initial free-fall time.
For example, in models calculated by Larson (1969a, 1972b) and
by Appenzeller and Tscharnuter (1975), the free-fall time is
2×10^5 yr and the time required for the accretion of 90% of the
envelope mass is also about 2×10^5 yr, whereas in a model of
Westbrook and Tarter (1975) which assumes a lower temperature
and hence allows a more rapid collapse of the outer envelope,
the corresponding accretion timescale is about an order of magnitude shorter.

1.3 Evolution of the Core

In the collapse models of Larson and of Appenzeller and Tscharnuter
(but not in that of Westbrook and Tarter), the material immediately
outside the core becomes effectively optically thin at an early
stage, so that the core is able to lose energy by radiation from
its surface like a normal star. The subsequent evolution of the
core then depends on the relative magnitudes of the accretion
time and the radiative cooling and contraction time for the core.
The ratio of these timescales is very different for low mass and
high mass protostars, which accordingly show qualitatively different types of core evolution.

For low mass protostars ($M \lesssim 5M_\odot$), the pre-main sequence contraction time for the core is long compared with the accretion
time, so throughout the accretion process the core maintains an
entropy and a radius that are characteristic of a pre-main sequence
star. Its surface is heated to a high temperature by infalling
matter and this produces most of the total protostellar luminosity,
but eventually accretion becomes unimportant and the surface
temperature of the core approaches that of a Hayashi track object.
During the later stages of the accretion process, radiative energy
losses from the outer layers of the core reduce its radius and
hence increase its contraction time until it becomes comparable
to the accretion timescale. Thus the accretion timescale
essentially determines the contraction time and hence the radius
and luminosity of the core when it becomes visible as a conventional pre-main sequence star. Since the temperature of the
Hayashi track is approximately constant, the contraction time can

be estimated as $\sim GM^2/RL \sim 10^7(R/R_\odot)^{-3}$ for a star of one solar
mass. If this is equated to an accretion timescale in the range
$\sim 10^4 - 10^6$ yr, a final radius in the range $\sim 2 - 10$ $R_\odot$ is predicted.
The radii of T Tauri stars are in fact observed to lie in this
range.

Thus the radii (or contraction times) of T Tauri stars
provide information about the accretion timescales for protostellar
envelopes and suggest that these are in the range $\sim 10^4 - 10^6$ yr.
The fact that these numbers are of the same order as the free-fall
times for condensations in molecular clouds suggests that the
free-fall time is in fact the relevant timescale for accretion,
despite the neglect so far of the effects of rotation and magnetic
fields.

For high mass protostars (M $\gtrsim$ 5$M_\odot$) the pre-main sequence
contraction time of the core becomes shorter than the envelope
accretion time, so the protostellar core can contract all the
way to the main sequence before it has accreted all of the envelope material. For massive stars the pre-main sequence contraction
time is given as a function of mass approximately by $t_c \sim 2 \times 10^7$
$(M/M_\odot)^{-2.5}$ yr; if we again adopt a range of accretion times $t_a \sim$
$\sim 10^4 - 10^6$ yr, we estimate that t_c becomes less than t_a for masses
greater than $\sim 3 - 20$ $M_\odot$. Stars of larger mass would then be predicted to complete their protostellar evolution by evolving up
the main sequence as they gain mass. Note that in this case the
distinction between "protostar" and "main sequence star" becomes
a semantic one, since a "protostar" in the present usage can contain as its core a main sequence star. A more meaningful distinction for the interpretation of observed objects is whether
the central stellar core is still accreting matter and growing
in mass.

There are a number of ways in which the core of a massive
protostar can react back on the envelope and retard or halt
further infall of matter when the core becomes very massive. The
most studied and perhaps most important effect is radiation
pressure (eg. Kahn, 1974; Westbrook and Tarter, 1975; Cochran and
Ostriker, 1977; Yorke and Krügel, 1977). Radiation pressure
acting on the dust grains in protostellar envelopes tends to concentrate the matter into dense shells or "cocoons", and the infall
of matter in these shells can be reversed by radiation pressure
when the core attains a mass of the order of 30 - 40 $M_\odot$. However,
this is not necessarily an upper limit on the core mass, since
Rayleigh-Taylor instabilities may break up the cocoon and allow
infall of more matter. A second possible effect is ionization,
which may heat up and blow off protostellar envelopes for core
masses above $\sim 30 - 60$ $M_\odot$ (Larson and Starrfield, 1971). A third
effect which might be important but has not yet been studied
quantitatively is the effect of a stellar wind from a hot stellar
core. Finally, a violent flareup of the core caused by rapid
heating of its outer layers as it approaches radiative equilibrium
can blow off the outer layers and the entire envelope; for example,

Appenzeller and Tscharnuter (1974) find that the envelope of a
60 $M_\odot$ protostar is ejected after the core attains a mass of only
18 $M_\odot$. In all cases, the result is sensitive to the initial conditions, and particularly to the density of the protostellar
envelope; the higher the density, the longer the dynamical
pressure of infalling matter can dominate over radiation pressure,
etc., and the larger is the mass that the core can attain. Further
quantitative exploration of these effects would be desirable.

1.4 Infrared Radiation from Protostars

Since the stellar core of a protostar remains completely hidden
until the envelope has nearly disappeared, the observed characteristics of a protostar depend on the properties of the envelope.
In particular, the luminosity from the core is absorbed by dust in
the envelope and reradiated at infrared wavelengths, so that the
emergent spectrum is determined by the transfer of infrared radiation through the envelope. Although there is no particular reason
to assume spherical symmetry, this is the only case for which detailed calculations have been made, so some results for predicted
infrared emission from spherical protostellar envelopes will be
summarized.
        In the simplest case, where thin dust shells are not an im-
portant feature, the density distribution in a spherical accretion
envelope has the form $\rho \propto r^{-3/2}$. In solving the radiative transfer
problem, it is usually also assumed, in the absence of more
detailed information, that the infrared emissivity of the dust
grains varies with wavelength as $\varepsilon_\lambda \propto \lambda^{-p}$ where $1 \leq p \leq 2$. In the
optically thin limit the temperature distribution is then $T \propto$
$\propto r^{-2/(4+p)}$; detailed calculations for the optically thick case
yield temperatures that are slightly lower (eg. Scoville and Kwan,
1976). The emergent spectra for models of this type have been
calculated in detail by Bertout (1976) and by Scoville and Kwan;
the results are similar to the earlier approximations of Larson
(1969b), and yield a spectrum not greatly different from that of
a blackbody with temperature equal to the temperature of the dust
at the radius where the infrared optical depth is equal to unity.
If spatial resolution of the protostellar envelope is possible,
the observed spectrum becomes cooler with increasing radius, since
the dust temperature decreases outward.
        As the amount of matter in the protostellar envelope declines,
the "photosphere" or surface of optical depth unity recedes inward
to smaller radii where the dust temperature is higher, so the
apparent temperature of the protostar increases. For a model
with a density distribution $\rho = \rho_0 r^{-3/2}$ and an opacity law
$\varepsilon_\lambda \propto \lambda^{-3/2}$, the apparent blackbody temperature T_b varies with ρ_0
and L according to

$$T_b \sim 300 \ (\rho_0/10^6)^{-2/5} \ (L/10^{35})^{1/10} \mathrm{K} \qquad (1)$$

(Larson, 1974). This yields apparent temperatures ranging from
≲50 K just after the core forms to ≳ 1000 K just before the
envelope becomes optically thin; most of the time is spent at
temperatures in the range ∿300 - 1000 K, so observed protostellar
temperatures should usually fall in this range. Note however
that quantitative predictions are sensitive to many uncertainties,
including the densities of protostellar envelopes; higher densities
would imply lower apparent temperatures.

    For massive protostars, detailed dynamical models including
the effects of radiation pressure have been calculated by Yorke
and Krügel (1977), and the infrared spectra of these models have
been calculated by Yorke (1977). These authors assume that at
temperatures < 100 K there are two types of dust grains: (1) bare
graphite particles, and (2) graphite cores with ice mantles; at
temperatures >150 K, the ices evaporate and only the refractory
cores remain. As a result, the protostellar envelope develops
a double cocoon structure: (1) an inner cocoon at a radius of
∿10^{14} cm consisting of refractory grains at a temperature of ∿1000
K, and (2) an outer cocoon at ∿10^{17} cm containing ice-coated particles at a temperature of ∿100 K. The outer cocoon is driven
outward by the pressure of near-infrared radiation from the
inner cocoon, and it is this effect which eventually halts further
infall of matter into the core and limits its mass.

    The spectrum of this model shows, as might be expected, two
peaks due to infrared emission from the two dust shells at
temperatures of ∿1000 K and ∿100 K (Yorke, 1977). Superimposed
on the shorter wavelength peak are narrow absorption features at
wavelengths of 3μ and 12μ due to maxima in the absorption
coefficient of ice at these wavelengths (silicates were not considered in these models). These results are in qualitative agreement with the infrared spectra of a number of objects thought to
be protostars or dust-embedded young stars; for example, the
Becklin-Neugebauer and Kleinmann-Low sources in Orion show, when
considered together, a double-humped spectrum which may be ex-
plainable by emission from an inner hot cocoon (the BN object)
and outlying cooler dust (the KL nebula) surrounding a massive
forming star or group of stars. The predicted existence of narrow
absorption features in the spectrum shows that such features
are naturally produced in a protostellar envelope with a temperature gradient (see also Finn and Simon, 1977), and do not require
additional extinction by foreground dust.

    While the observed infrared spectra of young objects can thus
be qualitatively understood, the detailed shape of the spectrum is
sensitive to the geometry of protostellar envelopes, which may
not be anything like the above idealized spherical models, and
to the properties of the dust grains, which may also be more complicated than has been assumed. For example, if there are two
types of dust grains, they will in general have different temperatures (Greenberg, this institute), and this effect can also produce
a broadened or double-peaked spectrum. Thus until such effects

are better understood, it may be difficult to interpret infrared
broad-band spectra in terms of detailed models of protostars; as
in the stellar case, more information about specific structural
properties of protostars may ultimately be obtained from the
study of narrow spectral features.

2. THREE-DIMENSIONAL COLLAPSE AND FRAGMENTATION

In order to understand better how protostars form and evolve, it
is necessary to study collapse with less restrictive assumptions
than spherical symmetry. We therefore consider next some results
of 2-dimensional collapse calculations made assuming axial symmetry,
and of 3-dimensional calculations assuming no symmetry. The
latter case is of particular interest because of the importance of
understanding the process of fragmentation of collapsing clouds,
which plays a crucial role in determining the stellar mass spectrum.

2.1 Summary of Two-Dimensional Collapse Results

If pressure forces are important, the collapse of an axisymmetric,
nonrotating prolate or oblate configuration generally yields
results similar to the spherical case, i.e. the development of a
strongly centrally peaked density distribution without large
deviations from spherical symmetry (Larson, 1972a). A problem of
wider interest has been the collapse of an axisymmetric rotating
cloud (see, e.g., Larson, 1977a). Many calculations agree in
finding that rotation does not alter the basic nonhomologous nature
of the collapse, and that infall of low-angular-momentum gas along
the axis of rotation leads to a runaway increase in density near
the center of the cloud, as in the spherical case. However, there
is still disagreement in detail as to whether a ringlike structure
forms near the center; calculations by Larson (1972a), Black and
Bodenheimer (1976), and Nakazawa et al. (1976) show the formation
of rings, while Tscharnuter (1975), Fricke et al. (1976), Deissler
(1976), and Kamiya (1977) find in general no tendency to form
rings. It may be significant that the most recent study by Kamiya
has, unlike all of the others, employed a Lagrangian grid which
should allow more accurate conservation of angular momentum. In
any case it is clear that the result is sensitive to subtle
differences in the computational techniques, and hence presumably
in the physical assumptions as well. Perhaps the assumptions of
axial symmetry and inviscid flow are so unrealistic that detailed
results are not very meaningful anyway. If a ring does form,
breakup of the ring is likely to lead to the formation of at least
a binary system, in which case the problem becomes 3-dimensional.
Clearly 3-dimensional collapse calculations are needed to pursue
these questions further.
        A different and perhaps more instructive type of 2-dimensional

calculation was made by Quirk (1973), who calculated the collapse
of a rotating thin disc of gas with a diameter of a few Jeans
lengths. He found fragmentation of the disc into several very
dense knots orbiting around each other, with a considerable
amount of leftover gas swirling around and slowly being either
accreted or dispersed. This result suggests that in the general
3-dimensional case a number of such dense condensations will form,
and that it will be necessary to follow the formation and orbital
motions of these condensations as well as the continuing accretion
of more material by them.

2.2 Simulation of 3-Dimensional Collapse by a Finite-Particle Scheme

Standard grid methods for 3-dimensional fluid dynamics are at
present limited to fairly coarse grids, which are not well suited
to following the formation and orbital motions of very small
dense condensations. A technique which seems better adapted to
this problem is to simulate the fluid by a number of "fluid particles", or fluid elements of finite mass whose motions are followed
individually as in an n-body calculation (Larson, 1978). Condensations are then represented by tight clusters of particles, and
there is in principle no limit to how small and dense such a cluster
can become.

In order to simulate gas dynamics it is necessary to introduce,
in addition to gravitational forces between all pairs of particles,
repulsive forces between neighboring particles to represent the
thermal pressure of the gas. To estimate the pressure force
between neighboring particles, consider two neighboring gas elements with density ρ and isothermal sound speed c, and suppose
that they have sizes comparable to their separation r and interact
via a pressure $P = \rho c^2$ over an interface of area $\sim r^2$. The repulsive force is then $\sim \rho c^2 r^2$ and, since the element mass is $\sim \rho r^3$, the
corresponding acceleration is $a_r \sim c^2/r$. In the calculations to be
described, this pressure force has been assumed to act only between
each particle and its nearest neighbor, although more complicated schemes have also been tried and yield similar results.

A second effect that must be included in simulating the fragmentation of collapsing clouds is some form of viscous dissipation,
since bound pre-stellar condensations cannot form unless the kinetic energy of collapse is somehow dissipated. Moreover, angular
momentum must be transferred outward in a protostar by some form of
effective viscosity. In the spherical collapse models, all of
the dissipation occurs in a spherical accretion shock at the
surface of the stellar core. In 3-dimensional collapse, shocks
will also occur but probably with a much less symmetrical spatial
structure, so the associated momentum transfer might be an important source of viscosity. The momentum transferred by shock
fronts can be estimated by considering two fluid particles of
separation r that are approaching each other with velocity u; if

$u \gtrsim c$, a shock will form between them and will overtake one of them after a time $\sim r/u$, imparting to it a velocity increment $\sim u$ corresponding to an average repulsive acceleration $a_r \sim u^2/r$. Combining this with the pressure effect, the total repulsive acceleration between neighboring particles is then $a_r \sim (c^2 + u^2)/r$, where the second term is included only if the particles are approaching.

This scheme attempts to include explicitly only one of the possible sources of viscosity in a collapsing cloud, i.e. that due to the propagation of shocks. In effect the scheme assumes a dissipation length for momentum transfer that is comparable to the particle spacing. A second effect that is implicitly included is the momentum transferred by gravitational torques in a non-axisymmetric mass distribution. Other sources of viscosity such as magnetic torques and turbulence may also be important, and the actual magnitude of the viscosity is quite uncertain, although it is unlikely to be much smaller than that provided by this simulation technique.

2.3 Tests of the Method

The ability of this scheme to simulate the pressure of a gas can be tested by enclosing the system in a reflecting sphere and comparing the measured pressure on the boundary with that predicted by adding up the forces between the particles (Larson, 1978); this yields agreement to an accuracy of $\sim 5\%$ with 150 particles. If the system is set oscillating radially, the measured period of oscillation also agrees to within $\sim 5 - 10\%$ with that predicted classically for an oscillating gas sphere. A more stringent test is to compare the behavior of a gravitationally collapsing system of particles with previous results for the isothermal collapse of a sphere; it is found that the particle system develops a centrally condensed structure with $\rho \propto r^{-2}$, as in previous calculations, and that the critical temperature required to prevent collapse for a system of fixed mass and radius agrees with previous results to an accuracy of $\sim 20\%$. Finally, the formation of an accretion disc can be simulated by following the collapse of a rotating cloud of particles around a central point mass, and from the timescale $\sim \mathcal{R}\omega^{-1}$ for the spiraling of particles into the central mass (Lynden-Bell and Pringle, 1974), an estimate for the effective Reynolds number of $\mathcal{R} \approx 50$ is obtained (Larson, 1978).

2.4 Results

To simulate the isothermal collapse of an approximately uniform rotating cloud, 150 particles have initially been scattered randomly in a sphere and given a solid-body rotation, and the subsequent evolution of the system has been calculated for various choices of the sound speed c and the initial angular velocity.

As anticipated, the system shows a strong tendency to form condensations, or small dense clusters of particles, that steadily
become denser and more massive as they accrete more particles.
To avoid the necessity of calculating the extremely rapid internal
motions in these dense clumps, which become increasingly tightly
bound, the particles have been merged when they come closer together than 1/200 of the initial cloud radius.

When the system is initially nearly supported against collapse
by thermal pressure and centrifugal force, the collapse is found to
proceed much as in the spherical case, with the formation of a
single dense condensation at the center. A central massive object
forms, surrounded by an accretion envelope whose rotational velocity and flattening increase toward the center. The central object may eventually accrete between ∿10 and 50 percent of the total
cloud mass through the action of viscosity and the spiraling in
of matter from the envelope. It is important that, because of
the centrally concentrated collapse and the effect of viscosity
in allowing a large amount of matter to accumulate at the center,
only a single condensed object forms; no successive fragmentation
is observed.

When the temperature and Jeans mass are reduced, more condensations are formed, the number varying approximately inversely
with the Jeans mass. For example, when the Jeans mass is about
half of the total mass, the typical result is the formation of a
binary system of two objects orbiting around each other. An
example of how binary formation typically occurs is shown in Figs.
1(a) - 1(c). At first the collapse proceeds roughly as before
with the formation of a single condensed object near the center,
but now the remaining envelope is less symmetrical and sometimes
shows a transient "spiral arm" structure (Fig. 1(a)); a second
satellite condensation forms in the spiral arm (Fig. 1(b)), and
it subsequently accretes almost as much mass as the first to form
the secondary component of a binary system (Fig. 1(c)).

As the temperature is further reduced, the early collapse
becomes less symmetrical, and the cloud develops an increasingly
lumpy structure. Most of the lumps collapse to form condensed
objects, but some of them merge with other lumps and some are
torn apart tidally. The same process of satellite formation
described above is observed to occur also on smaller scales, and
there is a strong tendency to form binary systems and multiple
systems with a hierarchial structure, such as triples in which
two of the objects form a close pair. However, as before there
appears to be little tendency for successive fragmentation after
the early stages of the collapse; basically what is observed is an
accumulation process, and the condensations end up with a wide
range of masses depending on how much mass each has been able to
accrete after its formation. The results of two calculations
made with a low initial temperature and small Jeans mass (about
1/16 of the total mass) are shown in Figs. 2(a) and 2(b); these
two cases differ only in the initial random positions of the

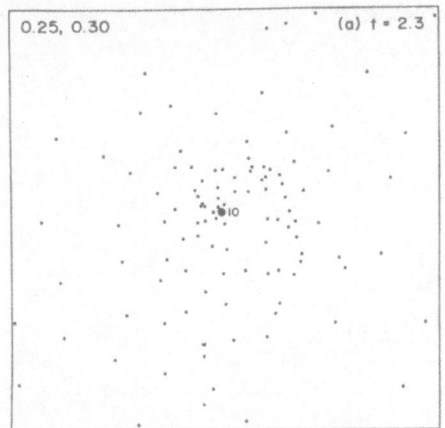 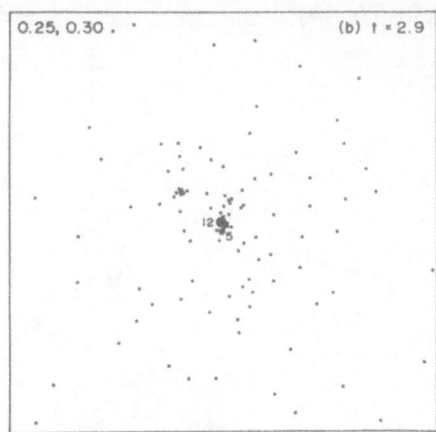

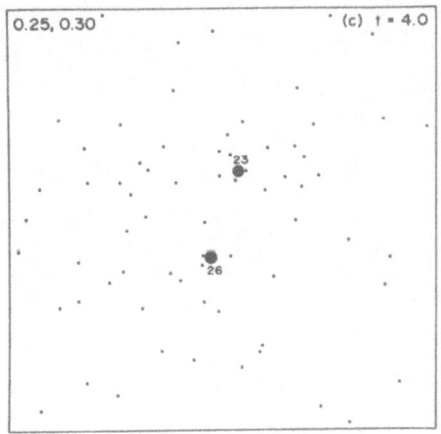

Fig. 1. Three stages in the collapse of an isothermal system of
150 particles, illustrating the formation of a binary system. The
particles are initially scattered randomly in a sphere of diameter 2.5 times the figure dimension and are given a counterclock-
wise rotation; all figures are projections along the rotation axis. The initial thermal and rotational energies are respectively
0.25 and 0.30 times the gravitational energy. The time is indicated in units of the free-fall time, and merged objects are represented by large dots whose mass is given in units of the particle
mass. Note that the two objects of masses 12 and 5 in Fig. 1(b)
eventually spiral together and merge to make the object of mass
26 in Fig. 1(c).

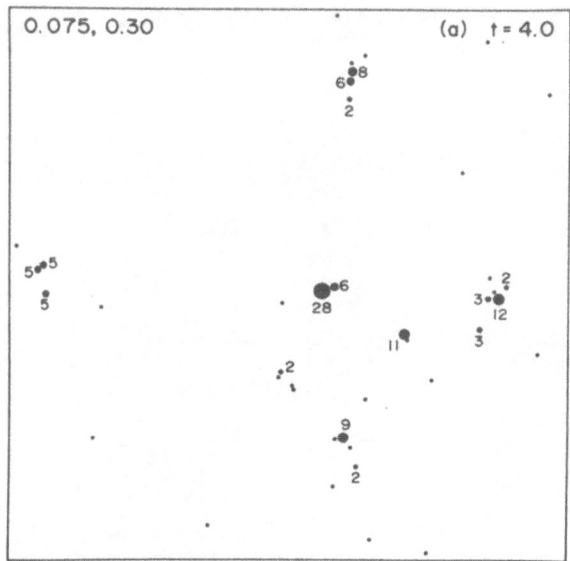

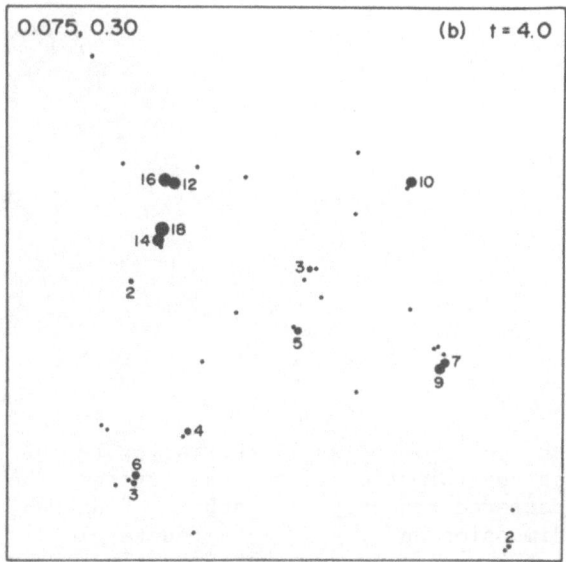

Fig. 2.   The results of two calculations with a much lower temperature and Jeans mass than in Fig. 1;   here the initial thermal
energy is about 0.075 times the gravitational energy.   The cases
shown in Figs. 2(a) and 2(b) differ only in the initial random positions of the particles.

particles. In these cases the frequent formation of hierarchial
multiple systems is evident; in addition, there is possibly a
tendency for the most massive objects to form near the center
of the cloud, and for peripheral objects to be less massive.

2.5 Implications for the Stellar Mass Spectrum

It is difficult to derive a complete mass spectrum from such
limited numbers of particles; however two parameters, the total
number of condensations formed and the mass of the largest one,
can readily be determined and their dependence on the initial conditions can be studied. It is found that the number of condensations depends almost entirely on the temperature or Jeans mass,
and not on the initial angular velocity of the cloud; in fact,
the number of condensations is almost exactly equal to the initial
number of Jeans masses in the cloud. Although this close coincidence is probably fortuitous, it is consistent with the observation that most of the fragmentation occurs during the earliest
stages of the collapse.
     While the number of objects formed thus depends on the Jeans
mass, their masses cover a wide range from considerably smaller
than the Jeans mass to many times larger. The formation of objects
less massive than the Jeans mass is possible because the collapse
of protostellar condensations produces dense cores of initially
much smaller mass, and these can survive as separate objects if
their envelopes are tidally stripped away. The formation of objects much larger than the Jeans mass is possible because some
favored condensations in the densest parts of the cloud are able
to accrete much more mass from the surrounding cloud. The mass
of the largest object that forms in this way is found to be almost
independent of the temperature or Jeans mass, but instead depends
strongly on the initial angular velocity of the cloud, increasing
with decreasing angular velocity. This is because a lower angular
velocity allows the cloud to become more condensed, and a larger
fraction of the material can then be accreted by the dense condensations, especially the most massive one.
     These results suggest that, in addition to the Jeans mass, a
second quantity that is relevant for the stellar mass spectrum
is the maximum mass that can be accreted onto a single object,
given the amount of angular momentum and the viscosity in a
collapsing cloud. A simple estimate shows that the fraction of
the cloud mass that can be accreted onto a central object through
the action of viscosity within one free-fall time is $\approx \mathcal{R}^{-1/2}$,
where $\mathcal{R}$ is the Reynolds number; substituting $\mathcal{R} \approx 50$ yields a predicted fraction of ≈ 0.15, which is of the same order as the observed fractional mass of the largest condensation. This result
emphasizes again the importance of understanding the sources of
viscosity in collapsing clouds.
     Evidently the resulting mass spectrum depends on the relative
numbers of objects that accrete various amounts of mass. The

calculations described above suggest that this may be determined
by the development of a hierarchy of accreting objects of different
masses, each with an associated accretion domain in which it is
the dominant accreting center. Each such object may accrete a
fraction of order $\mathcal{R}^{-1/2}$ of the matter in its domain, leaving
the rest to be accreted by lesser satellite objects. For example,
there might be a central most massive object whose accretion domain
is the whole cloud, around it may be several less massive objects
which accrete matter from smaller regions around them, and so on.
As long as the gas remains isothermal, it is possible that similar
processes occur on different scales and that different levels of
the hierarchy are related simply by scale factors. This suggests
the hypothesis of a self-similar accretion hierarchy, which in
general predicts a power-law mass spectrum for the accreting
objects.

To illustrate this, consider the following simple geometrical
example (Mandelbrot, 1975). Suppose that the cloud is represented
by an equilateral triangle, and suppose that the most massive condensation is represented by the inverted triangle constructed by
joining the midpoints of its sides. This leaves 3 smaller corner
triangles, each similar to the original triangle but having half
the linear dimension and one-quarter of the area. The same construction is then repeated for each of the 3 smaller triangles,
and so on. After n + 1 steps the number of inverted triangles (condensations) of the smallest size is $N = 3^n$, and the area (mass) of
each one is $m = 4^{-n}$ in units of the largest mass. We then have
$N = m^{-\log 3/\log 4} = m^{-0.79}$, or since N is the number of objects
in a given logarithmic mass interval, this is equivalent to

$$\frac{dN}{d \log m} \propto m^{-0.79}. \tag{2}$$

In the usual notation this corresponds to a power law mass spectrum
with exponent $x = 0.79$; for comparison, the Salpeter spectrum has
$x = 1.35$, and a representative value of x in stellar systems is
$x \sim 1$ (Tinsley, 1978).

In the limiting case of a hierarchy in which only a negligible
fraction of the mass available is accreted onto objects of each
mass, so that objects at all levels in the hierarchy have the same
total mass to accrete from, then the same amount of mass goes into
objects in each logarithmic mass interval, which corresponds to
$x = 1$. Apparently this concept predicts a reasonable first-order
approximation to the observed stellar mass spectrum.

The power law cannot continue to indefinitely small masses,
since there is a smallest fragment size determined by the Jeans
mass. This does not necessarily imply a sharp lower limit on
stellar masses because the Jeans mass may be considerably reduced
in the densest parts of a collapsing cloud, and because objects
smaller than the Jeans mass can form, as noted above. In the
numerical simulations, the smallest objects that form have masses

about one-fifth of the initial Jeans mass. An absolute lower
limit on the Jeans mass of the order of 10^{-2} $M_\odot$ is set by the
eventual effect of a finite opacity at high densities (eg. Low
and Lynden-Bell, 1976).

At the high mass end, an absolute upper limit is set by the
maximum fraction of the cloud mass that can be accreted onto a
single star; the simulations suggest that this fraction can be
substantial, and that the resulting maximum stellar mass increases
with increasing cloud mass and decreasing angular momentum. In
reality the upper mass limit may be set by other effects such as
radiation pressure, etc. as discussed earlier, but all effects
operate in the direction of increasing the mass limit in regions
of higher density. Observationally, this expectation seems to be
consistent with the fact that low-mass T Tauri stars are found
widely distributed through dark clouds, while the most massive
young stars and protostars seem to form only in the densest
concentrations of molecular gas.

3. STAR FORMATION AND GALACTIC EVOLUTION

3.1 Large-Scale Aspects of Star Formation

If the fragmentation of collapsing clouds and the properties of
the resulting stars are largely determined by the initial conditions existing when gravitational collapse begins, it is important
to understand how molecular clouds are formed and what determines
their properties. This must evidently involve the larger scale
dynamics of the interstellar medium, and ultimately the evolution
of the galaxy as a whole. Many properties of galaxies depend in
turn on star formation processes, which control their gas contents
and chemical evolution. The detailed state of the gas also depends
on star formation through effects such as ionization and supernova
explosions. Even the basic structure of galaxies depends on the
efficiency of star formation during the initial protogalactic
collapse; this determines how much mass ends up in a spheroidal
component and how much in a disc. Thus it is important to understand the processes that act to control star formation on a
galactic scale (Larson, 1977b).

3.2 Conditions Required for Star Formation

On the smallest scale, i.e. the scale of collapsing clouds and
protostars, the most important limiting condition for star formation is the Jeans criterion, i.e. the condition that self-gravity
must dominate over the thermal pressure (plus any effects of rotation or magnetic fields) of a collapsing cloud or protostar. As
already noted, this condition appears to be amply satisfied in many
molecular clouds; however, if the much less dense "standard clouds"
are to collapse gravitationally, the Jeans criterion is more

restrictive and requires a minimum cloud mass of several thousand solar masses.

On the scale of spiral arms, the main effect tending to counteract the self-gravity of the gas and prevent collapse is not thermal pressure but differential galactic rotation, or the tidal force of the galactic gravitational field. The condition that self-gravity exceed tidal forces is essentially the classical Roche condition, and requires that the gas density exceed a value approximately equal to the mean density of the system (in this case, the whole galaxy) interior to the point considered.

On the largest scale, i.e. that of a collapsing protogalaxy, the most important condition limiting the occurrence of star formation may be the requirement that the gas be able to cool within a free-fall time from the high temperatures generated by high velocity collisions and shock fronts (eg. Rees and Ostriker, 1977). The cooling time varies inversely with the gas density, so here again the basic requirement for star formation is a high enough gas density. We consider below some processes that may be important on various scales in compressing gas to a high enough density for star formation to occur.

3.3 Mechanisms of Gas Compression

Considering first the smallest scales, there are two ways in which an interstellar cloud can be compressed to a high enough internal density for individual protostars to fragment out and begin collapsing. The first is the gravitational collapse of an isolated cloud whose mass exceeds the Jeans mass, which is a few times 10^3 $M_\odot$ for "typical" cloud parameters. The collapse of such a cloud would raise the internal density (and perhaps also lower the temperature) to the point where the Jeans mass becomes $\lesssim 1$ $M_\odot$ and protostars of small mass can form. However, this may be an inefficient way to form stars, if only a relatively small core region of the cloud collapses to a high density and if not much fragmentation occurs after the earliest stages of the collapse, as the calculations suggest for the isothermal case. Thus some process of rapid external compression of a cloud may be a more efficient way of producing the conditions required for star formation.

External compression can be produced by a collision with other material which drives a shock front into the cloud, thus compressing the cloud gas. For a highly supersonic shock the compression results from a near balance between the dynamical pressure ρu^2 of the gas entering the shock and the thermal pressure ρc^2 of the post-shock gas after it has cooled (generally very rapidly) to a new equilibrium temperature. Thus the density is increased by a factor $\sim(u/c)^2$, where u is the shock speed and c is the sound speed in the cooled gas; this factor can be quite large. For example, if two clouds collide at a velocity $u \sim 10$ km s^{-1} and the shocked gas cools to ~ 10 K, then $c \sim 0.2$ km s^{-1} and $(u/c)^2 \sim 3000$; a cloud of initial density 10 cm^{-3} would then be compressed to a density $\sim$3 x 10^4 cm^{-3}, producing directly the conditions needed for a protostar of one solar mass to collapse. A possible example of star formation associated with colliding clouds in NGC 1333 has been described by Loren (1976). It should be noted that shock compression may also be important inside a collapsing cloud; for example, a typical collapse velocity u $\sim$ 1 km s^{-1} is still quite supersonic and can produce a compression factor of $\sim$30, sufficient to reduce the Jeans mass by a factor of $\sim$5.

Other situations where shock compression may play a role in inducing star formation involve compression of interstellar clouds by supernova blast waves or shocks driven by ionization fronts (eg. Elmegreen and Lada, 1977).

On the intermediate scale of spiral arms, there can again be both internal and external mechanisms for compressing the gas. If the gas in the galactic disc becomes dense enough for its self-gravity to overcome tidal forces, gravitational instabilities can occur and bring about the collapse of parts of the gas layers. A detailed stability analysis by Goldreich and Lynden-Bell (1965) shows that the gas density required for gravitational instability in the solar neighborhood is $\sim$2 - 3 cm^{-3}, a value not greatly in excess of the observed density; these authors suggest that dissipation of turbulent motions and settling of the gas into a thinner and denser layer could trigger gravitational instabilities and hence star formation. The most unstable wavelength is of the order of 1 kpc, and the amount of mass in a region of this size is of the order of 10^6 - 10^7 M$_\odot$, similar to the observed masses of the largest star-forming molecular cloud complexes.

A much more widely discussed possibility is that an underlying density wave in the stellar disc induces a spiral compression wave or shock front in the gas layer. Several types of density waves are possible: bar-like distortions are often found in numerical simulations, and may be quite long-lived; spiral density waves can also exist, but are likely theoretically to be more transient; and purely transient disturbances of many forms can be induced by tidal encounters between galaxies. For a bar- or spiral-shaped density wave, the induced response in the gas is a trailing spiral compression wave, which can become a shock front if the amplitude is large enough. In the "classical" two-phase model of the interstellar medium, a shock in the intercloud gas can compress and trigger the collapse of embedded clouds (Woodward, 1976), but this picture is contingent on the existence of a general intercloud medium of temperature 10^4 K, which is now in doubt. Other possibilities are that the compression wave may take the form of a region of enhanced cloud collisions, or that a density wave may serve simply to enhance the gas density and thereby trigger gravitational instabilities in a broad spiral shaped region with irregular small-scale structure.

Observations of spiral galaxies have so far failed to yield clear-cut evidence that star formation takes place in a wave that

propagates with respect to the gas and has a well-defined front,
as postulated in at least the simpler versions of density wave
theory; neither the colors nor the surface brightness profiles
generally show the predicted effects (Schweizer, 1976). It has
in any case always been evident that star formation often occurs
in regions or in galaxies where there is no large-scale spiral
pattern, and hence presumably no density wave; thus local phenomena, such as gravitational instabilities, must also play a role
in initiating star formation in some circumstances.

Detailed multicolor surface photometry of galaxies can help
to clarify the mechanisms of star formation. For example, data
for M 83 reported by Talbot (1977) show evidence for two modes
of star formation in this galaxy. The inner region, dominated by
a strong bar, shows a classical two-armed spiral pattern with young
stars and HII regions distributed in relatively well-defined arms,
as might be expected for star formation driven by a density wave.
However, the outer half of the galaxy lacks a regular spiral
pattern, yet has many disconnected spiral-shaped patches in which
star formation is proceeding just as vigorously as in the main
spiral arms. This suggests that star formation in the outer part
of this galaxy is driven not by a density wave but by the Goldreich-
Lynden-Bell gravitational instability mechanism.

Finally, we note that star formation in collapsing protogalaxies may also be triggered by the shock compression produced
as different parts of an irregularly collapsing protocloud collide
with each other at high velocity. If the density is high enough
for cooling to be effective, very large compression factors can
result. For example, with a velocity $u \sim 100$ km s^{-1}, the compression factor is $\sim 10^2$ if the gas rapidly cools to 10^4 K, and
$\sim 10^6$ if it cools to 10 K. In the first case, the gas can for
example be compressed from 10^{-2} to 1 cm^{-3}, while in the second
case, clouds of density 1 cm^{-3} can be compressed to $\sim 10^6$ cm^{-3}.
Such processes could be important in producing the rapid early
star formation required in most models for the formation of
spheroidal stellar systems.

3.4 Empirical Star Formation Rates in Galaxies

Information about the large-scale processes responsible for star
formation in galaxies can be obtained by studying the star formation rates in various types of galaxies and looking for correlations with morphological features indicative of the large scale
dynamics. For example, if supersonic gas motions and shock compression are of general importance for star formation, the observed star formation rates might be particularly high in systems
showing evidence for disturbed gas motions or shock fronts - eg.,
colliding or tidally interacting galaxies. Similarly, high star
formation rates might be expected in galaxies that are very young
and still collapsing, if such systems can be found. Thus it may
be instructive to study star formation rates in peculiar or

distorted galaxies and see if correlations can be found with
evidence for collisions, etc.

The total star formation rate in a galaxy can be estimated
from its UBV colors, which indicate the relative number of young
and old stars present. For morphologically normal galaxies from
the Hubble Atlas of Galaxies (Sandage, 1961), it is found that the
colors form a narrow sequence in the two-color (U-B, B-V) diagram
and that this sequence is well reproduced by synthetic population models which assume star formation rates declining monotonically with time at different rates (Larson and Tinsley, 1978,
hereafter LT). The position of a galaxy on this sequence is a
unique measure of the relative amount of star formation during the
past $\sim 10^8$ yr, and the observed colors range from those of a system
with a constant star formation rate to those of a system in which
all of the star formation occurred $\sim 10^{10}$ yr ago. The star formation rates as inferred from the colors show at least a broad correlation with the gas contents of the galaxies as determined from
21 cm observations.

By contrast, a sample of galaxies taken from the Atlas of
Peculiar Galaxies (Arp, 1966) shows a much greater scatter in the
two-color diagram, both above and below the sequence of normal
galaxies. Also, they extend to much bluer colors than the normal
galaxies. These results cannot be explained by models with monotonically declining star formation rates, but can be explained if
many of the peculiar galaxies have experienced recent strong bursts
of star formation (LT). A very recent burst produces hot stars
that make a galaxy bluer in U-B than the normal (U-B,B-V) relation,
but as the burst ages the U-B color rapidly becomes redder, and
after $\sim 5 \times 10^7$ yr it becomes redder than the normal (U-B, B-V)
relation. With the help of a grid of models for systems with
bursts of star formation, LT were able to show that the color distribution of the Arp galaxies can be explained if many of them
experience bursts with durations as short as $\sim 2 \times 10^7$ yr, involving
up to a few percent of the total mass.

A further significant finding of LT is that nearly all of the
large color scatter of the Arp galaxies is produced by galaxies
showing evidence of collisions or tidal encounters with companions;
the non-interacting Arp galaxies, even though peculiar in other
ways, show almost as narrow a distribution as the normal galaxies.
Thus there is evidence that bursts of star formation are often
triggered by interactions between galaxies. This conclusion is
strengthened by dividing the interacting galaxies into early and
late stages of interaction on the basis of whether they possess
tidal tails; numerical simulations show that tidal tails do not
appear until the interacting systems have passed the point of
closest approach, typically $\sim 5 \times 10^7$ yr after tidal distortions
have first become evident (Toomre and Toomre, 1972). In agreement
with this prediction, the interacting systems without tails tend
to have bluer U-B colors than those with tails, and correspondingly
shorter burst ages of $\leq 5 \times 10^7$ yr. Independent evidence for an

association between interactions and high rates of star formation has been shown by Rieke (this institute), who finds that
interacting galaxies have systematically higher infrared luminosity than noninteracting ones.

The above results for the integrated UBV colors of galaxies
do not indicate in detail the mechanisms responsible for the high
star formation rates in interacting systems. It is possible, for
example, that intense star formation could be triggered by a strong
transient density wave induced by the interaction; an example of
this may be the "Cartwheel" ring galaxy studied by Fosbury and
Hawarden (1977), where intense star formation is apparently
occurring in a ring-shaped disturbance generated by a collision
with another galaxy. A second possibility is a strong shock
front caused by a direct collision of the gas components of the two
galaxies; an example may be an interacting system studied by C.R.
Lynds (private communication) which has an apparent sharp front
of active star formation at one edge. Finally, it is possible that
the interaction causes disturbed gas to be dumped into the nucleus
of one or both galaxies, where it becomes highly compressed in the
deep potential well and generates a burst of star formation. The
high infrared fluxes found by Rieke are in fact based on observations of the nuclear regions only, and this suggests that much of
the enhanced star formation activity occurs in the nucleus. Similar processes may also account for much of the nuclear activity seen
in more active galaxies; for example, it is now beginning to be
recognized that many Seyfert galaxies are in interacting systems
(eg. Adams, 1977).

We conclude by mentioning a particularly interesting class of
objects for the study of star formation: galaxies like M 82 and NGC
253 in which there is abundant evidence for intense star formation
activity, yet nearly all of the action is hidden from view at
visible wavelengths by dust. Consequently UBV colors fail to
reveal the anomalously high star formation rates. Most of the
bolometric luminosity of these galaxies (at least in the inner
regions) is emitted at infrared wavelengths by heated dust that is
almost certainly closely associated with young stars or protostars,
as in regions of active star formation in our galaxy (Rieke, this
institute). This makes clear the importance of studying infrared
radiation from galaxies: apparently the most intense bursts of
star formation, such as those that might be expected to occur
during the formative phases of galaxies, occur in regions so deeply
embedded in dust that they are not seen optically and can be observed only at infrared wavelengths.

REFERENCES

Adams, T.F., 1977, Astrophys. J. Suppl. 30, 19.
Appenzeller, I. and Tscharnuter, W., 1974, Astron. Astrophys. 30,
423.

Appenzeller, I. and Tscharnuter, W., 1975, Astron. Astrophys. 40, 397.
Arp, H.C., 1966, Atlas of Peculiar Galaxies. California Institute of Technology. Also in Astrophys. J. Suppl. 14, 1.
Bertout, C., 1976, Astron. Astrophys. 51, 101.
Black, D.C. and Bodenheimer, P., 1976, Astrophys. J. 206, 138.
Bodenheimer, P. and Sweigart, A., 1968, Astrophys. J. 152, 515.
Cochran, W.D. and Ostriker, J.P., 1977, Astrophys. J. 211, 392.
Deissler, R.G., 1976, Astrophys. J. 209, 190.
Elmegreen, B.G. and Lada, C.J., 1977, Astrophys. J. 214, 725.
Finn, G.D. and Simon, T., 1977, Astrophys. J. 212, 472.
Fosbury, R.A.E. and Hawarden, T.G., 1977, Mon. Not. Roy. Astron. Soc. 178, 473.
Fricke, K.J., Möllenhoff, C. and Tscharnuter, W., 1976, Astron. Astrophys. 47, 407.
Goldreich, P. and Lynden-Bell, D., 1965, Mon. Not. Roy. Astron. Soc. 130, 125.
Kahn, F.D., 1974, Astron. Astrophys. 37, 149.
Kamiya, Y., 1977, preprint.
Larson, R.B., 1969a, Mon. Not. Roy. Astron. Soc. 145, 271.
Larson, R.B., 1969b, Mon. Not. Roy. Astron. Soc. 145, 297.
Larson, R.B., 1972a, Mon. Not. Roy. Astron. Soc. 156, 437.
Larson, R.B., 1972b, Mon. Not. Roy. Astron. Soc. 157, 121.
Larson, R.B., 1974, Fund. Cosmic Phys. 1, 1.
Larson, R.B., 1977a, Star Formation, IAU Symposium No. 75, ed. by T. De Jong and A. Maeder, p. 249. D. Reidel Publishing Co., Dordrecht and Boston.
Larson, R.B., 1977b, The Evolution of Galaxies and Stellar Populations. ed. B.M. Tinsley and R.B. Larson, P. 97. Yale University Observatory, New Haven, Connecticut.
Larson, R.B., 1978, J. Comput. Phys., in press.
Larson, R.B. and Starrfield, S., 1971, Astron. Astrophys. 13, 190.
Larson, R.B. and Tinsley, B.M., 1978, Astrophys. J. 219, in press.
Loren, R.B., 1976, Astrophys. J. 209, 446.
Low, C. and Lynden-Bell, D., 1976, Mon. Not. Roy. Astron. Soc. 176, 367.
Lynden-Bell, D. and Pringle, J.E., 1974, Mon. Not. Roy. Astron. Soc. 168, 603.
Mandelbrot, B., 1975, Les Objets Fractals. Flammarion, Paris.
Nakazawa, K., Hayashi, C. and Takahara, M., 1976, Prog. Theor. Phys. 56, 515.
Quirk, W.J., 1973, Bull. Am. Astron. Soc. 5, 9.
Rees, M.J. and Ostriker, J.P., 1977, Mon. Not. Roy. Astron. Soc. 179, 541.
Sandage, A., 1961, The Hubble Atlas of Galaxies. Carnegie Institution of Washington.
Schweizer, F., 1976, Astrophys. J. Suppl. 31, 313.
Scoville, N.Z. and Kwan, J., 1976, Astrophys. J. 206, 718.

Talbot, R.J., 1977, The Evolution of Galaxies and Stellar Popu-
lations, ed. B.M. Tinsley and R.B. Larson, P. 128, Yale University
Observatory, New Haven, Connecticut.
Tinsley, B.M., 1978, Structure and Properties of Nearby Galaxies,
IAU Symposium No. 77. D. Reidel Publishing Co., Dordrecht and
Boston, in press.
Toomre, A. and Toomre, J., 1972, Astrophys. J. 178, 623.
Tscharnuter, W., 1975, Astron. Astrophys. 39, 207.
Westbrook, C.K. and Tarter, C.B., 1975, Astrophys. J. 200, 48.
Woodward, P., 1976, Astrophys. J. 207, 484.
Yorke, H.W., 1977, Astron. Astrophys. 58, 423.
Yorke, H.W. and Krügel, E., 1977, Astron. Astrophys. 54, 183.

INFRARED EMISSION OF THE GALACTIC CENTER AND EXTRAGALACTIC SOURCES

George Rieke

Steward Observatory and Lunar and Planetary Laboratory, University of Arizona, Tucson, Arizona

1. INFRARED EMISSION OF THE GALACTIC CENTER

The basic rationale for studying galactic nuclei, including our own, is to search for phenomena fundamentally different from those we can study nearby. In some cases, such as Seyfert galaxies, such processes are clearly occuring, but we are frustrated by the lack of detail that we can see in what is spatially an unresolved point. The Galactic Center would probably not hold our attention for long if we observed it from beyond the Local Group, but its proximity allows study on a scale completely unattainable for any other galactic nucleus. Thus, in the study of galaxies, it holds a unique place analogous to that of the sun in stellar astronomy.

The interstellar obscuration toward the Galactic Center is so heavy that the region has never been detected at wavelengths shorter than 1μ (until the X-ray region). The position of this source was therefore first determined with reasonable accuracy ($\sim 1°$) only 60 years ago and by indirect means. Infrared observations have been essential to refining this position, which is probably known to $\sim 5''$ through study of 1) the distribution of extended 2μ emission due to late-type stars in the central bulge (Becklin and Neugebauer, 1968); 2) the location of a cluster of unique 2μ and 10μ sources (Rieke and Low, 1973; Becklin and Neugebauer, 1975); and 3) the velocity-rotation pattern from the 12.8μ Ne$^+$ line (Wollman, Geballe, Lacy, Townes and Rank, 1977).

The interstellar extinction plays such a dominant role in many of the observed properties of the Galactic Center that it must be understood before discussing the Galactic Center sources in detail. The extinction was first estimated at $A_V \sim 27$ (assuming a normal extinction law) by Becklin and Neugebauer (1968), who identified the origin of the extended 2-μ emission by showing it to have the

159

the same physical scale as the distribution of late-type stellar emission in the nucleus of M31, and then applied the appropriate color corrections. The luminous red supergiants in the central cluster show variations in color that indicate non-uniform extinction with $23 \leq A_v \leq 30$ (Rieke, Telesco, and Harper, 1977). On the other hand, the silicate absorption feature at 10μ (Gillett and Woolf, 1973), which is measured for a different group of sources than the red supergiants, is of a strength correcponding to $A_v \sim 40$ to 50 if the ratio of silicate absorption to A_v is the same as is observed toward VI Cyg No.12 (Rieke, 1974; Gillett et al. 1975) and o Sco (unpublished). The molecular clouds delineated by formaldehyde absorption appear as shadows against the extended 2μ source (Rieke, Telesco and Harper, 1977; unpublished observations).

Two pieces of evidence indicate that the extinction law toward the Galactic Center may differ from the "normal" law observed in the solar neighborhood: 1) The 10-μ silicate absorption strength in the spectrum of one source (No.7) is about 1.5 times stronger than would correspond to the $A_v \sim 30$ seen in the near infrared (Becklin et al. 1977); however, note that 1) this source has a 10μ excess, making interpretation of its spectrum somewhat uncertain; and 2) the position angle of polarization for many of the sources rotates by $\sim 60°$ between 2 and 11μ (Knacke and Capps, 1977; unpublished observations); if the polarization is due to absorption or scattering out of the beam by aligned dust grains along the line of sight, two types of grains with different alignment would be needed to explain this behavior.

In summary, the extinction toward the Galactic Center is roughly $A_v \sim 30$. However, there are significant variations over angular distances of 10-20" and very large variations over 1-2'. The extinction law may differ substantially from that in the solar neighborhood. Clearly, further understanding of the extinction is important to progress studying this region.

As already mentioned, the extended 2-μ emission from the Galactic Center is convincingly identified with the distribution of late-type stars. The following will discuss the nature of the other types of source observed in this region - compact near-infrared objects, far infrared emission, and compact middle-infrared sources. The compact sources will be identified in the nomenclature introduced by Rieke and Low (1973) and extended by Becklin and Neugebauer (1975).

High-resolution maps show a number of compact (in fact, spatially unresolved), near infrared sources clustered within $\sim 30"$ of the Galactic nucleus (Becklin and Neugebauer, 1975). CO bands have been detected in the spectrum of the brightest of these sources, No.7 (Treffers et al. 1976), and they are inferred to occur in the spectra of sources 11 and 12 from a general depression seen near 2.35μ at moderate spectral resolution (Neugebauer et al. 1976). These sources are therefore thought to be red supergiants. The broadband near-infrared colors of sources 9, 14, 15, 16, 17 and 19 are similar to those of sources 7, 11 and 12, indicating that the former six sources are probably also stars or clusters of stars.

The spectral distribution of the far-infrared emission strongly indicates that it arises from heated dust. It has been thought that the energy is generated by the cool stars responsible also for the extended 2-μ source, but arguments below indicate that hot stars may play an important - possibly dominant - role.

The hottest dust, seen as extended emission at 10μ, coincides closely with the thermal radio emission, while the cooler dust seen at 60μ coincides roughly with the radio sources but appears to be of somewhat larger angular extent (Rieke, Telesco, and Harper, 1977). These characteristics are similar to those of many HII regions. Within 2' of the center, the dominant far-infrared source has many other properties similar to an HII region like M17 - e.g. total luminosity, electron density, mass of ionized gas, IR luminosity vs. Ly-α luminosity, and projected distance from the radio peaks to nearby molecular clouds. Aitken, Griffiths, and Jones (1976) have shown that the excitation of the 12.8-μ Ne$^+$ line from this region can be satisfactorily explained if the ultraviolet flux is provided by stars at 40,000° K and if the abundance of neon relative to hydrogen in both stars and gas is about 3 times the solar value. If such stars provide the ionizing flux, their total luminosity is $\gtrsim 5 \times 10^6$ L$_\odot$, in agreement with the observed infrared luminosity. The late-type stars in the same volume have a total luminosity of the same order, but the above arguments (particularly the spatial coincidence of infrared and thermal radio emission) indicate that the hot stars make the dominant contribution to the energy ultimately radiated in the far infrared.

On the whole, the correspondence between the spatial distribution of far infrared and thermal radio emission holds over at least the central 20' of the galaxy, indicating that hot stars play an important - possibly dominant - role in the far infrared emission from this whole region (Rieke, Telesco, and Harper, 1977). A few notable exceptions (see, e.g. Gatley et al. 1977) may represent particularly dusty regions containing cooler stars.

Of particular interest are a group of compact sources in the Galactic Center which cannot be readily classified as stars or as any other form of infrared source that is common in other parts of the Galaxy (Rieke, Telesco, and Harper, 1977).

One example is source 16, which is 2'-3' in diameter, lies near the centroid of the distribution of compact sources, and has 1.6- -3.6μ colors similar to those expected of stars but has an excess at 4.9μ and no depression at 2.3μ, corresponding to CO bands. Assuming that this source is a cluster of stars of sufficiently early type or low luminosity that CO bands are not expected, Becklin et al. (1977) demonstrate that the density within the cluster exceeds 10^6 M$_\odot$ pc^{-3}. This density is at least an order of magnitude greater than is found in the nucleus of dense globular clusters.

Source 3 has unique properties for this region: even after

correction for $A_v \sim 30$, it has strong silicate absorption and a
spectrum corresponding to a 400° K blackbody (outside the absorption feature). It has a diameter <1" and a luminosity ~ 5 x
x 10^4 $L_\odot$; no compact radio source lies at its position, nor is any
OH emission detected there (Rieke, Telesco, and Harper, 1977; Becklin et al. 1977). These characteristics suggest that it may be a
very young star, similar to objects detected in and near other
HII regions.

Sources 1, 2, 5, 6, 9, and 10 have a number of properties in
common. Their spectra show two components, particularly after
correction for extinction of $A_v \sim 30$ (Rieke, Telesco, and Harper,
1977). These components suggest the presence of a near-infrared
source which is dominant for $\lambda < 5\mu$ and a middle-infrared excess
which dominates the spectrum for $\lambda > 5\mu$. High resolution maps
(Rieke, Telesco, and Harper, 1977; Becklin et al. 1977) show that
the diameters of these sources grow with increasing wavelength -
from <1".5 at 3.6 and 4.9μ to 3"-4" at 20μ. There is some evidence
(based on the possibly abnormal extinction law toward the Galactic
Center) that these sources also have 10μ "silicate" emission excesses (Becklin et al. 1977). All of these characteristics
suggest the presence of compact near-infrared luminosity sources
embedded in and heating a thin cloud of dust. There is some
evidence, although weak, that the required luminosity is supplied
by stars which are seen directly for $\lambda < 5\mu$, but with their colors
modified by additional extinction compared with other stars in the
same region (Rieke, Telesco, and Harper, 1977).

In addition to the fascinatingly broad variety of sources already detected in the Galactic Center, the possibility of as yet
unobserved phenomenon must be kept in mind. Infrared observations
of the rotation in this region, and thus of the mass, will play
a particularly important role in searching for massive objects at
the core of the Galactic nucleus; already, such measurements demonstrate that the dominant mass within the central 30" of the
Galaxy is that of the stars, ~ 4 x 10^6 $M_\odot$ (Wollman et al. 1977).

2. INFRARED EMISSION OF NORMAL GALAXIES

About ten years ago, it was discovered that a number of external
galaxies were even more powerful infrared sources than the Galactic Center (e.g. Pacholczyk and Wisniewski, 1967). Shortly thereafter, infrared surveys (Kleinmann and Low, 1970; Rieke and Low,
1972) had detected emission from representatives of virtually
every type of extragalactic object, including dwarf ellipticals
(M32), dwarf galaxies with young stellar populations (II Z w 40),
many normal spirals, interacting galaxies (NGC 5195), giant
ellipticals (NGC 1052), many Seyfert galaxies, radio galaxies
(Cygnus A) and QSO's (3C 273). The most radical implication of
these measurements was that many galaxies contain previously unseen
energy sources whose output dwarfs that detected in other wavelength

regions. At the same time,the mechanism for generating this energy has been controversial.

The following discussion of recent progress in this area will concentrate on a few individual galaxies that have been studied in detail in the infrared and on the broad categories of extragalactic infrared source for which sufficient examples have been observed to define the general properties of the class – normal spirals, interacting galaxies, Seyfert galaxies, and QSO's (particularly of the BL Lac type).

Many of the features of the infrared emission of normal spiral galaxies can be seen most clearly in the Galaxy. The preceding discussion of the Galactic Center has concentrated on the inner few parsecs, a region that falls within the finest single resolution element attainable on the nearest external infrared-emitting galaxies. This region is surrounded by a strong far infrared source of diameter $\sim$400 pc, with a structure that appears to be determined largely by the early-type stars that are also responsible for the thermal radio sources in the same region. This source is presumably responsible for strong emission at 10μm; the available very large beam observations are subject to calibrational uncertainties, but do show a very extended source in this region (Houck et al., 1971).

Outside the central 400 pc, the infrared appearance of the Galaxy can be deduced from recently available all-sky surveys at 10-30 μm (Heintz, Jereki, Low, and Rieke, 1977) and from observations with a balloon-borne telescope in the 60-300 μm region (Low, Kurtz, Poteet, and Nishimura, 1977). Giant HII regions within $\sim$7 kpc of the Galactic Center are the dominant radiation source over this entire spectral range; at the longer wavelengths, there is an additional low-surface-brightness component probably associated with dust heated by relatively cool stars or by hot stars outside HII regions. Discrete sources of the former type have also been observed at 11 and 20μ – e.g. CRL 437 and CRL 1855 (a part of the ρOph cloud)(Price and Walker, 1976).

Current techniques could detect a galaxy like our own at 60μ out to $\sim$6 Mpc and 10μ to $\sim$10 Mpc (assuming that the ratio of far infrared to 10μ flux is about the same for the Galactic Center as for other spiral galaxies). Only very limited information about its structure and spectrum at 10μ could be obtained unless it were closer than $\sim$4 Mpc. The more extended component of the infrared flux, corresponding to the region in the Galaxy out to 7 kpc from the center, would be undetectable with existing instruments, except that the flux from a few of the most powerful HII regions could be observed. However, the total flux from this region should be comparable to or greater than that from the central 400 pc, and cooled telescopes in space observing with large beams, such as the planned sky survey IRAS, would respond to that component of the total flux.

The nuclei of M31 and M33 are not strong infrared emitters; in fact, M31 has been detected at 10 μm, but there is no excess above an extrapolation of the stellar flux distribution (Rieke and Lebof-

sky, 1977). Therefore, the Galaxy is the best example for studies
of the general properties of a spiral exhibiting a moderate infrared output, as well as being the only galactic nucleus whose infrared emission can be studied on scales well below 1 pc.

NGC 253 is the normal spiral most thoroughly studied in the
infrared after our own (Becklin et al., 1973; Harper and Low, 1973;
Rieke and Low, 1975; Gillett et al., 1975; Hildebrand et al., 1977).
It generates a luminosity of $1.5 \times 10^{10} L_\odot$ (about 20 times that of the
Galactic Center) in a region of diameter $\sim$300 pc, and a luminosity
$\gtrsim 2 \times 10^9 L_\odot$ from a 45 pc region. The mass within the 300 pc region
was estimated by Rieke and Low (1975) to be $\leq 6 \times 10^8 M_\odot$; on the basis
of better spectroscopic material, Ulrich (1977) has recently estimated the mass of this same region to be $\leq 3 \times 10^7 M_\odot$. Thus, M/L for
the infrared source is ≤ 0.002, about 3 orders of magnitude below
the value for the Galactic Center, and in a range that can only be
sustained by stellar populations for a short time compared with
the age of the universe.

The emission mechanism for NGC 253 is undoubtedly thermal radiation by dust. There is abundant evidence for dust near the
nucleus from the visible colors, from the general appearance of
the nuclear region, and from the presence of a broad dip in the
spectrum at 10μ which is probably in part absorption by silicates.
In addition, the shape of the far infrared spectrum is similar to
that for Galactic thermal sources, the 10-μ spectrum contains
features found in other thermal sources (which probably contribute to the spectral dip through excess emission near 8 and 12μ),
and the physical extent of the emitting region is similar to that
in the Galactic Center, where thermal reradiation appears to
account for virtually all the infrared emission.

The very low value of M/L places strong constraints on the
nature of the ultimate energy source, particularly since there are
presumably many low-luminosity stars in the nucleus that are not
part of the current energetic episode and which contribute substantially to the nuclear mass but negligibly to the luminosity.
Thus, the value M/L $\sim 2 \times 10^{-3}$ may be an upper limit for the true
energy source; it is, however, already closely comparable with the
value for Salpeter's (1959) "young cluster" luminosity function,
extrapolated to a maximum $M_V = -7$. A further constraint on any
hypothetical stellar luminosity function is placed by the relatively low level of free-free emission from the nucleus. Although
the infrared peak coincides with a compact radio source, this
source has a power law spectrum, and an upper limit to the free-
free emission an order of magnitude below the observed radio flux
can be deduced from the strength of the 12.8 μm Ne$^+$ line (Gillet
et al., 1975). Ulrich (1977) has deduced that 10^3 06 stars, with
a total luminosity of $10^9 L_\odot$, are needed to excite the emission
lines from the nucleus; however, the apparent low level of free-
free emission implies that the bulk of the nuclear luminosity is
generated by some cooler source.

Wherever the pertinent observations are available, the properties of other strongly infrared-emitting spiral galaxies are simi-

lar to those of NGC 253, although the luminosities tend to be smaller and M/L larger. For example, NGC 2903, 5236, and 6946 have far infrared fluxes in about the same proportion to 10μ fluxes as NGC 253 (M51 may have an even stronger relative far infrared flux) (Harper, 1977). Absorption features at 10μ have been observed in the spectra of NGC 2903 and 5236 (Lebofsky and Rieke, 1977). The physical diameters of the 10-μ sources in NGC 2903, 3504, 4536, 5236, and 6946 all fall in the range 150-600 pc (Rieke, 1976).

This similarity of infrared properties can be used together with observations of a larger sample of spiral galaxies at 10 μ to derive an "average" M/L ratio for these galaxies.

There are 41 spiral galaxies (excluding NGC 1068) sufficiently northerly to be observed from Arizona and with integrated magnitudes brighter than 11. 35 of these have been observed at 10μ, and 16 plus the Galactic Center have been detected (Rieke and Lebofsky, 1977). In all cases except M31, the 10 μm flux lies well above the stellar continuum. The total luminosity of each galaxy was estimated from the 10 μ flux under the assumptions that the spectral distribution was similar to that for the galaxies detected in the far infrared and that the observed flux would grow linearly with aperture out to a diameter projected on the galaxy of 300 pc. Upper limits to the masses of the same regions were computed wherever spectroscopic observations were available, by assuming spherically symmetric mass distributions; more realistic models would have yielded nuclear masses smaller by a factor of 1.5 - 2. Some representative M/L ratios were: NGC 2903, 0.05; NGC 4528, 0.04; NGC 5055, 0.6; NGC 4826, 1.7; Galaxy, 1.9. For galaxies where the nuclear mass could not be measured directly, the average of $2.5 \times 10^8 M_\odot$ found for 16 bright spirals was assumed. Under these assumptions, the average M/L ratio for the entire sample (detected and undetected) is ≲0.2.

It is implied by this average ratio as well as the individual ratios for specific galaxies that spiral galaxy nuclei frequently contain much more powerful energy sources than have been detected outside the infrared spectral region. Typical values for the mass to light ratio in these regions have been estimated to be an order of magnitude higher than the values of M/L estimated here. The 10 μ absorption features found even in some relatively face-on galaxies are probably caused in part by strong "silicate" absorption in dense interstellar clouds, which obscure the stars or other energy sources that generate the infrared luminosity.

Complete thermonuclear burning of a primordial abundance of hydrogen and helium can release ∼0.8 percent of the total mass as energy. Thus, M/L ratios ≲0.1 cannot be sustained by thermonuclear processes for more than 10^{10} years. A more realistic limit on M/L would be substantially higher than 0.1, since real stars are not likely to achieve complete burning, both on observational and theoretical bases. At the same time, most evolutionary theories would predict that the average M/L of a galactic nucleus would tend to decrease with time to the current epoch, so that the

effective M/L of an average spiral galaxy nucleus over the life of
the universe may be substantially less than 0.2.

If it is assumed that stars account for the bulk of the energy
emitted by spiral galaxy nuclei, it may be required that the material within these regions be replenished to avoid exhaustion of
the available hydrogen. The very low values of M/L found for the
nuclei of some galaxies imply that luminous star formation occurs
in bursts in these regions. A period of star formation might
lead to a subsequent dramatic increase in the rate of supernova
explosions, thereby accounting for some of the "active" phenomena
occuring in galactic nuclei.

The conclusions drawn from the infrared measurements in many
respects support arguments by Larson and Tinsley (1977) from UBV
observations and model galaxy calculations that galaxies with
morphological peculiarities and very blue U-B colors are undergoing bursts of star formation. However, probably because of
reddening in the nuclei of many galaxies with young stellar populations, the infrared measurements indicate that this phenomenon
is much more widespread than they suggested. A further prediction
from their work can also be tested in the infrared: that tidal
interaction triggers episodes of star formation.

M. Tarenghi and I (1977) have observed a number of interacting
systems at $10\,\mu m$, partly selected from a list provided by W.G.
Tifft and chosen on the basis of condensed nuclei and strong
emission lines. Detected galaxies include NGC 2992, 3690 (Markarian 171), 4922N, 4933NE, 5195, 5953, and 7714, IC694, K161NE,
K466SW, and K466NE, where K designates galaxies in the list of
Karachentsev (1972). The luminosities of these galaxies tend to
range to significantly higher values than those for the sample of
normal spirals; under the same assumptions discussed previously,
most of the detected interacting galaxies have luminosities between 10^{10} and 10^{11} $L_\odot$. In addition to the list of detected double galaxies, a number of other strong $10-\mu$ emitting galaxies such
as NGC 1614 and 4194 are morphologically very peculiar and therefore provide further support for the arguments of Larson and Tinsley (1977); these objects might be interacting systems where one
member is projected on top of the other.

3. INFRARED EMISSION OF ENERGETIC EXTRAGALACTIC SOURCES

As first pointed out by Oke et al. (1970), the most variable QSO's
have spectra rising steeply into the infrared. Among the most
extreme examples are the BL Lac-type objects, which have received
the bulk of attention in infrared studies of QSO's.

These sources are characterized by power law infrared spectra,
with spectral indices $\alpha \sim 1$. Some examples are known with much
steeper optical spectra, but in these cases a break occurs in the
infrared spectrum, and toward the long wavelength side of this
break the spectral properties are similar to those of other members

of the class. Examples include BL Lac itself, PO735+178, and
A00235+164 (Rieke et al. 1977). In the latter two cases, the
break appears to result from absorption in the material responsible for high redshift absorption lines in the spectra of the
sources or to be intrinsic to the object itself. Since normal
absorption-line QSO's are not as red as these sources, the second
interpretation is more likely. It implies that the lack of
emission lines may result from a deficiency of ultraviolet flux.

The spectral distributions of these sources do not change
appreciably as the objects vary even by an order of magnitude
(Rieke and Kinman, 1974), except possibly during the most rapid
variations. Changes on timescales of 24 hours are seen in both
the optical and infrared (see e.g. Rieke et al. 1976).

The polarimetric properties of these sources are also note-
worthy. The strength of polarization does not change between the
optical and infrared (Knacke, Capps, and Johns, 1976), although
in some cases a position angle rotation has been reported (Rieke
et al. 1977).

An infrared excess is a nearly universal characteristic of
Seyfert galaxies. Of 52 observed at 10μ, nearly 90% have been
detected (Rieke, 1977). The undetected galaxies are among the
fainter and more distant of the sample, indicating that additional
detections would be expected if greater sensitivity had been
available (1 - σ levels of 10 - 15 mJy were typical).

A broad range of spectral distributions is found. If the slope
of the spectrum is characterized by the spectral index α in a fit
of the form $F = C\nu^{-\alpha}$ between 3.6μ (where the fluxes have been
corrected for the contribution from stars) and 10μ, values of α
range from ≤ 0.5 (Markarian 10, 376, 506, 509) to ≥ 2.5 (NGC 1068,
3227, 6764, Markarian 3, 198). The steepness of the infrared
spectrum is correlated with the spectroscopic type: those of type
1 (broad Balmer lines) have an average value of $\alpha \sim 1.0$, while
those of type 2 have an average of $\alpha \sim 2.2$.

Although a large percentage of known, bright Seyfert galaxies
have been observed in the infrared, it is likely that this sample
is biased and conclusions about the general properties of this
type of source may reflect this bias. At redshifts greater than
3000 Km/sec (57 Mpc if H_O = 53 Km/sec Mpc), the space density
of known Seyfert galaxies goes down rapidly with increasing
distance. Most of the more distant Seyfert galaxies were discovered through their inclusion on the lists of Markarian. As a
result, the known sample is biased toward ultraviolet-bright
galaxies. For example, application of standard tests for completeness indicates that the 30 brightest known Seyfert galaxies
in the ultraviolet represent 50-60% of the total down to the
brightness of the faintest of the 30; the corresponding percentage
for the infrared sample is only 10-20%. The incompleteness of the
sample is also evident from the decrease in relative numbers of
type 2 galaxies with increasing distance. The strength of the ultraviolet excess in type 1 galaxies seems to grow as the infrared

excess becomes smaller (those with $\alpha \gtrsim 1.5$ have an average U-B of ~ 0.1, those with $1.2 < \alpha < 1.5$ have U-B ~ -0.4, and those with $\alpha \leq 1.2$ have U-B ~ -0.8). As a result, the observed decrease in α with increasing distance (for redshifts <10,000 Km/sec, $\langle \alpha \rangle \sim 1.4$; for larger redshifts $\langle \alpha \rangle \sim 0.9$) is probably a result of the same sort of selection that discriminates against type 2 galaxies.

Within the limits imposed by the biases of the sample, the density of Seyfert galaxies as a function of infrared luminosity is roughly: $1-10 \times 10^9$ L_Θ $- \sim 3 \times 10^{-24}$/pc^3; $10 - 100 \times 10^9$ L_Θ $- \sim 6 \times 10^{-25}$/pc^3; $100 - 1000 \times 10^9$ L_Θ $- \sim 3 \times 10^{-25}$/pc^3. (In converting from flux between 8 and 13µ to total flux a factor of 8 has been used, since recent far infrared measurements indicate that the spectra of a number of Seyfert galaxies rise relatively slowly into that spectral region (Harper, 1977). The large number of low luminosity Seyferts may indicate a continuity between the properties of these galaxies and those of high-luminosity "normal" spirals.

Limited surveys of the infrared properties of Seyfert galaxies have been carried out by Stein and Weedman (1976), who measured 30 galaxies at 3.5 µm only, and by Neugebauer, Becklin, Oke, and Searle (1976), who measured 8 galaxies, mostly between 1.6 and 10.6 µm. Both groups of authors argue that Seyfert galaxies of type 1 have non-thermal infrared emission sources, while those of type 2 have thermal sources. This conclusion is brought into question, however, by more extensive measurements, including infrared spectral distributions from 1.25 µm to 10.6 µm for 36 Seyfert galaxies (Rieke, 1977), more detailed photometry of individual galaxies (e.g. Rieke, 1976), and infrared polarimetry (Kemp et al. 1977).

For example, it was argued by both groups that type 1 Seyfert galaxies have smooth, approximately power-law spectra extending through the visible and infrared. Yet the more extensive observations demonstrate that many of these sources, including the class prototype NGC 4151, have a spectral inflection near 1µ that suggests that the visible and infrared sources may be distinct.

On the basis of the same type of broad-band photometry available for other type 1 Seyferts, Markarian 231 appeared to have a unified visible-to-infrared power law spectrum. However, more detailed observations (Rieke, 1976) indicated a strong absorption at 10µ and a probable emission feature at 3.4µ (later confirmed: Soifer, 1977). Between 2 and 25 µm, the spectrum of this galaxy is similar in many respects to that of M82. Both spectra have a strong resemblance to Galactic thermal sources. The infrared polarization of both galaxies also exhibits very similar behavior, decreasing slowly and monotonically from $\sim 3\%$ in the visible to $\sim 0.5\%$ at 2.2µ (Kemp et al.,1977; Lebofsky, Kemp and Rieke, 1977). This trend probably results from scattering in a thick cloud of dust. Although it is exceptional in a number of respects, Markarian 231 demonstrates that at least some type 1 Seyfert galaxies emit thermally.

The generally accepted diameter and gas density for the region producing the forbidden lines in type 1 Seyferts would imply a substantial optical depth in the visible and ultraviolet from absorption by dust, unless the dust is substantially depleted relative to the normal dust to gas ratio of 1% (by mass). Thus, a thermally reradiated component would be expected in the spectrum of these galaxies. This component might be detected through observation of spectral turnovers near 2μ and in the far infrared, or of a 10-μ emission feature.

The most convincing demonstration of non-thermal emission would be detection of rapid fluctuations in emission. Extensive monitoring of a limited number of Seyfert galaxies has failed to reveal any fluctuations larger than a factor of 1.5. The photometric surveys described earlier might also detect variations where the same galaxy has been observed in different surveys. However, repeated measurements of cases where discrepancies of a factor of 1.5 or more were found have revealed no additional changes, and it seems likely that the original discrepancies arise from photometric errors. Fluctuations smaller than a factor of 1.5 have been reported in a number of cases. Widespread acceptance of the reality of these smaller changes will probably require application of more accurate measurement techniques. However, the available observations already apparently establish that the infrared variability is of smaller amplitude than that found in the optical (see e.g. Pacholczyk and Weymann, 1968), and that infrared variations do not occur synchronously with optical ones (Penston et al. 1974).

Strong infrared polarization might be expected from a non-thermal source, as is found for QSO's. However, NGC 4151 is virtually unpolarized at 2.2μ (Kemp et al. 1977). The residual polarization that is observed could result from scattering by dust (the wavelength dependence is similar to that for interstellar extinction) or to the dilution of infrared flux from the polarized source seen in the visible by a stronger, unpolarized infrared source.

From the evidence cited above, it seems likely that the ultraviolet and infrared sources in type 1 Seyfert galaxies are distinct. The infrared sources differ in a number of important respects from those in QSO's - including extent of variability, correlation of infrared variations with optical/UV ones, and polarimetric behavior.

The most thoroughly studied type 2 Seyfert galaxy is the class prototype, NGC 1068. From a number of points of view, it has been argued that the infrared emission is thermal reradiation by dust. A more definitive test of this suggestion is made possible by the recent calculations by Jones et al. (1977), who predict the spectrum over a broad spectral range under the assumption that all of the mid-infrared flux is radiated by a thick dust cloud. Their model was able to account for the available measurements of the spectrum between 4 and 34μ (e.g. Rieke and Low, 1975; Kleinmann, Gillett, and Wright, 1976) and gives a plausible explanation for the angular extent of the source measured by Becklin et al. (1973).

The calculations also predicted that silicates, seen weakly in absorption at 10μ, would appear in emission at 20μ; this prediction has been confirmed by recent observations (Lebofsky, Rieke, and Kemp, 1977). The close correspondence between theory and observation establishes the thermal origin of the 4 - 40μ flux from NGC 1068 on a much firmer basis than before, although the possibility of a small contribution (≲30%) from some other mechanism remains.

The transition from absorption at 10μ to emission at 20μ can occur only if the total optical depth in the cloud falls within a restricted range, $0.3 \lesssim \tau_{10} \lesssim 1.2$ (Leung, 1976), corresponding to $5 \lesssim A_v \lesssim 20$. Therefore, it seems likely that any source at the center of the dust cloud is inaccessible to direct optical study but can be observed in the near infrared. Short of 4μ, the calculations of Jones et al. cannot fit the observed spectrum without a source component in addition to the dust cloud. Therefore, the spectral range 1 - 4μ will be of great importance in studies of the central source.

In that spectral range, the polarization of NGC 1068 exhibits a remarkable behavior, not yet observed in any other extragalactic source. After dropping smoothly from 7% at θ = 103° in the ultraviolet to ∿1% at 94° in the red (Angel et al. 1976), it rises to ∿4% at 2.2μ and rotates to 125° by 3.45μ (Lebofsky, Rieke, and Kemp, 1977). The most straightforward explanation for these observations is that a central, power-law source, with wavelength-independent polarization of 2.5% at ∿126°, is embedded in the dust cloud which modifies the polarization of the energy by scattering from aligned grains. A similar model, with a spectral index α ∿ ∿ 0.9, is successful in fitting the 1 - 4μ spectrum if allowance is made for absorption and scattering within the central cloud corresponding to $A_v \sim 15$. Extrapolated to 10μ, this power law would fall below the limits on an unresolved nuclear source.

A test of this model would entail extending the polarization measurements to 13μ, including observations within and outside of the silicate absorption, and searching for circular polarization in the near infrared.

Detailed study of a number of galaxies of each type will be necessary before drawing general conclusions about the distinctions between the two classes of Seyfert galaxy. Of particular interest will be NGC 1275, with a relatively flat infrared specturm for type 2 (α ∿ 1.7) and many differences in other respects from NGC 1068; Markarian 6, which may represent an intermediate type (see e.g. Neugebauer et al. 1976); Markarian 78, which has a relatively flat infrared spectrum and large ultraviolet excess for type 2; NGC 4151; NGC 7603, with a relatively steep infrared spectrum for type 1 (α ∿ 1.7) and small ultraviolet excess, and Markarian 509, with a particularly flat spectrum (α ∿ 0.4).

REFERENCES

Aitken, D.K., Griffiths, J. and Jones, B., 1976, Mon. Not. Roy.
Astr. Soc. 176, 73P.
Angel, J.R.P., Stockman, H.S., Woolf, N.J., Beaver, E.A. and Mar-
tin, P.G., 1976, Astrophys. J. (Letters) 206, L5.
Becklin, E.E. and Neugebauer, G., 1968, Astrophys. J. 151, 145.
Becklin, E.E. and Neugebauer, G., 1975, Astrophys. J. (Letters)
200, L71.
Becklin, E.E., Fomalont, E.M. and Neugebauer, G., 1973, Astropys.
J. (Letters) 181, L27.
Becklin, E.E., Matthews, K., Neugebauer, G. and Willner, S.P.,
1977, preprint.
Becklin, E.E., Matthews, K., Neugebauer, G. and Wynn-Williams,
C.G., 1973, Astrophys. J. (Letters) 186, L69.
Gatley, I., Becklin, E.E., Werner, M.W. and Wynn-Williams, C.G.,
1977, Astrophys. J. 216, 277.
Gillett, F.C. and Woolf, N.J., 1973, in Woolf, I.A.U. Symp. No.
52, editors - Greenberg and van de Hulst, p. 485.
Gillett, F.C., Kleinmann, D.E., Wright, E.L. and Capps, R.W.,
1975, Astrophys. J. (Letters) 198, L95.
Gillett, F.C., Jones, T.W., Merrill, K.M. and Stein, W.A., 1975,
Astron. and Astrophys. 45, 77.
Harper, D.A., 1977, private communication.
Harper, D.A. and Low, F.J., 1973, Astrophys. J. (Letters) 182,
L89.
Heintz, J., Jereki, P., Low, F.J. and Rieke, G.H., 1977, in pre-
paration.
Hildebrand, R.H., Whitcomb, S.E., Winston, R., Stiening, R.F.,
Harper, D.A. and Moseley, S.H., 1977, Astrophys. J. 216, 698.
Houck, J.R., Soifer, B.T., Pipher, J.L. and Harwit, M., 1971,
Astrophys. J. (Letters) 169, L31.
Jones, T.W., Leung, C.M., Gould, R.J. and Stein, W.A., 1977, As-
trophys. J. 222, 52.
Karachentsev, I.D., 1972, Comm. Spec. Astrofiz. Obs. Acad. Nauk
USSR, AS 7, 3.
Kemp, J.C., Rieke, G.H., Lebofsky, M.J. and Coyne, G.V., 1977,
Astrophys. J. (Letters) 215, L107.
Kleinmann, D.E. and Low, F.J., 1970, Astrophys. J. (Letters) 159,
L165.
Kleinmann, D.E., Gillett, F.C. and Wright, E.L., 1976, Astrophys.
J. 208, 42.
Knacke, R.F. and Capps, R.W., 1977, Astrophys. J. 216, 271.
Knacke, R.F., Capps, R.W. and Johns, M., 1976, Astrophys. J.
(Letters) 210, L69.
Larson, R.B. and Tinsley, B.M., 1977, preprint.
Lebofsky, M.J., Kemp, J.C. and Rieke, G.H., 1977, unpublished.
Lebofsky, M.J., Rieke, G.H. and Kemp, J.C., 1977, in preparation.
Leung, C.M., 1976, Astrophys. J. 209, 75.

Low, F.G., Kurtz, R.F., Poteet, W.M. and Nishimura, T., 1977, Astrophys. J. (Letters) 214, L115.
Neugebauer, G., Becklin, E.E., Oke, J.B. and Searle, L., 1976, Astrophys. J. 205, 29.
Neugebauer, G., Becklin, E.E., Beckwith, S., Mathews, K. and Wynn-Williams, C.G., 1976, Astrophys. J. (Letters) 205, L139.
Oke, J.B., Neugebauer, G. and Becklin, E.E., 1970, Astrophys. J. 159, 341.
Pacholczyk, A.G. and Weymann, R.J., 1968, Astronom. J. 73, 850.
Pacholczyk, A.G. and Wisniewski, W.A., 1967, Astrophys. J. 147, 394.
Penston, M.V., Penston, M.J., Selmes, R.A., Becklin, E.E. and Neugebauer, G., 1974, Mon. Not. Roy. Astr. Soc. 169, 357.
Price, S.D. and Walker, R.G., 1976, AFGL-TR-76-0208.
Rieke, G.J., 1974, Astrophys. J. (Letters) 193, L81.
Rieke, G.H., 1976, Astrophys. J. (Letters) 206, L15.
Rieke, G.H., 1976, Astrophys. J. (Letters) 210, L5.
Rieke, G.H., 1977, in preparation.
Rieke, G.H. and Kinman, T.D., 1974, Astrophys. J. (Letters) 192, L115.
Rieke, G.H. and Lebofsky, M.J., 1977, Astrophys. J. (Letters) submitted.
Rieke, G.H. and Low, F.J., 1972, Astrophys. J. (Letters) 176, L95.
Rieke, G.H. and Low, F.J., 1973, Astrophys. J. 184, 415.
Rieke, G.H. and Low, F.J., 1975, Astrophys. J. 197, 17.
Rieke, G.H. and Low, F.J., 1975, Astrophys. J. (Letters) 199, L13.
Rieke, G.H., Telesco, C.M. and Harper, D.A., 1977, Astrophys. J., in press.
Rieke, G.H., Lebofsky, M.J., Kemp, J.C., Coyne, G.V. and Tapia, S., 1977, Astrophys. J. (Letters), in press.
Rieke, G.H., Grasdalen, G.L., Kinman, T.D., Hintzen, P., Wills, B.J. and Wills, D., 1976, Nature 260, 754.
Salpeter, E.E., 1959, Astrophys. J. 129, 608.
Soifer, B.T., 1977, private communication.
Stein, W.A. and Weedman, D.W., 1976, Astrophys. J. 205, 44.
Tarenghi, M. and Rieke, G.H., 1977, in preparation.
Treffers, R.R., Fink, U., Larson, H.P. and Gautier, T.N., 1976, Astrophys. J. (Letters) 209, L115.
Ulrich, M.H., 1977, in preprint.
Wollman, E.R., Geballe, T.R., Lacy, J.H., Townes, C.H. and Rank, D.A., 1977, preprint.

INFRARED ASTRONOMICAL BACKGROUND RADIATION

Martin Harwit

Cornell University, Ithaca, New York, U.S.A. and
MPI für Radioastronomie, Bonn, West Germany

Measurements of astronomical background radiation are interesting
because when they succeed they tell us the energy budget of the
universe -- at least that part of the budget that gives rise to
electromagnetic radiation.   Neutrino and gravitational radiation
background measurements would be required before a complete led-
ger could be set up.
   Peter Mezger's lectures have shown a composite chart of background components that are incident on the earth's atmosphere
from outside. His figure covers the range from 10 MHz in the radio
wavelength range to 10^{21} Hz or roughly 10 Mev in the gamma ray domain.   Gamma and X-ray background measurements have to be obtained
from above the atmosphere; ultraviolet background radiation also
must be obtained at high - rocket or satellite - altitudes; visible background radiation can be obtained from the ground, but the
results are much more reliable when telescopes are placed in spacecraft. At the other extreme of the wavelength range, radio background observations from 10 MHz to higher frequencies can be carried out rather well from the ground.   Peter Clegg's contribution
shows how observations are carried out in the microwave and submillimeter range where the 2.7 K cosmic background radiation finds
its peak.
   In this lecture I will describe measurements that tell us the
strength of the background radiation incident on the atmosphere in
the infrared part of the spectrum. To do that properly, the discussion will also have to concern itself with visible radiation
at one end of the scale, and the microwave background at the other,
since most of the natural sources of radiation cover extended spectral range.
   The information we have about infrared background radiation today, is still provisory. While some rocket observations have been

173

G. Setti and G. G. Fazio (eds.), Infrared Astronomy, 173-180.

possible, the detectors that were available at the time were
orders of magnitude less sensitive than those available right now.
A cosmic background radiation explorer satellite would therefore
be able to yield data of much greater cosmological significance
than information currently available.

The chief sources of infrared background radiation that we now
recognize are these:

1. Nearest to us, in the ecliptic plane and stretching out to the
orbits of asteroids, lies a thin sheet of interplanetary dust.
In the visible part of the spectrum this dust scatters sunlight
and gives rise to the zodiacal glow. At 10μ the dominant component of the zodiacal radiation is the thermal emission from these
grains.

2. A second near infrared component can be attributed to Galactic
sources. There is a concentration of this radiation toward the
Galactic plane. A contribution from the Galaxy's spherical distribution of population II stars is also expected, but its strength
still is uncertain.

3. In the far infrared, beyond about 20μ, HII regions emit power-
fully. The radiation comes partly from dust embedded in the ionized gas, and partly from dark clouds surrounding the hot plasma.
The dust grains in these cooler regions absorb much of the visible
and ultraviolet radiation escaping the HII domain, and convert the
energy into far infrared radiation. Since the ionized gases in
the Galaxy are found predominantly in the plane, the far infrared
flux reaches us from a narrow Milky Way band that becomes increasingly bright toward the Galactic center.

4. We expect an isotropic radiation component that can be attributed to the integrated flux reaching us from distant galaxies.
At low spatial resolution the individuality of the nearer sources
will become indiscernable; and the more distant sources would
appear to blend into a continuum even at higher spatial resolution.
In the far infrared, around 100μ where currently available telescopes are diffraction limited at roughly half a minute of arc,
this limit should be reached even for moderately distant galaxies.
At present, however, we have no observational evidence for the
existence of such a component. We can only estimate its strength
on the basis of far infrared observations obtained on individual
nearby galaxies.

5. At the longest wavelengths, bordering the millimeter range, the
isotropic 2.7 K cosmic background spills over into the far infrared.

Observational Difficulties

In the infrared spectral range diffuse sources are difficult
to detect because radiation from most sources tends to be observed
by differential measurements. Radiation incident on the detector
is "chopped" between fields of view containing a source and ad-

jacent fields that are suspected to be empty. If the 'empty'
field, however, is not truly empty, then its signal is subtracted
from that of the field of view containing the 'source'. An isotropic source of radiation can therefore never be detected by
means of a chopping technique.

One might ask oneself why chopping should have come into use
at all in that case? The reason is quite compelling: The emission
from the atmosphere can be enormous, and so highly variable in many
parts of the infrared spectrum, that its brightness in the telescope's field of view often is many tens of thousand times brighter than the source to be viewed. A small chopper 'throw' is an
advantage since atmospheric brightness varies across the sky and
is smallest when immediately adjacent regions are compared to each
other. Such a small throw, however, also makes detection of slightly extended sources difficult. Since chopping techniques are used
not only in ground based observations, but also with airborne and
balloon telescopes, the study of diffuse sources has not progressed
as rapidly as have observations of compact infrared sources.

Thus far the only astronomical observations that have not made
use of chopping devices have been rocket observations and a number of selected balloon studies specifically designed to make background measurements in the submillimeter range.

Units

Various units can be used to express sky brightness or surface
brightness of a diffuse source, for example the intensity per unit
wavelength band or frequency band, respectively $I(\lambda)$ or $I(\nu)$, measured over unit solid angle. The measurements are also conveniently normalized to a frequency bandwidth $\Delta\nu=\nu$ or wavelength
bandwidth $\Delta\lambda=\lambda$. This allows a comparison of the total sky brightness in the visual, infrared, and radio domains, for example.
The units used therefore are watt meter^{-2}sterad^{-1} measured at the
the telescope and express quantities $\lambda I(\lambda)$ or $\nu I(\nu)$.

Observations in the Visual Range

In the visible and near infrared, some background observations
have been possible from the ground. Roach and Smith (1968) for
example made direct observations of the zodiacal dust component,
and by subtraction derived a component that could be extragalactic.
The zodiacal light, in this wavelength range is well known (cf.
Allen, 1973) to have a brightness of 2 x 10^{-4} W m^{-2}sr^{-1} at 10°
elongation from the sun, along the ecliptic plane. Going further
along the plane, this brightness rapidly drops to 10^{-6} W m^{-2}sr^{-1}
at elongations around 90°. At greater elongations, there is a
further drop of some 15% or so in intensity; but in the antisolar direction, the sky brightness again picks up. Here one is

dealing with the Gegenschein which produces an antisolar direction brightness roughly equal to the surface brightness at 90°
elongation in the ecliptic. The minimum zodiacal light intensity
found near the ecliptic pole is roughly half of that same flux.
Roach and Smith (1968) thought they could attribute roughly $\lesssim$ 5%
of this minimum zodiacal light flux to extragalactic sources, and
Lillie (1968) using rocket observational data came up with similar
results. Clearly, however, substantial subtractions of zodiacal
and Galactic components dominate such a computation and the results
of these authors must be treated with caution.

Better results by far can be obtained from spacecraft. Hanner
et al. (1974) and Weinberg et al. (1974) observed diffuse radiation in two spectral ranges, 3900-5000 Å and 5950-7200 Å from the
Pioneer 10 spacecraft. At elongations of about 100° where the
ecliptic component of the zodiacal light is at a minimum, they
observed a drop by a factor of 10 in going from 1 A.U. out to
2.4 A.U. Beyond 3.3 A.U. a furthur drop made the zodiacal glow
invisible at a brightness level roughly 6×10^{-9} W $m^{-2}sr^{-1}$ in the
blue filter.

A puzzling result due to Mattila (1976) should be mentioned
here. He observed the surface brightness of a high galactic la-
titude dark cloud, L 134 at two different wavelengths, and compared this to the sky brightness adjacent to the cloud. Use of
the two wavelengths permitted subtraction of scattered star light
or light from stars at large distances within our galaxy, on the
basis of the sharply rising stellar spectrum expected in the 4000 Å
range where the two wavelengths were centered.

The extragalactic background is expected to be flat, and Mattila
therefore concludes from his observations that his high (10^{-7}w
m^{-2} sr^{-1}) relatively flat spectrum flux must be extragalactic.
This might be reconcilable with cosmological models in which galaxy formation took place relatively recently, at an era corresponding to a red-shift of around 3 to 5.

We have gone into so much detailed description of these visual
domain observations, because a similar historical development may
be anticipated in the infrared in the next decade. Mattila's
technique, even though his results are controversial, also helps
to point out that differential observations do have a place even
in the determination of highly diffuse radiation, since they can
act to separate out Galactic foreground emission in a manner re-
mininiscent of the normal infrared chopping techniques mentioned
above. One important difference however is the use of color dif-
ferentiation as a second dimension in the differential measurement. It emphasizes the probable future need for a combination
of diffuse measurements with spectral techniques which should help
to untangle the contributions from diffuse sources lying along one
and the same line of sight.

Infrared Observations

There have been relatively few direct infrared observations of background radiation. Zodiacal light observations have been carried out at 2.4μ from balloons. At this wavelength the upper atmospheric OH glow which emanates from altitudes above 60 km is at its lowest, and the view is therefore relatively unconfused. At 5, 13 and 20 microns the Cornell University group obtained data consistent with grains having an albedo around 0.3 and a temperature of 230 to 350°K, along a viewing direction somewhat more than 90° from the sun. The ecliptic plane surface brightness is found to be $\sim 10^{-5}$ W m^{-2}sr^{-1}.

A variety of observations have also been carried out at angles far closer to the sun. Both ground based and airplane studies were involved.

The brightness of the Galaxy and the Milky Way plane has been observed by several groups in Japan and at Heidelberg, at wavelengths of 2.4μ. At longer wavelengths, between 5 and 120μ early measurements were made by the Cornell University group which found a number of previously unrecognized sources and also measured the diffuse 100μ radiation at larger distances from the center, along the Galactic plane. More recently, balloon observations have been providing confirming data on a larger scale.

Extragalactic fluxes in the infrared can, at present, only be given upper limits. These observations again carried out from rockets by the Cornell group still have a long way to go, and will probably have to await launch of a satellite probe designed to observe the infrared background flux.

In the submillimeter region a wide variety of observations have been carried out from airplanes, balloons and rockets to determine the short wavelength spectrum of the cosmic microwave background. Details of these observations are found in the talk by Peter Clegg, in this same volume.

Acknowledgements

The author is grateful to the Alexander von Humboldt Stiftung of the Federal Republic of Germany for a travel grant to the NATO Advanced Study Institute on Infrared Astronomy, and in particular for a U.S. Senior Scientist Award that enabled him to spend a year at the Max-Planck-Institute for Radioastronomy in Bonn. He thanks Prof. Peter Mezger, director of the Institute for Radioastronomy, for his hospitality during that year.

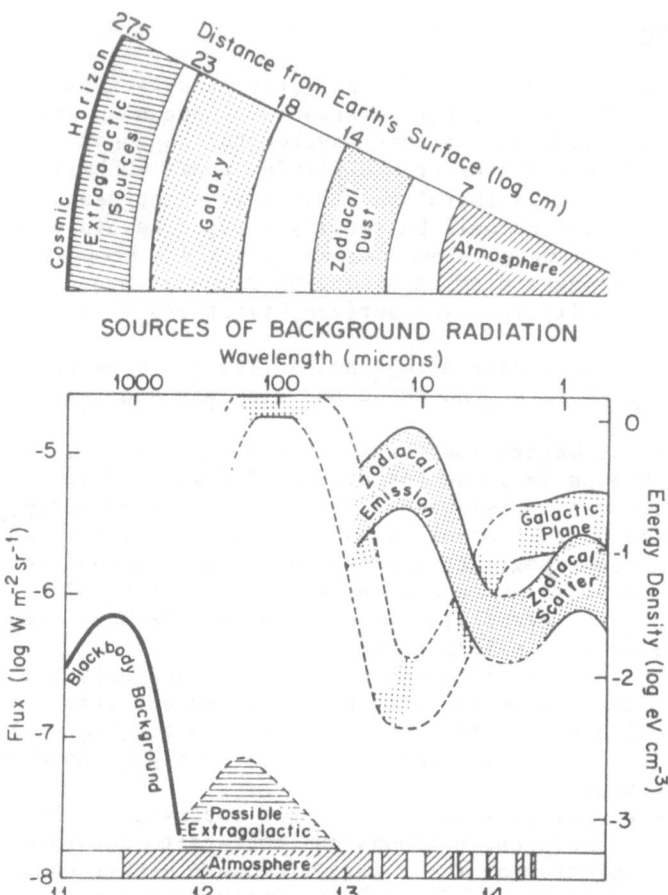

Sources of Background Radiation. The top part of the figure shows
the distance from an Earthbound observer to some of the more important sources. The bottom half of the diagram shows the spectral
character and intensity of the sources. Except in the shorter, submillimeter ranges the blackbody cosmic background radiation has been
measured with the greatest precision. The zodical glow is quite
varied; in general it peaks toward the sun and in the ecliptic
plane. The range shown here is characteristic of portions of the
night sky. The galactic plane contribution also is quite variable,
and has been measured reliably only in the 2.4µ region of the infrared. Measurements also exist in the 100µ region; but the contributions from individual sources are more difficult to identify
and isolate from a general diffuse background, at these longer wavelengths. The extragalactic source contribution is least well understood among infrared sources. The possible surface brightness shown
was estimated by Stecker et al. (1977).

REFERENCES

General Reference

Harwit, M., 1970, Revista del Nuovo Cimento $\underline{II}$, 253.
This review covers observations up to the late 1960's and includes
wavelengths between 1μ and 1 mm. The main changes since that review have been in firm determinations of the zodiacal dust emission
around 10μ, the Galactic plane emission at 100μ, and the microwave cosmic background radiation just short of 1 mm. These topics
are covered in more recent references given below.

Special Topics

Zodiacal Dust

Allen, C.W., 1973, Astrophysical Quantities, (London: Athlone Press)
This book tabulates the 5000 Å sky brightness due to zodiacal
light at different solar elongations and elevation angles above
the ecliptic.
Lillie, C.F., 1968, University of Wisc. Ph. D. Thesis.
This deals with rocket determination of the zodiacal light inten-
sity at 4000 Å.
Roach, F.E., and Smith, L.L., 1968, Geophys. J. Roy. Astr. Soc. $\underline{15}$,
227.
This article shows ground-based determination of the visual dif-
fuse glow.
Soifer, B.T., Houch, J.R., and Harwit, M., 1971, Ap. J. $\underline{168}$, L 73.
This article contains Rocket observations of the zodiacal glow at
5, 13 and 20μ.
Weinberg, J.L., Hanner, M.S., Beeson, D.E., DeShields, L.M. II,
and Green, B.A., 1974, J. Geophys. Res. $\underline{79}$, 3665.
Pioneer 10 observations find that the visual zodiacal light drops
to an unmeasurably low value beyond 3.3 A.U. from the sun.

Galaxy

Hanner, M.S., Weinberg, J.L., DeShields, L.M. II, Green, B.A., and
Toller, G.N., 1974, J. Geophys. Res. $\underline{79}$, 3671. Pioneer 10 observa-
tions at large distances from the sun are described in this article.
A comparison is made of the observed brightness in fields of view
subtending roughly 10 sq. deg., and the predicted brightness of
known stars in the field.
Houck, J.R., Soifer, B.T., Pipher, J.L., and Harwit, M., 1971,
Ap. J. $\underline{169}$, L 31. 5, 13, 20 and 100μ observations of the central
portions of the Galaxy are made.
Ito, K., Matsumoto, T., and Uyama, K., 1977, Nature $\underline{265}$, 517.
This paper describes the Galactic center and plane brightness dis-
tribution at 2.47μ.
Low, F.J., Kurtz, R.F., Poteet, W.M., and Nishimura, T., 1977,
Ap. J. $\underline{214}$, L115.
Balloon observations at 60 to 300μ show a distribution of extended
sources in the Galactic plane.

Pipher, J., 1973, I.A.U. Symposium No. 52, 559.
Diffuse brightness at 100μ, at low latitudes are estimated on the
basis of rocket observations.

Extragalactic Radiation

Matilla, K., 1976, Astron. + Astrophys. 47, 77.
This article shows a subtraction technique that makes use of dark
clouds at high Galactic latitudes that leads to a surprisingly
high estimate of the extragalactic flux.
Stecker, F.W., Puget, J.L., and Fazio, G.G., 1977, Ap. J. 214, L51.
The extragalactic flux from known sources is estimated for the
spectral range between 10μ and 1 mm.

COSMIC BACKGROUND: MEASUREMENTS OF THE SPECTRUM

P.E. Clegg

Queen Mary College, University of London, England

1. INTRODUCTION

Assume the existence of Planckian relict radiation at a present temperature of ∿2.9 K.
- Reasonable assumptions about the matter density of the universe lead to the conclusion that the photons now entering radiometer horns or spectrometers last interacted with matter at a redshift of ∿1500, i.e. only ∿200,000 y after the "origin" of the universe.
- In contrast, redshifts of visually identified objects in the universe are ≤4 corresponding to epoch ∿10^9 y after creation. Antiquity of the relict radiation therefore provides a unique opportunity of direct investigation of the early universe.
  The importance of verifying the existence of a thermal microwave background is evident. Many observations have been made resulting in statements such as "There is now little doubt that the microwave background does not have a black-body shape" (Davis, 1976). I maintain that there is considerable doubt and that more measurements are required:
- to confirm the generally Planckian shape of the spectrum,
- to determine the shape of the spectrum precisely and thereby extract the maximum historical information.
  I shall present:
- a brief theoretical background,
- a review of observations,
- possibilities for the future.

G. Setti and G. G. Fazio (eds.), Infrared Astronomy, 181-198.
All Rights Reserved. Copyright © 1978 by D. Reidel Publishing Company, Dordrecht, Holland.

2. THEORETICAL BACKGROUND

The theory of the early universe is presented in more detail by
Longair in the volume and in many standard texts (e.g. Peebles,
1971; Weinberg, 1972). Briefly we consider a standard Friedmann
cosmology, expanding uniformly with scale factor $R(t)$.
    Redshift z of a photon between epochs t and t_o (present) is
given by

$$1 + z = R(t_o)/R(t) . \qquad (2.1)$$

For a universe filled with photons alone, of number density n_γ,
photon conservation gives

$$n_\gamma \propto R^{-3} . \qquad (2.2)$$

For Plankian photons at a temperature T_γ

$$n_\gamma = \frac{4aT^3}{3k} , \qquad (2.3)$$

so that

$$T_\gamma \propto R^{-1} . \qquad (2.4)$$

If matter is also present with number density n_H, there will be
thermal interaction between matter and radiation. What effect does
this have?

Heat capacity of radiation	$\sim 4aT^3$
Heat capacity of ionised hydrogen	$\sim 3n_Hk$

$$\text{Ratio } \sigma = \frac{4aT_\gamma^3}{3n_Hk} = \frac{n_\gamma}{n_H} . \qquad (2.5)$$

For $T_{\gamma o} \sim 2.9$ K, $n_{Ho} \lesssim 3 \times 10^{-6}$ g cm^{-3}, $\sigma_o \sim 10^9$ where subscript
zero denotes present value.
$\therefore$ the heat capacity of radiation is easily able to compensate any
attempt by matter to change its temperature and, consequently,
relation (2.4) remains valid. Furthermore, because σ is independent
of $R(t)$ (n_γ and n_H both proportional to R^3 because photons and
baryons are conserved), this result always applies. The behaviour
of T_γ as a function of R is indicated in Fig. 1; also shown is the
corresponding age for an Einstein-de Sitter universe.
    At sufficiently early epochs the average photon energies will
be sufficiently high to ionise the hydrogen. Eventually, as the
photons are redshifted with epoch, there will be insufficient
photons of high enough energy to maintain ionisation and matter
will recombine. With current estimates of photon and matter density
this occurred at

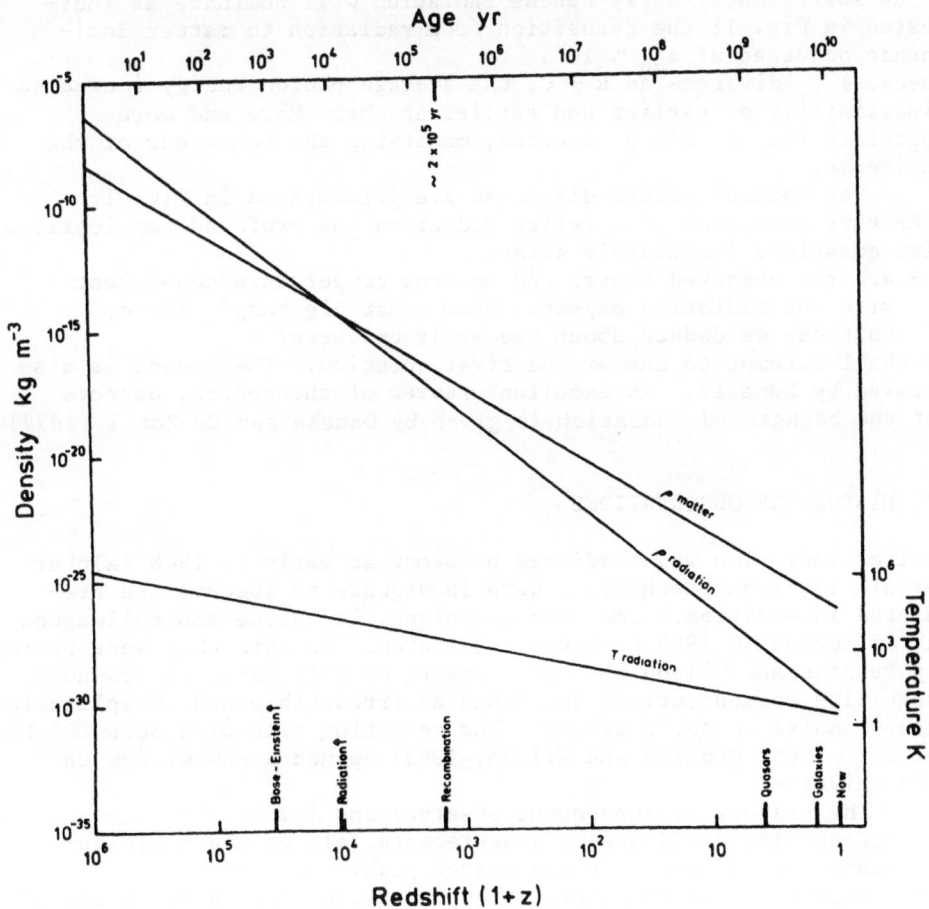

Fig. 1. Matter density, radiation density and radiation temperature as a function of redshift. Various indicated epochs are
discussed in the text. The epoch corresponding to z for an
Einstein-de Sitter universe is indicated on the upper axis.

$$z_{rec} \sim 1500 \qquad\qquad\qquad (2.6)$$

What are the consequences of all this?

The <u>dynamics</u> of the universe is determined by the dominant
energy density:

Radiation $u_r = aT^4 \propto R^{-4}$

Matter $u_m = \rho c^2 = n_H m_H c^2 \propto R^{-3}$

∴ at sufficiently early epochs radiation will dominate as indicated in Fig. 1; the transition from radiation to matter dominance occurred at $z_{eq} \sim 10^4$.

Because T_γ diverges as $R \to 0$, the average photon energy increases indefinitely at earlier and earlier epochs. More and more particle fields will be created, modifying the behaviour of the universe.

The various points discussed are illustrated in Fig. 1. The very existence of a relict radiation has profound implications. Two questions immediately arise:
- are the observed fluxes and antenna temperature consistent with the radiation expected from a hot big bang? If so, what can we deduce about the early universe?

I shall attempt to answer the first question. The second is discussed by Longair. An excellent review of theoretical aspects of the background radiation is given by Danese and De Zotti (1977).

3. HISTORY OF OBSERVATIONS

Relict radiation was predicted by Gamow as early as 1948 (Alpher et al., 1948) but techniques were inadequate to observe the predicted intensities. Improved techniques led Dicke and colleagues at Princeton in 1965 to propose a search. In this they were beaten by Penzias and Wilson who were working at Bell Labs. on low-noise satellite communications and found an irreducible and inexplicable excess noise in their system. The resulting pair of papers (Dicke et al., 1965; Penzias and Wilson, 1965) opened a new window on the universe.

The history of subsequent observations is:
- Monochromatic radiometer measurements, all on the Rayleigh-Jeans side of the supposed Planck peak.
- Interstellar thermometry, using rotational levels of CN and CH.
- Broadband measurements from rocket and balloon platforms.
- Spectral measurements from balloon platforms.

These observations, and their results interpreted on the hypothesis of a Planck spectrum, are summarised in Fig. 2. The overall agreement with a thermodynamic temperature of ~ 2.9 K is at first sight remarkably good. Many uncertainties and ambiguities remain however. A review of the current position is given by Robson and Clegg (1977).

3.1 Monochromatic Radiometry

The principle of this method is very simple:
- Measure the total power in a (small) bandwidth using a horn feeding a low-noise radiometer.
- Subtract the contribution from all known extraneous sources, such as the atmosphere, the Galactic background, losses in feeds, side-lobe response from the ground, etc.

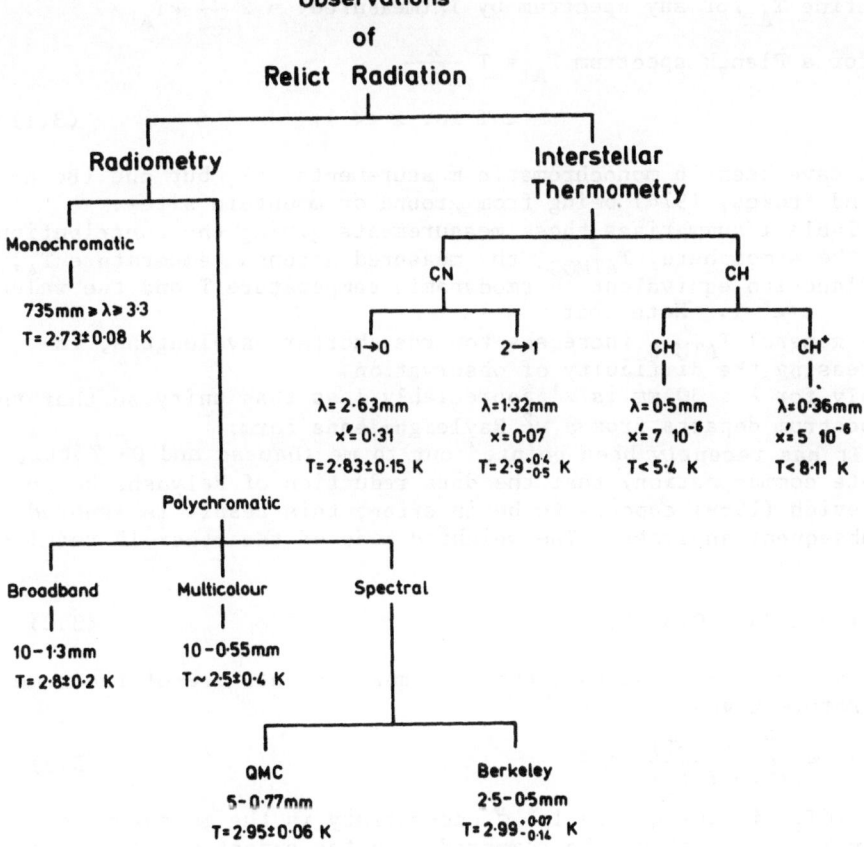

Fig. 2. Summary of observations of the microwave background radiation.

- What is left is the cosmic background!

   The practice is exceedingly difficult because it involves subtracting comparable and relatively uncertain quantities to leave a small remainder. For example, only 1/3 of the power originally measured by Penzias and Wilson at 73.5 mm was attributable to the cosmic background. It is often convenient to discuss results in terms of antenna temperature T_A because
- it is frequently the quantity directly measured,
- it is easy to see the departure from a Rayleigh-Jeans spectrum.

Antenna temperature:

For a Planck spectrum $I_\nu = \dfrac{2\nu^2}{c^2} kT \dfrac{x}{e^x - 1}$ where $x = h\nu/kT$.

For high T and low ν, $x \ll 1$ and $I_\nu \approx 2 \dfrac{\nu^2}{c^2} kT$.

Define T_A for any spectrum by I_ν(measured) $= 2 \frac{\nu^2}{c^2} kT_A$

$\therefore$ for a Planck spectrum $T_A = T \frac{x}{e^x-1}$

$$\approx T \text{ for } x \ll 1. \qquad (3.1)$$

There have been 16 monochromatic measurements, all but one (Boynton and Stokes, 1974) being from ground or mountain sites.

Table 1 summarises these measurements giving the contribution from the atmosphere, T_{ATMOS}, the measured antenna temperature T_A, the Planckian equivalent thermodynamic temperature T and the value of $x' = x/e^x-1$. Note that:

- In general T_{ATMOS} increases towards shorter wavelengths, increasing the difficulty of observation.
- Only for $\lambda \leq 30$ mm is x' appreciably less than unity so that the spectrum departs from a ν^2 Rayleigh-Jeans form.

It has recently been pointed out to me (Danese and De Zotti, private communication) that the data reduction of Pelyushenko and Stankevich (1969) appears to be in error; this result is ignored in subsequent analyses. The weighted mean of the other 15 results gives

$$T = 2.74 \pm 0.08 \text{ K.} \qquad (3.2)$$

A χ^2 test of the hypothesis that all measurements are of the same temperature gives

$$\chi^2 = \sum_{i=1}^{n} \frac{(T_i - \overline{T})^2}{(\delta T_i)^2} \approx 9 \qquad (3.3)$$

where (δT_i) is the quoted $1 - \sigma$ uncertainty in the measurement. This value of χ^2 should be compared with the expected value of

$$(n-1) \pm [2(n-1)]^{1/2} \approx 14 \pm 5 . \qquad (3.4)$$

These measurements are therefore consistent with a confidence of 75%.

Fig. 3 plots the measured intensities as a function of λ and illustrates the good fit with a 2.74 K Planck spectrum (solid line). But, apart from the 3.3 mm measurements, the results are almost equally consistent with a Rayleigh-Jeans spectrum (dashed line). If the measurement by Howell and Shakeshaft (1967), which is well below the Galactic background is also ignored, the other 11 results are best fitted with a $\nu^{2.12\pm0.05}$ spectrum, with a χ^2 of 9.

The first clear indication of departure from a Rayleigh-Jeans spectrum was obtained by the Princeton group (Stokes et al., 1967; Wilkinson, 1967); the reader is referred to these papers to gain appreciation of the difficulties of the shorter wavelength measurements. A $2\frac{1}{2}\sigma$ departure from ν^2 was observed, confirmed by χ^2 of 7 compared with 2 ± 2 for the ν^2 hypothesis. The most precise monochromatic confirmation of the non-Rayleigh-Jeans nature comes from

Table 1

Monochromatic Measurements

λ(mm)	T_{ATMOS}	T_A	T	x'	Reference
735 492	10.4±0.7	3.7±1.2	3.7±1.2	1.00	Howell and Shake-shaft (1967)
212	2.3±0.2	3.2±1.0	3.2±1.0	0.99	Penzias and Wilson (1967)
207	2.2±0.2	2.8±0.6	2.83±0.6	0.99	Howell and Shake-shaft (1966)
300 209 150		2.5±0.37[x]	2.53±0.37[x]	0.99	Pelyushenko and Stankevich (1969)
73.5	2.3±0.3	3.5±1.0	3.6±1.0	0.97	Penzias and Wilson (1965)
32	3.03±0.04	2.78±0.5	3.00±0.50	0.93	Roll and Wilkinson (1966)
32	1.38±0.02	$2.47^{+0.16}_{-0.25}$	$2.69^{+0.16}_{-0.25}$	0.92	[†]Stokes et al. (1967)
15.8	3.99±0.25	$2.35^{+0.12}_{-0.17}$	$2.78^{+0.12}_{-0.17}$	0.85	Roll and Wilkinson (1966)
15	∿6	2.0±0.4	2.45±0.4	0.82	Welch et al. (1967)
9.24	4.62±0.36	2.45±0.26	3.17±0.27	0.77	Ewing et al. (1967)
8.56	1.38±0.02	$1.81^{+0.17}_{-0.21}$	$2.56^{+0.17}_{-0.22}$	0.71	[†]Wilkinson (1967)
8.2	17±0.8	2.89±0.7	$3.7^{+0.71}_{-0.72}$	0.78	Puzanov et al. (1968)
3.58		0.95±0.58	$2.43^{+0.69}_{-0.81}$	0.39	Kislyakov et al. (1971)
3.3	11.53±0.24	0.90±0.32	$2.47^{+0.40}_{-0.43}$	0.36	Boynton et al. (1968)
3.3	11.9±1.5	1.01±0.20	$2.61^{+0.25}_{-0.26}$	0.39	Millea et al. (1971)
3.3	1.21±0.37	0.92±0.4	$2.50^{+0.49}_{-0.55}$	0.37	Boynton and Stokes (1974)

[x]Error analysis is suspect.
[†]Note that tables are transposed between these papers.

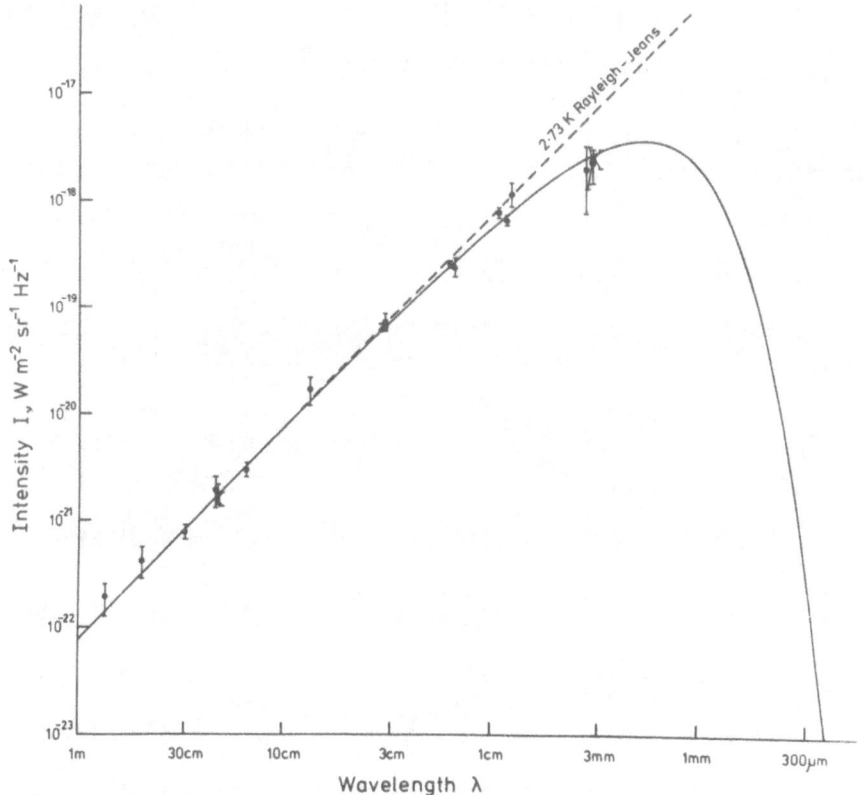

Fig. 3. Monochromatic radiometric measurements (References as in Table 1). The solid curve is a Planckian for the mean thermodynamic temperature of 2.73 K. The dashed curve is the equivalent Rayleigh-Jeans spectrum.

the airborne 3.3 mm experiment of Boynton and Stokes (1974) with comparatively low atmospheric contribution at the zenith of 1.21 ± ± 0.37 K, but the experiment is extremely difficult.

The conclusion from the monochromatic measurements? Almost certainly not a ν^2 spectrum, but the shorter wavelength of observation is below the expected Planck peak at ∿2 mm.

3.2 Interstellar Thermometry

The methods are:
- Observe the rotational splitting of the optical absorption of starlight in interstellar CN and CH.
- Assign an excitation temperature to the rotational level population.
- Eliminate unlikely excitation mechanisms.

- Attribute the excitation to the background radiation.
    The excitation was observed in 1941 (McKellar, 1941) but only later were new spectra by Herbig of ζ Oph and ζ Per interpreted by Field and Hitchcock (1966) in terms of a cosmic background radiation temperature of between 2.7 and 3.4 K.
    The method has since been refined and is reviewed by Thaddeus (1972). Although other excitation mechanisms are possible (e.g. collisional excitation):
- arguments against are persuasive,
- in any case, the excitation must almost certainly represent an upper limit to the background temperature.
The results are summarised in Fig. 4.

3.3 Broadband and Multicolour Measurements

    The peak of the intensity I_ν of a Planck spectrum occurs at a wavelength λ_m given by $\lambda_m T = 5$ mm K.

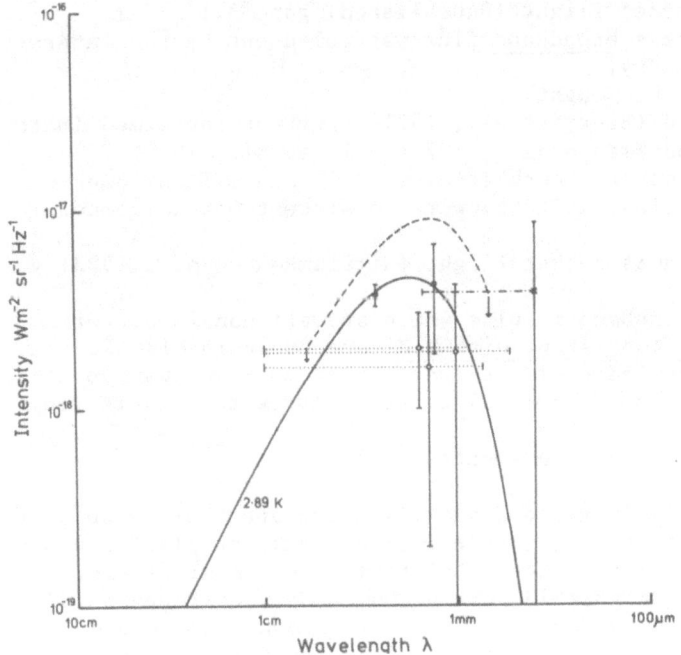

Fig. 4. Broadband observations around the Planckian peak. The solid curve is a Planck spectrum for a temperature of 2.89 K (see text). References: |·····Ī··•|, Muehlner and Weiss, 1973 a,b; |—·—·—Ī|, Houck et al., 1972; ┌────┐Williamson et al., 1973. Also shown are the most precise results of interstellar thermometry using CN (Borbolot et al., 1969; Heygi et al., 1974).

for T = 2.7 K, $\lambda_m \sim 2$ mm.
Atmospheric emission makes it impossible to observe this peak from
the ground or at aircraft altitudes.

There are two possibilities. Observe from:
- a balloon platform: Long observation time but a residual atmosphere,
- a rocket platform: Almost no "atmosphere" but limited time.

The first experiment was by the Cornell group (Shivanandan
et al., 1968) covering a range 1.3 - 0.4 mm. The result:
- The measured flux, <u>interpreted as coming from a Planck spectrum</u>
  gave a temperature of $8^{+2.2}_{-1.3}$ K! Because this was around the peak,
  where flux $\propto T^4$, this corresponds to ~75 times more flux than
  expected.

Muehlner and Weiss (1970) at MIT, choosing the alternative of
a balloon platform carrying a multicolour broadband photometer also
obtained an equivalent temperature of ~8 K.

The wild goose chase was on and a review is given by Blair
(1974). Briefly:
- A Los Alamos rocket flight (unpublished) gave $3.1^{+0.5}_{-1.0}$ K.
- A universal excess <u>broadband</u> flux was ruled out by the inter-
  stellar thermometry.
  ∴ a line was to be sought.
- A line was found (Beery et al., 1971), but not confirmed (Mather
  et al., 1971; Beekman et al., 1972; Nolt et al., 1972).
- Later Cornell rocket (Houck et al., 1972) and MIT balloon
  (Muehlner and Weiss, 1973a,b) were consistent with a tempera-
  ture of 2.7 K.
- A second Los Alamos rocket flight (Williamson et al., 1973) gave
  $3.8^{+10}_{-3.8}$ K.

The latest broadband results which are all consistent with
2.7 K, <u>but equally consistent with 0 K</u>, are shown in Fig. 4. At
this point, out knowledge of the spectrum around 1 mm was unsatisfac-
tory at the very least. Spectral measurements were clearly required.

3.4 Spectrophotometric Measurements

Measurement of a millimetre and submillimetre spectrum, even at the
low resolving power of ~10, requires time. Balloon platforms are
therefore the preferred choice. The first successful flight,
after a balloon burst in 1973, was by Queen Mary College in 1974
(Robson et al., 1974; Robson, 1976). A liquid-helium cooled
polarising Michelson interferometer, with spectral resolution of
1 cm^{-1}, fed two InSb hot electron bolometers with NEP's of 10^{-12}
WHz$^{-1/2}$ (cf. Lena, this volume, for nomenclature). The interferometer is shown in Fig. 5, and the cryostat in Fig. 6. Radiation
entered the cryostat through a 25 μm thick polyethelene window,
the thermal emission from which was used to calibrate the spectrum.
The package was flown at a pressure altitude of 2.7 + 3.0 mb (~40
km). Fig. 7 shows the raw spectrum obtained whilst Fig. 8 shows the

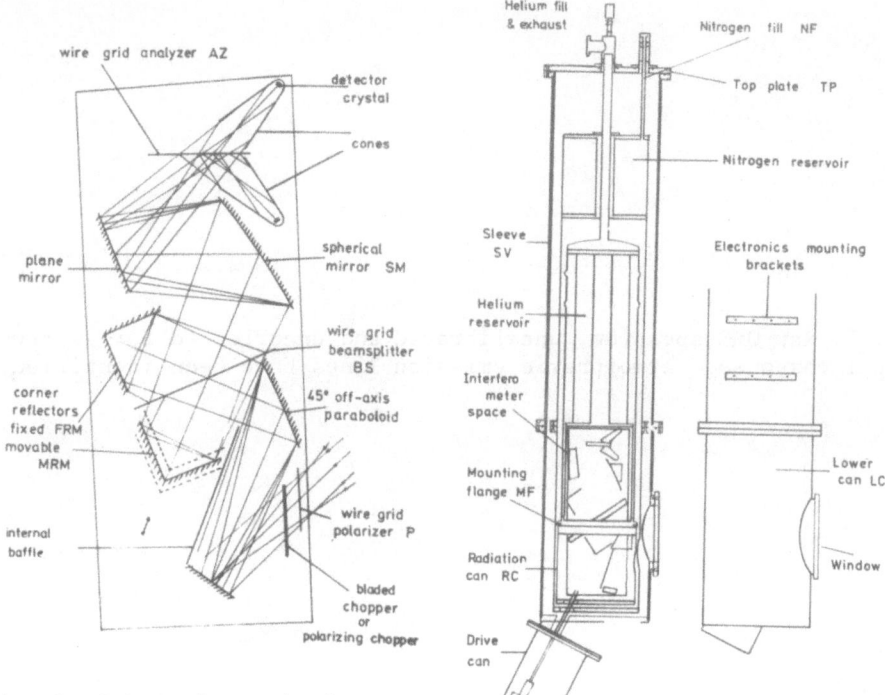

Fig. 5. Schematic optical arrangement of QMC liquid-helium cooled polarizing Michelson interferometer.

Fig. 6. QMC flight cryostat.

calibrated spectrum with atmospheric lines removed. In both spectra, the background spectra below ∿10 cm^{-1} appears obvious. The spectrum above 12 cm^{-1} is attributed mainly to window emission. The dashed line in Fig. 7 indicates the predicted window emission, based on (difficult) laboratory measurements. There is a discrepancy between predicted and observed window emission. Because we have more confidence in the higher frequency, higher absorption region, the spectrum has been calibrated there. The interferometer measures the difference between radiation entering the cryostat and the radiation within the cryostat cavity. A theoretical 2.7 K minus 1.4 K (the cavity temperature) curve is therefore shown in Fig. 8.

Results: The best fit to the 12 independent spectral points is
        T = 2.94 ± 0.06 K with a χ^2 of 7 (cf. 11 ± 5).
Situation: Most satisfactory.

        Later in 1974 Berkeley (Woody et al. 1975) flew a very similar experiment, to 3.2 – 3.4 mb (∿38 km) altitude. Main differences between the QMC and Berkeley instruments are:
 - Berkeley had no window.
 - Coupling to the radiation. To avoid diffraction of hot objects

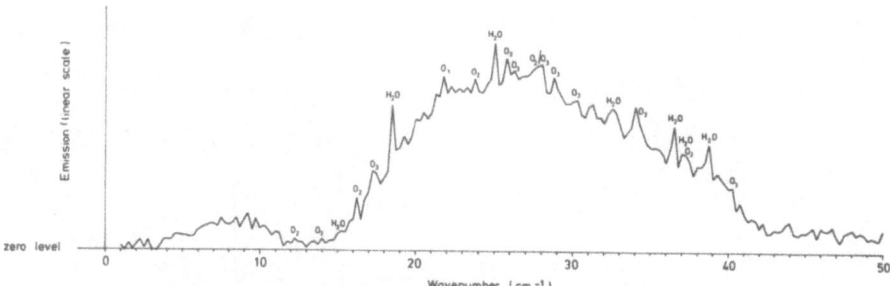

Fig. 7. Raw QMC spectrum, uncalibrated and uncorrected for instrumental response. Atmospheric emission lines have been identified.

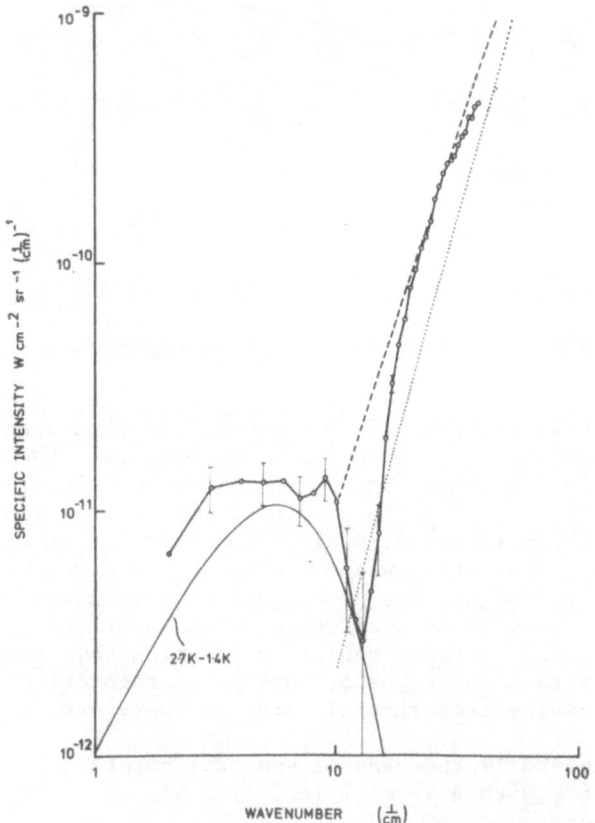

Fig. 8. Calibrated QMC spectrum with atmospheric lines removed. The
dashed curve indicates expected window emission (see text). Error
bars are 1-σ internal errors. The smooth solid curve is the spectrum expected from a 2.7 K black body (see text).

into the beam, QMC focused on an imaginary aperture, far from any real aperture. Berkeley used a specially designed horn.

The Berkeley spectrum showed no turn-over near 10 cm^{-1} and was dominated by the atmosphere. They deduced a background spectrum by fitting a model atmosphere with an isothermal, exponential pressure profile, a constant mixing ratio, and they published line strengths and widths for the atmospheric lines. The adjustable parameters (which they claim to be approximately orthogonal) were: column densities of O_2, O_3, H_2O and thermodynamic temperature of a Planckian spectrum. The residuals of the fit look impressively low (Woody, 1975) but are not altered dramatically by fitting other continuum spectra or by omitting a background altogether.

Result: $T = 2.99^{+0.07}_{-0.14}$ K.

Why is there a disturbing discrepancy between the Berkeley and QMC spectra? Different flight altitudes? A second Berkeley result from a higher altitude (private communication) appears to

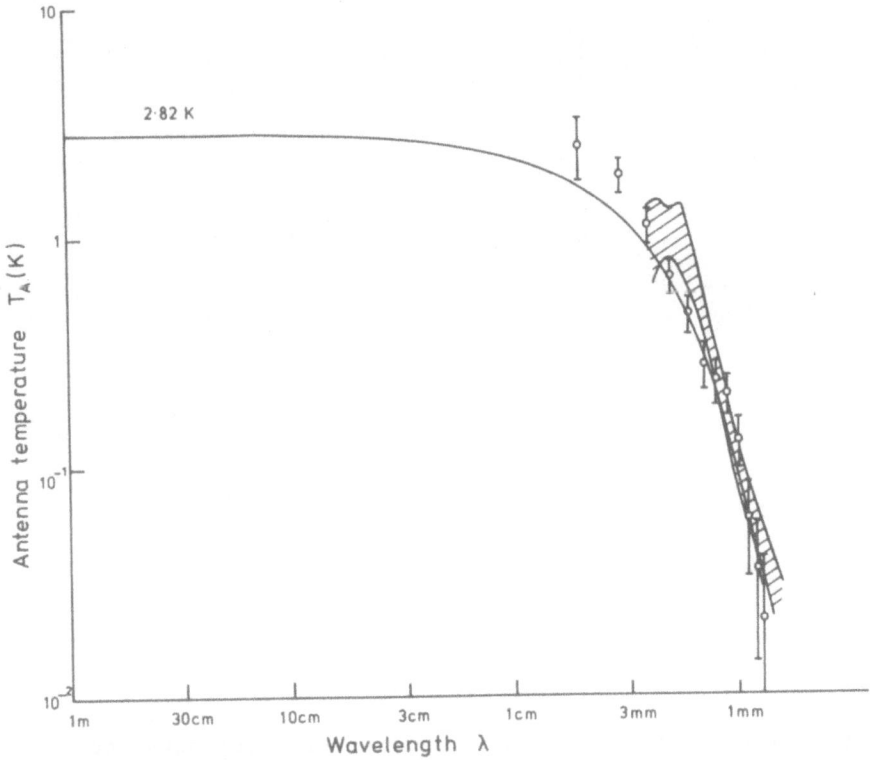

Fig. 9. Comparison of QMC antenna temperature spectrum (open circles, 1 - σ error bars) with that of Berkeley (shaded region, 2 - σ limits). The solid curve is a 2.82 K Planck spectrum.

show a turn-over above 10 cm^{-1}. A fault in QMC resolution? Oxygen
lines at 11 and 12 cm^{-1} appear weaker than they should.

    Data from a recent flight by Leeds University are still being
reduced.

    Fig. 9 shows the QMC and Berkeley results together, the shaded
region indicating Berkeley 2 - σ limits whilst the QMC error bars
are 1 - σ. Taken at face value, the results are in very good agreement.

4. CONCLUSIONS

Fig. 10 shows all monochromatic radiometer results together with
the QMC spectral points. Also shown is an upper limit derived
from a recent very difficult pseudo-monochromatic measurement at

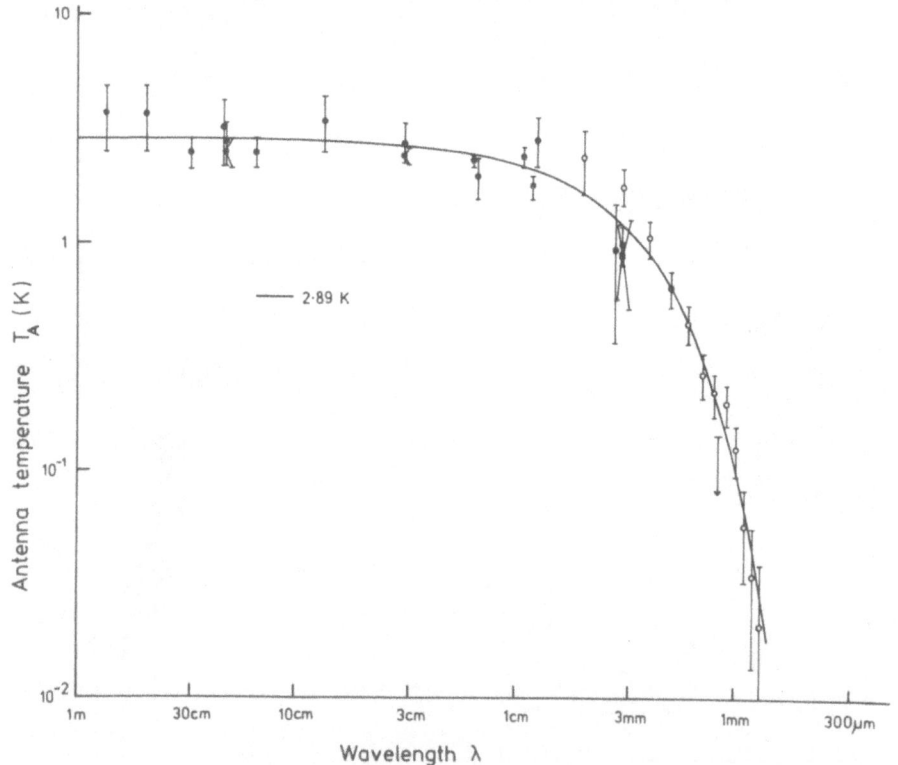

Fig. 10. Antenna temperature as a function of wavelength for all
monochromatic radiometric measurements (fitted circles) and QMC
spectrophotometric points (open circles). The upper limit at
1.2 mm is a pseudo-monochromatic measurement (Dall'Oglio et al.,
1976). The solid curve is a 2.89 K Planck spectrum (see text).

at 1.2 mm (Dall'Oglio et al., 1976); because of the difficulties
of interpretation of the result of this measurement, it has not
been included in the following.

The weighted mean of all the monochromatic results, excluding
that of Pelyushenko and Stankevich (1969), the QMC spectral points
and the Berkeley temperature is:

$$T = 2.89 \pm 0.04 \text{ K.}$$

$$\chi^2 = 22 \quad (cf. \ 27 \pm 7)$$

(4.1)

Omission of either the QMC or the Berkeley results does not significantly affect this value.

Although this may be considered very satisfactory, I believe
caution is indicated by the history of these observations. I
suggest the following conclusions:
 - There is clear evidence of continuum energy in the wavelength
   range 735 - 1 mm.
 - The spectrum of this energy is certainly not Rayleigh-Jeans.
 - The spectrum probably turns over shortward of 1 mm.
 - The spectrum is possibly Planckian.
More precise measurements are needed!

5. FUTURE OBJECTIVES

I believe that unambiguous verification of the Planckian nature of
the background spectrum is still required; this may be possible
from balloon platforms and experiments are in preparation.

More, however, could be asked. A Planck spectrum is not
necessarily expected. Galaxy formation, for example, might be
expected to transfer energy to the photons (cf Longair in this
volume). Many authors have discussed this (e.g. Chan and Jones,
1975a).

The distortions produced in the background spectrum depend
upon the magnitude and epoch of energy injection. In particular,
for heating by a hot plasma before an epoch z_a, Compton scattering
and bremsstrahlung can act together to produce a Bose-Einstein (see
Fig. 1) rather than a Planck spectrum. After z_a, bremsstrahlung
cannot produce photons rapidly enough to establish an equilibrium
between the plasma and the photons. A fairly extreme distortion
produced under the latter condition is shown in Fig. 11 (adapted
from Chan and Jones, 1975b). Current observations cannot distin-
quish the distorted curve from a Planck spectrum.

Much greater discrimination could be obtained from space platforms. The solid rectangles in Fig. 11 indicate the limits which
could be achieved in an experiment proposed for Spacelab by D.H.
Martin with the present author.

We should therefore welcome the proposed satellites COBE and
IRSAT discussed by Moorwood (this volume). Even more will be

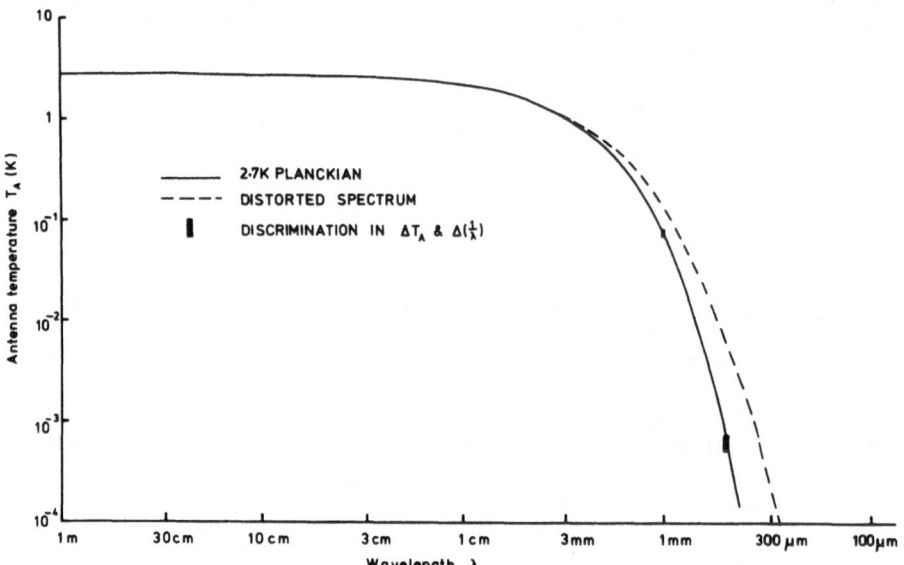

Fig. 11. Distortion of a Planck spectrum produced by heating of the photons by a hot plasma post z_a (see text). Solid rectangles denote the wavelength and antenna temperature discrimination of a proposed Spacelab experiment.

revealed if these satellites can also measure micro-anisotropy to the required precision (Longair, this volume).

REFERENCES

Alpher, R.A., Bethe, H.A. and Gamow, G., 1948, Phys. Rev. 73, 803.
Beckman, J.E., Ade, P.A.R., Huizinga, J.S., Robson, E.I., Vickers, D.G. and Harries, J.E., 1972, Nature 237, 154.
Beery, J.G., Martin, T.Z., Nolt, I.G. and Wood, C.W., 1971, Nature 230, 36.
Blair, A.G., 1974, IAU Symposium No. 63, edited by M.S. Longair, p. 143. D. Reidel Publishing Co., Dordrecht.
Bertolot, V.J., Clauser, J.F. and Thaddeus, P., 1969, Phys. Rev. Letters 22, 307.
Boynton, P.E. and Stokes, R.A., 1974, Nature 247, 528.
Boynton, P.E., Stokes, R.A. and Wilkinson, D.T., 1968, Phys. Rev. Letters 21, 462.
Chan, K.L. and Jones, B.J.T., 1975a, Astrophys. J. 195, 1.
Chan, K.L. and Jones, B.J.T., 1975b, Astrophys. J. 198, 245.
Dall'Oglio, G., Fonti, S., Melchiorri, B., Melchiorri, F., Natale, V., Lombardi, P., Trivero, P. and Sivertsen, S., 1976, Phys. Rev. D 13, 1187.

Danese, L. and DeZotti, G., 1977, La Rivista del Nuovo Cimento, in press.
Davis, M., 1976, Frontiers of Astrophysics, edited by E.H. Avrett, p. 472. Harvard University Press.
Dieke, R.H., Peebles, P.J.E., Roll, P.G. and Wilkinson, D.T., 1965, Astrophys. J. 142, 414.
Ewing, M.S., Burke, B.F. and Staelin, D.H., 1967, Phys. Rev. Letters 19, 1251.
Field, G.B. and Hitchcock, J.L., 1966, Astrophys. J. 146, 7.
Heygi, D.S., Traub, W.A. and Carleton, N.P., 1974, Astrophys. J. 190, 543.
Houck, J.R., Soifer, B.T., Harwit,M., and Pipher, J.L., 1972, Astrophys. J. 178, L29.
Howell, T.F. and Shakeshaft, J.R., 1966, Nature 210, 1318.
Howell, T.F. and Shakeshaft, J.R., 1967, Nature 216, 753.
Kislyakov, A.G., Chernyshev, V.I., Lebskii, Yu. V., Mal'tsev, V.A., and Serov, N.V., 1971, Soviet Astron. 15, 29.
Mather, J.C., Werner, M.W., and Richards, P.L., 1971, Astrophys. J. 170, L59.
McKellar, A., 1941, Publs. Dominion Astrophys. Obs., Victoria, B.C. 7, 251.
Millea, M.F., McColl, M., Pedersen, R.J., and Vernon, F.L., 1971, Phys. Rev. Letters 26, 919.
Muehlner, D. and Weiss, R., 1970, Phys. Rev. Letters 24, 742.
Muehlner, D. and Weiss, R., 1973a, Phys. Rev. D7, 326.
Muehlner, D. and Weiss, R., 1973b, Phys. Rev. Letters 30, 757.
Nolt, I.G., Radostitz, J.V. and Donelly, R.J., 1972, Nature 236,444.
Peebles, P.J.E., 1971, Physical Cosmology, Princeton University Press.
Pelyushenko, S.A. and Stankevich, K.S., 1969, Soviet Astronomy 13, 223.
Penzias, A.A. and Wilson, R.W., 1965, Astrophys. J. 142, 419.
Penzias, A.A. and Wilson, R.W., 1967, Astrophys. J. 72, 315.
Puzanov, V.I., Salomonovich, A.E. and Stankevich, K.S., 1968, Soviet Astron. 11, 905.
Robson, E.I., 1966, Far Infrared Astronomy, edited by M. Rowan-Robinson, p. 115. Pergamon Press.
Robson, E.I., Vickers, D.G., Huizinga, J.S., Beckman, J.E. and Clegg, P.E., 1974, Nature 251, 591.
Robson, E.I. and Clegg, P.E., 1977, IAU Symposium No. 74, edited by M.S. Longair, D. Reidel Publishing Co. (in press).
Roll, P.G. and Wilkinson, D.T., 1966, Phys. Rev. Letters 16, 405.
Shivanandan, K., Houck, J.R. and Harwit, M.O., 1968, Phys. Rev. Letters 21, 1460.
Stokes, R.A. and Wilkinson, D.T., 1966, Phys. Rev. Letters 16, 405.
Stokes, R.A., Partridge, R.B. and Wilkinson, D.T., 1967, Phys. Rev. Letters 19, 1199.
Thaddeus, P., 1972, Ann. Rev. Astron. Astrophys. 10, 305.
Weinberg, S., 1972, Gravitation and Cosmology, J. Wiley and Sons Publishing Co.

Welch, W.J., Keachie, S., Thornton, D.D. and Wrixon, G., 1967,
Phys. Rev. Letters 18, 1068.
Wilkinson, D.T., 1967, Phys. Rev. Letters 19, 1195.
Williamson, K.D., Blair, A.G., Catlin, L.L., Hiebert, R.D., Loyd,
E.G. and Romero, H.V., 1973, Nature 241, 79.
Woody, D.P., 1975, Ph.D. Thesis, University of California, Berkeley.
Woody, D.P., Mather, J.C., Nishioka, N.S. and Richards, P.L., 1975,
Phys. Rev. Letters 34, 1036.

COSMOLOGICAL ASPECTS OF INFRARED AND MILLIMETRE ASTRONOMY

M.S. Longair

Mullard Radio Astronomy Observatory, Cavendish
Laboratory, Cambridge, England

1. INTRODUCTION AND PLAN

Past experience has shown that every time a new waveband is opened
up, fundamental discoveries of importance to cosmology are made.
I have every expectation that the infrared and submillimetre wavebands will be no exception to this rule. The aim of this short
set of lectures is to discuss some of the more obvious aspects of
particular interest to cosmology. Naturally, such predictions of
what one expects to find in an unexplored waveband are hazardous
but the apologia for such an exercise is that, if they encourage
infrared astronomers to make the appropriate observations, they
are likely to discover even more exciting things than one could
have predicted.
I will divide the material into two separate parts, the first
devoted to the Hot Model of the Universe and the second to Observations of Discrete Infrared Sources at cosmological distances. In
the first part we will develop a mode for the Universe in three
successive approximations, the aim being to produce a more and more
complete picture of how observed structures in the Universe came
about. In the first approximation we develop the canonical Hot
Model of the Universe under the assumption that it remains isotropic
and homogeneous throughout its evolution. In the second approximation we include small perturbations into the model, the hope being
that these may develop into real objects such as galaxies, clusters
of galaxies, etc. This analysis will reveal a number of fundamental
problems in galaxy formation. In the third approximation we follow
these perturbations into the non-linear stages of collapse in
which any complete theory should explain the detailed properties
of galaxies, clusters, etc. None of the theories are sufficiently
secure for such detailed predictions to be made. In view of these

G. Setti and G. G. Fazio (eds.), Infrared Astronomy, 199-230.

uncertainties, it is equally valuable to work backwards and ask
what young galaxies might look like and whether or not they are
observable.

It should be emphasised that only the first approximation in
this scheme has gained any general acceptance by astronomers and
cosmologists. There is no consensus concerning approximations two
and three and none of the arguments is decisive. The reasons for
this situation will be described below.

In the second part we discuss the theoretical treatment of
observations of objects at cosmological distances. In this part
we will present a set of useful relations with which to evaluate
intrinsic properties of objects such as proper diameters, space
densities and intrinsic luminosities from a knowledge of their
observed properties and their redshifts. We will also look at
some of the problems which have been encountered in other wavebands and the morals to be drawn from the various controversies
and mistakes which have been made. Hopefully, this catalogue of
errors may avoid their repetition in the infrared waveband.[†]

It is worthwhile summarising the points of contact with infrared and millimetre astronomy:

  i) The spectrum of the microwave background radiation. Distor-
     tions of the spectrum of the microwave background radiation
     from a pure Planck spectrum are expected in all three approxi-
     mations described in the first part. These are expected in
     the centimetre, millimetre and submillimetre wavebands.

 (ii) Fluctuations in the intensity distribution of the microwave
     background radiation on the celestial sphere. The observa-
     bility of those expected due to collapsing protogalaxies,
     proto-clusters and large scale systems are perhaps of the
     most importance. We must also investigate other sources of
     fluctuations in the background radiation to assess whether
     those due to galaxy formation may be swamped by other unre-
     lated effects.

(iii) The detectability of young galaxies. We will make a case that
     the infrared waveband is probably the most important for the
     detection of galaxies soon after their epoch of formation.

 (iv) How to treat cosmological observations of discrete sources
     in the infrared waveband.

PART I - THE HOT MODEL OF THE UNIVERSE

2. FIRST APPROXIMATION - THE ISOTROPIC HOT MODEL

The isotropic hot model of the Universe is based upon one observation and one fundamental postulate. The observation is the isotropy

---

[†]At this point Professor Giancarlo Setti commented pessimistically,
"Cosmology is like love: everyone wants to make his own mistakes!"

of the microwave background radiation which has been shown to be
isotropic to better than one part in 10^3 on all angular scales
$\theta \gtrsim 1'$ arc.   The postulate is known as the cosmological principle
according to which the large scale properties of the Universe
observed by us are typical of what would be observed by any observer at the same cosmological epoch. Thus isotropic world models
are the obvious starting point for the construction of model Universes.

An important simplifying feature of isotropic models is that
all the large scale properties are also "local" properties and most
of the important features can be derived from consideration of the
properties and dynamics of any representative piece of it. For
example, for the purposes of treating radiation, we may consider
a small portion of the Universe to be a box with perfectly reflecting walls. Fig. 1 shows what happens when this box expands
with the Universe. The wavelength of the radiation expands in proportion to the size of the box. This may be shown either by considering the Doppler shifts of the radiation in collisions with the
walls of the expanding box or by considering the adiabatic expansion
of a gas of photons. The size of the Universe at any time is
defined by the scale factor R(t) which is just the relative distance between two points which partake in the isotropic expansion
of the Universe. It is convenient to normalise R(t) to unity at
the present epoch. Therefore,

$$R(t) \propto \lambda \propto \nu^{-1} \tag{2.1}$$

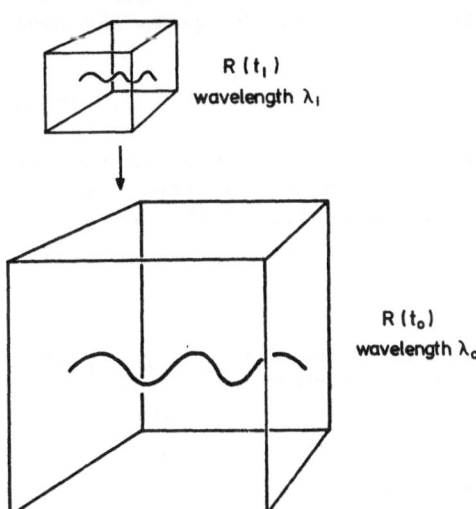

$R(t_1)$
wavelength λ_1

$R(t_0)$
wavelength λ_0

Fig. 1. The Universe as an expanding box with perfectly reflecting
walls. The scale factor R(t) measures the relative size of the
Universe at time t.

Redshift Z is defined to be

$$Z = \frac{\lambda_{obs} - \lambda_{em}}{\lambda_{em}} \qquad (2.2)$$

where λ_{obs} is the observed wavelength of a spectral feature and λ_{em} is the wavelength with which it was emitted by the source. Now

$$\frac{\lambda_{obs}}{\lambda_{em}} = \frac{R(t = present\ epoch)}{R(t)} = \frac{1}{R(t)} . \qquad (2.3)$$

Therefore

$$R(t) = \frac{1}{(1+Z)} . \qquad (2.4)$$

This illustrates the cosmological interpretation of the redshift. The redshift of a source indicates the size of the Universe when the source emitted the radiation. However, we do not know when the radiation was emitted because we do not yet know the dynamics, the acceleration or deceleration, of the Universe between the time the radiation was emitted and received.

We can now work out how the energy of a photon changes with epoch in an expanding Universe.

$$E = h\nu \propto (1+Z) . \qquad (2.5)$$

Since this result is applicable for photons of all energies, it is evident that the radiation temperature T_r must change in the same way

$$T_r \propto (1+Z) . \qquad (2.6)$$

As the box expands, assuming no creation or annihilation, the number densities of objects – protons, galaxies, photons – decrease as $R^{-3}(t)$, and hence the mean density of matter in the Universe decreases as

$$\rho_{matter} \propto R^{-3} \propto (1+Z)^3. \qquad (2.7)$$

However, for photons, their inertial mass density $\rho_{radiation}$ is $\Sigma_\nu h\nu n_{photon}$, where n_{photon} is the photon number density and the summation over all frequencies of photons in unit volume. $n_{photon} \propto R^{-3}$ and $\nu \propto R^{-1}$, and hence

$$\rho_{radiation} \propto \Sigma h\nu\ n_{photon} \propto R^{-4} \propto (1+Z)^4 \qquad (2.8)$$

This is what would be expected because $T_r \propto (1+Z)$, and hence from the Stefan-Boltzmann Law, $\rho_{radiation} \propto T_r^4 \propto (1+Z)^4$.

As yet we have not prescribed the dynamics of the world model. Because of the cosmological principle, in isotropic models global properties are local properties and we can obtain the correct results by considering only a small bit of Universe and applying the Newtonian theory of gravitation. In this case we consider, a spherically symmetric region of radius r centered on the Earth. Consider the gravitational deceleration of a galaxy at distance r due to the mass within this radius; in this approximation we use Gauss's law and neglect the rest of the Universe outside r. Then, if m_g is the mass of the galaxy

$$m_g r'' = - \frac{4\pi}{3} \frac{\rho G r^3 m_g}{r^2} \qquad (2.9)$$

where ρ is the mean density of the Universe. Therefore,

$$r'' = - \frac{4\pi}{3} \rho G r . \qquad (2.10)$$

Notice that all trace of the galaxy has disappeared from this equation which is telling us about the dynamics of the Universe as a whole. Let us normalise all quantities to their values at the present epoch, t_0. Then

$$r = R(t) r_0 \quad ; \quad \rho = \rho_0 R^{-3}(t) . \qquad (2.11)$$

Therefore,

$$R'' = - \frac{4\pi}{3} \frac{G\rho_0}{R^2} . \qquad (2.12)$$

It is convenient to write this relation in dimensionless form by introducing H_0, Hubble's constant. Conventionally, H_0 is the constant in the local velocity distance relation for galaxies

$$\nu = H_0 r . \qquad (2.13)$$

We will adopt a value of $H_0 = 50$ km s^{-1} Mpc^{-1} throughout these notes. Evidently, if the Universe expanded at constant velocity, H_0^{-1} defines the age of the Universe. We can define a critical density ρ_{crit} by the combination of constants

$$\rho_{crit} = \frac{3H_0^2}{8\pi G} \qquad (2.14)$$

and then a density parameter Ω by

$$\Omega = \rho_0 / \rho_{crit} \qquad (2.15)$$

where ρ_0 is the mean density of matter in the Universe at the
present epoch. ρ_{crit} will be found to be "critical" in the sense
defined below. Rewriting the equation for R, we find

$$\overset{\shortparallel}{R} = -\frac{\Omega H_0}{2}\frac{1}{R^2} .$$

(2.16)

A result of the same form is found in a full general relativistic
treatment of the problem. The only differences are the inclusion
of relativistic corrections to the inertial mass density for matter
with finite pressure p and the proper inclusion of the curvature
of the space.

The interpretation of Ω may be understood as follows. If we
consider an expanding sphere of radius r, the ratio of the gravitational potential energy of the system to its kinetic energy of
expansion is just Ω, i.e.

$$\Omega = \frac{\lvert\text{ Gravitational Potential Energy }\rvert}{\text{Kinetic Energy of Expansion}} .$$

(2.17)

This result enables the exact solutions of equation (2.16) shown
in Fig. 2 to be understood. The case $\Omega = 1$ is critical in the
sense that in this model the Universe has just got sufficient
kinetic energy to make it reach infinite radius at zero velocity.
In the case $\Omega > 1$, the mean matter density is sufficiently great
to bring the Universe to rest in a finite time and it then re-
collapses; the gravitational potential energy exceeded the kinetic
energy of expansion. If $\Omega < 1$, the Universe expands forever and
reaches infinite radius with a finite velocity, i.e. the Universe
exceeds its own escape velocity.

These results are often referred to as the dynamics of dust-
filled Universes. I hesitate to use this expression at a summer
school in which dust has played such a central role; in the cosmological context, it is taken to mean non-interacting particles

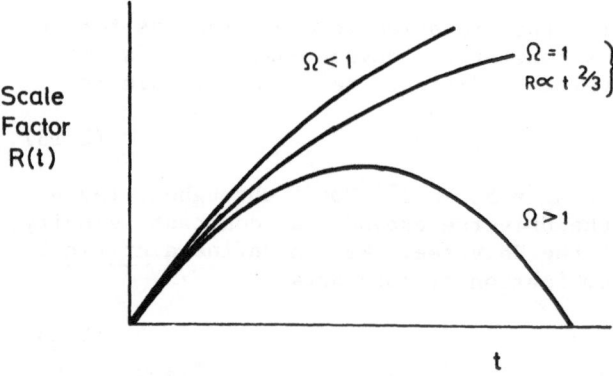

Fig. 2. The dynamics of isotropic world models as parameterised
by the scale factor R(t).

with zero pressure.

Two final notes may be useful. First, it is clear that the density parameter is closely related to the deceleration of the Universe. From equation (2.16), the dimensionless deceleration parameter $q_o = -(\ddot{R}/\dot{R}^2)$ is just

$$q_o = \Omega/2 \ . \tag{2.18}$$

Very often, world models are characterised by q_o and then the "critical" value, separating "oscillating" from "ever-expanding" models, corresponds to $q_o = 1/2$. The model $\Omega = 1$ ($q_o = 1/2$) is known as the Einstein-de Sitter model, for which

$$R = (\tfrac{3}{2} H_o t)^{2/3} \ . \tag{2.19}$$

Second, as a rough approximation, at redshifts $\Omega Z > 1$, the dynamics of the world model behave similarly to the Einstein-de Sitter model $R \approx \Omega^{1/3}(3/2\ H_o t)^{2/3}$; at redshifts $\Omega Z < 1$, the dynamics approximate those of the empty world model $\Omega = 0$, for which $R = H_o t$, i.e. undecelerated expansion. The model having $\Omega = 0$ is known as the Milne model.

The above results are sufficient to develop most of the important results which we will use in what follows.

The temperature history of the Universe follows directly by extrapolating the present Universe backwards in time using the above results. As described in the lectures by Clegg (this volume), the spectrum of the microwave background radiation follows closely a Planck distribution at radiation temperature $T_r \approx 2.7$ K. At earlier epochs the radiation temperature was $T(Z) = 2.7\ (1+Z)$. We can therefore identify certain epochs which will be significant in the temperature history of the Universe.

At $Z \approx 1500$, $T_r \approx 4000$ K and there are sufficient ionising photons in the tail of a Planck distribution of this temperature to ionise a neutral intergalactic medium. Therefore at earlier epochs, the intergalactic gas must be fully ionised.

At $Z \approx 10^{8-9}$, $T_r \approx 3 \times 10^{8-9}$ and the photons of the background radiation are energetic enough to dissociate nuclei. At earlier epochs neutrons, photons, electrons, etc. must all be in equilibrium with the radiation field.

At $Z \approx 10^{12}$, baryon pair production becomes important and so on.

It is also important that at large redshifts, most of the inertial mass density is in the form of radiation rather than matter. The ratio of energy densities in radiation and matter are

$$\frac{\varepsilon_{radiation}}{\varepsilon_{matter}} = \frac{\varepsilon_{rad}(Z=o)\ (1+Z)^4}{\Omega \rho_{crit} c^2\ (1+Z)^3} = \frac{10^{-4}}{\Omega}\ (1+Z) \ . \tag{2.20}$$

Thus at redshifts $Z \gtrsim 10^4 \Omega$, most of the mass density in the Universe is in the form of radiation. In this case the relativistic

"corrections" to the mass appearing in equation (2.16) are of the
same magnitude as the density term itself, and the differential
equation for the case of a "radiation-dominated" Universe is

$$\ddot{R} = - \frac{8\pi GR}{3} \rho_{radiation} \; . \tag{2.21}$$

If ρ_o is the present mass density in radiation and $\rho_{rad} \propto R^{-4}$,

$$R = (\frac{32\pi G\rho_o}{3})^{1/4} t^{1/2} \; . \tag{2.22}$$

A further important parameter is the ratio of number densities
of photons to baryons (or electrons).

$$\frac{n_\gamma}{n_{matter}} \approx \frac{10^8}{\Omega} \; . \tag{2.23}$$

The CANONICAL HOT BIG BANG MODEL OF THE UNIVERSE is what results if one starts an isotropic model Universe in equilibrium at
dense, high-temperature early state and then lets the model evolve
forwards in time using the best available knowledge of nuclear
interactions, etc. The temperature history of this model for the
$\Omega = 0.1$ world model is shown schematically in Fig. 3. This value
of Ω has been selected since a number of independent lines of
evidence favour (but do not prove!) the hypothesis that the Universe is of low density. The various epochs mentioned above are
indicated on the diagram. It should be noted that in the $\Omega = 0.1$
model the epoch of recombination more or less coincides with the
epoch at which the Universe changes from being radiation to matter
dominated. The details of the thermal history of the Universe in
this approximation have been thoroughly described by a number of
authors. We will only note a few points of significance for the
future development.

(i) The total energy density in mass-less particles which dominate the dynamics at $Z \gtrsim 10^4 \Omega$ consists not only of the
contribution of photons but also of all sorts of neutrinos
which decouple at redshifts $Z \gtrsim 10^{10}$. Thus the dynamics
of the radiation dominated phase at $Z \lesssim 10^9$ is given by
equation (2.22) with

$$\rho = \varepsilon_{massless \atop particles} /c^2 = \chi(t) a T_{rad}^4 \tag{2.24}$$

where T_{rad} is the radiation temperature and $\chi(t) = 1.7$ for
$Z < 10^9$.

(ii) At epochs $Z > 10^9$, the thermalisation times for the radiation and matter distributions are very short because the
process of electron-positron pair production at these temperatures brings the electrons and positrons into equilibrium
with the radiation field, i.e. $n_{e^+e^-} \approx n_\gamma$. There is, therefore, no problem in understanding why the spectrum of the

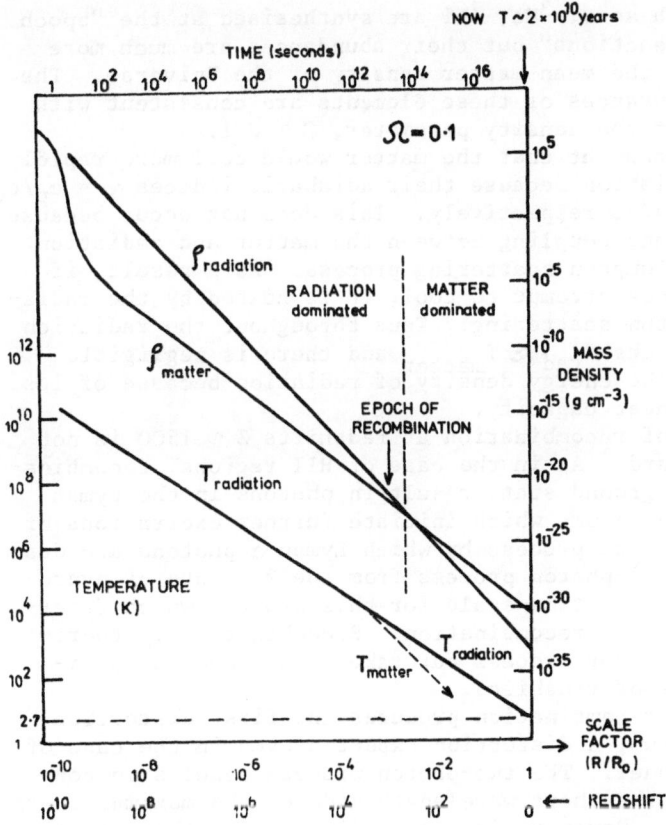

Fig. 3. The temperature history of the Universe.

microwave background radiation is of Planckian form. However, when the electrons and positrons annihilate, at $Z \approx$ $\approx 10^9$ the number densities of electrons and positrons falls by a factor of $\sim 10^8 - 10^9$. It is found that there is still time to set up an equilibrium spectrum but not to change significantly the numbers of photons present by the bremsstrahlung process.

(iii) For a wide range of parameters of the canonical model, about 25% of the mass of the Universe is converted into helium at $Z \sim 10^9 - 10^8$. This is in remarkably good agreement with the abundance of helium wherever it is observable in the Universe. Helium is very difficult to synthesise in large quantities by normal astrophysical nucleosynthesis in stellar interiors. This may be regarded as one of the triumphs of the hot model of the Universe since this provides an independent check upon the validity of the extrapolations back to early epochs. In addition, small amounts of light elements such as D, ^{3}He, ^{7}Li are synthesised at the "epoch of nuclear reactions" but their abundances are much more sensitive to the mean matter density of the Universe. The observed abundances of these elements are consistent with low values of the density parameter, $\Omega \sim 0.1$.

(iv) It might be thought that the matter would cool more rapidly than the radiation because their adiabatic indices $\gamma = c_p/c_v$ are 5/3 and 4/3, respectively. This does not occur because there is strong coupling between the matter and radiation through the Compton scattering process. As a result, if the matter does attempt to cool, it is heated by the radiation by Compton scattering. Thus throughout the radiation dominated epochs $T_{rad} \approx T_{matter}$ and there is negligible effect upon the energy density of radiation because of its much higher heat capacity.

(v) The process of recombination at redshifts $Z \sim 1500$ is not straightforward. As in the case of HII regions, recombinations to the ground state result in photons in the Lyman series or continuum, which initiate further excitations or ionisations. The process by which Lyman-α photons are destroyed is the 2-photon process from the 2s state of hydrogen and it is the time-scale for this process which determines the rate of recombination. Recombination is therefore not an abrupt process but takes place over a considerable range of redshifts.

The process of recombination produces the first distortion of the Planck spectrum, a distortion expected even in the case of the isotropic hot model. The two-photon process results in continuum radiation to the short wavelength side of the maximum of the Planck distribution. However, it can easily be seen that it is of very small energy density in comparison with that of the microwave background radiation. The energy deposited per Lyman-α photon corresponds only to ~ 10 eV compared with an energy density in radiation which is of the same order as the total rest mass energy of the matter. In addition, recombination lines are expected at lower energies but their intensities are estimated to be $\lesssim 10^{-7}$ of the intensity of the continuum radiation and hence are negligible.

The end result of the canonical model is very dull. The Universe consists of cold matter ($T \sim 0.01$ K) and radiation. There is no structure and certainly no astronomers to study it. The next step is to include some embryonic structure into the model in the form of small fluctuations and investigate whether they can develop into known types of object.

3. SECOND APPROXIMATION - GROWTH OF SMALL FLUCTUATIONS

First, we ask what scales will collapse under self-gravitation in an expanding Universe. The instability criterion turns out to be essentially the same as that found in the collapse of interstellar

gas clouds - i.e. the Jeans' stability criterion. The condition
for collapse is that the self-gravitational forces are stronger
than the internal pressure forces which support the cloud against
collapse. If we consider the forces acting on a cubic centimetre
of material of density ρ at a typical point in the cloud, this
condition may be written

$$\frac{GM\rho}{R^2} \gtrsim \frac{p}{R} \quad \text{i.e.} \quad R \gtrsim \left(\frac{p}{G\rho^2}\right)^{\frac{1}{2}}. \tag{3.1}$$

M is the mass of the cloud, R its radius and p the internal
pressure. An alternative way of writing this result is to note
that the time it takes a sound wave to cross the cloud is $\sim R/c_s$,
where c_s is the sound speed and the collapse time is $(G\rho)^{-1/2}$.
If the former is greater than the latter, collapse occurs. This
is identical with the first result because $c_s \approx (p/\rho)^{1/2}$.

Second, we ask what is the growth rate of the instability
under the influence of self-gravitation alone. This basic result
was first developed by Lifshitz. We can derive the essential
result by considering again the local Euclidean model, but now
we consider a spherically symmetric perturbation which has density
ρ' slightly greater than the mean density of matter in the Universe. This region of enhanced density will behave just like a
Universe of slightly higher density than the background model.
The evolution of the perturbation is illustrated schematically in
Fig. 4. The growth rate may be worked out in the approximation

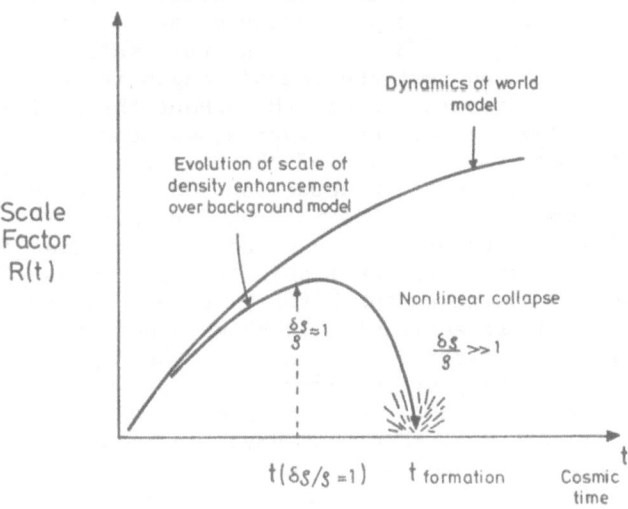

Fig. 4. A schematic diagram illustrating the evolution of density
perturbations in matter dominated Universes. $t(\delta\rho/\rho = 1)$ means
the epoch at which the density perturbation has $\delta\rho/\rho = 1$ and t(formation) is the epoch at which the object forms.

in which the perturbation is small. Provided $\Omega Z > 1$, i.e. the
background model behaves like an Einstein-de Sitter model, the
small perturbation must lie on an evolutionary track which eventually leads to collapse. A simple calculation shows that the perturbation in density, expressed as a density contrast $\Delta\rho/\rho$, grows
as

$$\frac{\Delta\rho}{\rho} \propto (1+Z)^{-1} \, . \tag{3.2}$$

If $\Omega Z < 1$, the growth rate of the perturbation is very small because
the background model expands forever and a small perturbation will
not shift the perturbed region onto a collapsing world line.

The importance of this result is that the perturbations do not
grow exponentially with time. The growth rate is only algebraic
and hence the fluctuations cannot develop out of statistical fluctuations in the distribution of matter in the Universe. The perturbations must appear with finite amplitude at some epoch.

Let us apply these results to the Universe. Immediately after
recombination, the matter and radiation decouple and the temperature
of the matter is ~ 4000 K. At a redshift of $\sim 10^3$, the Jeans' mass
is $M_J \sim 10^{5-6} M_\odot$, i.e. all masses greater than those of globular
clusters are unstable. Thus there is nothing to prevent fluctuations on the scale of galaxies and clusters collapsing to form
bound systems from the epoch of recombination onwards.

However, prior to this epoch, the Universe is radiation dominated and the matter and radiation are in close thermal contact.
In the limit $\varepsilon_{rad} \gg \varepsilon_{matter}$, the appropriate sound speed is that
of a fully relativistic gas, $c_s = c/\sqrt{3}$. Inserting this value into
the Jeans' criterion, it is found that the Jeans' length is just
smaller than the size of the Universe itself throughout the radiation dominated phase. By the size of the Universe, we mean the
"horizon scale", $r \approx ct$ where t is the age of the Universe. Thus,
during the radiation dominated epochs, fluctuations on the scales
of galaxies and clusters do not grow. The internal pressure of
the photon gas is sufficiently great to prevent collapse. This
means that the fluctuations from which real objects form can only
grow after recombination according to the Lifshitz rule. The behaviour of fluctuations on different scales is shown schematically
in Fig. 5. This result means that the amplitudes of the "initial"
fluctuations must be large if they are to grow into real objects
having $\Delta\rho/\rho \gg 1$ by the present epoch. For example, if we require
objects to have $\Delta\rho/\rho \gg 1$ at $Z = 5$, they must have had $\Delta\rho/\rho = 1$
at some earlier epoch, $Z \approx 10$. Then, at recombination, $Z \approx 10^3$,
since $\Delta\rho/\rho \propto (1+Z)^{-1}$, $\Delta\rho/\rho$ must have been 10^{-2} and prior to this
epoch the fluctuations must have been of roughly constant amplitude.

What types of fluctuation in the radiation and matter distribution could have existed prior to recombination? Conventionally,
perturbations of three types are considered.

(i) <u>Adiabatic fluctuations</u> are sound waves prior to recombination. The matter and radiation are strongly coupled and the density

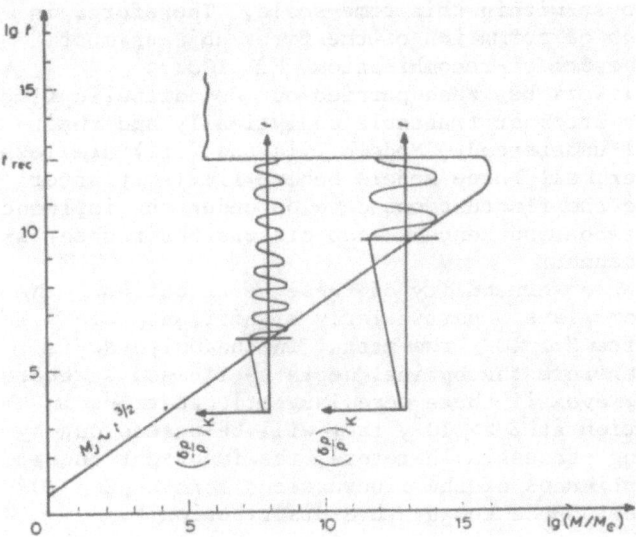

Fig. 5. A schematic diagram illustrating the evolution of perturbations on different scales as a function of cosmic time t. Before
the epoch of recombination, adiabatic perturbations on scales less
than the horizon are sound waves. After recombination, the Jeans'
mass drops to $M \sim 10^{5-6}$ M_Θ and perturbations on all greater scales
are unstable.

and pressure of the perturbation change according to the adiabatic
law as it oscillates. The origin of these fluctuations is not
known and it is assumed that they are part of the primordial structure of the Universe. In computations, a primaeval fluctuation
spectrum is selected which reproduces the observed distribution
of objects such as galaxies and clusters of galaxies at the present
day.

(ii) Entropy or isothermal fluctuations are fluctuations in the
matter distribution but not in the distribution of radiation. They
are thus essentially stationary prior to recombination. Again,
the fluctuation spectrum and its origin are unknown.

(iii) Whirl or vortex theories invoke turbulence to produce the
fluctuation spectrum. Before recombination, the sound speed is
$c/\sqrt{3}$ and therefore all turbulent motions are subsonic. Hopefully,
the results of laboratory experiments on subsonic turbulence may
be applicable. In this theory, it is assumed that a Kolmogorov
spectrum of whirl perturbations is set up. This theory has the
attraction that it is the only model which makes any definite prediction of the fluctuation spectrum. After recombination, the
sound speed drops abruptly and the turbulent motions become highly
supersonic. Most workers agree that these supersonic eddies will
dissipate in about one turn-over time and hence if real objects are

to form, they must do so within this time-scale. Therefore, in this theory, the epoch of formation of the first objects must occur soon after the epoch of recombination, $Z \sim 100$.

Most theoretical work has been carried out on adiabatic fluctuations because they are most tractable analytically and their behaviour is now well understood. Models (ii) and (iii) are less well defined. However, all three models behave similarly after recombination because the fluctuations develop under the influence of self-gravitation alone and hence we can discuss their observability in a similar manner.

In discussing the observability of these perturbations, the epoch of recombination plays a particularly significant role. This is because at redshifts $Z > 10^3$, the matter in the Universe is fully ionised and therefore the optical depth to Thomson scattering is very large. Thus, even if there were large fluctuations in the temperature distribution at $Z \gg 10^3$, they will be washed out by the Thomson scattering process. Therefore, the important questions are "What are the amplitudes of the fluctuations $\Delta\rho/\rho$ at $Z \simeq 10^3$ and what is the effect on the temperature distribution?".

We have shown above that the fluctuations must have amplitude at least $\Delta\rho/\rho = 10^{-3}$ at $Z = 10^3$ and that in the case of whirl perturbations, they must be very much larger, $\Delta\rho/\rho \approx 0.1 - 1$. A question which has provoked considerable interest is "What fluctuations can survive from the earliest epochs to the epoch of recombination?". Silk was the first to demonstrate that, in the case of adiabatic fluctuations, the coupling between the matter and radiation is not perfect and that there is "friction" between them which causes damping of all masses less than $\sim 10^{12}$ $M_\odot$ by the epoch of recombination. Thus, in the adiabatic picture, the only perturbations which survive to recombination are those on the scale of the most massive galaxies and greater. Furthermore, the dissipated energy goes into heating the matter and this may result in observable distortions of the spectrum of the microwave background radiation (see later). Similar damping occurs in the Whirl Models but regeneration of small scale eddies takes place continuously through the decay of larger eddies. In both cases, information is lost about the spectrum of primaeval fluctuations on scales $M < 10^{12}$ $M_\odot$. No damping occurs in the case of isothermal perturbations because there is no motion of the perturbations.

For adiabatic fluctuations, density fluctuations result in temperature fluctuations $\Delta T/T = 1/3 \; \Delta\rho/\rho$. However, more important for all three types of perturbation are the temperature fluctuations which result from Doppler scatterings of the background radiation by collapsing proto-objects. As soon as pressure support is removed by the decoupling of the matter and radiation, all three types of fluctuation start condensing immediately. From the fact that the density contrast is increasing, there must be velocities in the matter distribution as a result of the continuity equation

$$\frac{\partial}{\partial t} \left(\frac{\Delta\rho}{\rho}\right) = dw \; u \; . \tag{3.3}$$

To find the magnitude of the predicted intensity fluctuations, statistical averages are taken over a spectrum of density fluctuations
designed to result in the formation of real objects.

Two further complications must be introduced to make proper
predictions. First, strong temperature fluctuations are only
observed if the optical depth of the fluctuations to Thomson
scattering is greater than unity. If they are not, the photons
can diffuse out of the fluctuation and the fluctuations are much
reduced in strength. This is an important effect for masses
$M \lesssim 10^{15} M_0$. Second, the process of recombination is not instantaneous and there is a finite optical depth to Thomson scattering
from redshifts $Z \approx 1500$ to $Z \approx 750$ which produces further damping
of temperature fluctuations on all angular scales. Detailed calculations of these effects were first performed by Sunyaev and Zeldovich and the results of various computations for adiabatic fluctuations are shown in Fig. 6. The model which probably corresponds
most closely to reality is that for which $\Omega = 0.1$ and $\Delta\rho/\rho = 1$ at
$Z = 10$. It can be seen that the predictions of theory are not far

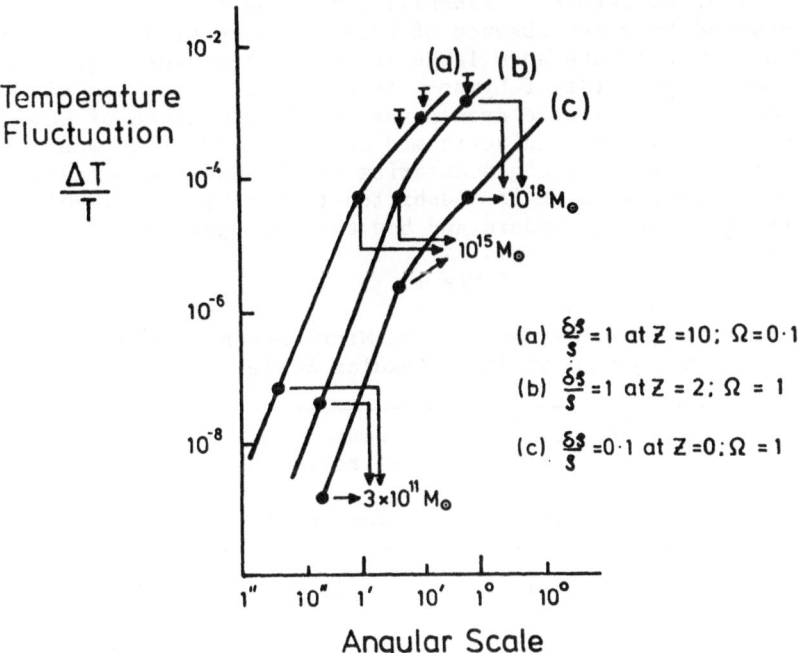

Fig. 6. The expected amplitude of fluctuations in the microwave
background radiation as a function of angular scale for an assumed
spectrum of irregularities to produce observed objects in the Universe. The curves correspond to different assumptions about the
world model. Some of the observed upper limits are indicated on
the diagram.

below the existing upper limits. Also shown on the curves are
the masses corresponding to different angular scales. According
to these calculations, there is most hope of detecting primordial
fluctuations for masses $M \gtrsim 10^{15}$ M_Θ on angular scales $\theta \gtrsim 1'$.
For reference, in Table 1, we give a recent compilation of limits
to intensity fluctuations in the microwave background radiation.

Similar computations have been carried out for "whirl" theories
and in these cases, even taking account of the two damping effects
mentioned above, the predicted fluctuations exceed the upper limits
in Table 1. The reason for this is that in this theory, objects
form soon after recombination and hence $\Delta\rho/\rho$ must be much larger
at $Z \sim 1500$ leading to larger velocities and fluctuations.

The results for isothermal fluctuations are similar to those
for adiabatic fluctuations.

There is, however, a further complicating factor and this is
that we have neglected any further scatterings of the background
radiation between the epoch of recombination and the present epoch.
Evidence that such scatterings may be important is derived from
the observation that the intergalactic gas is almost certainly
very highly ionised between a redshift $Z = 3$ and the present epoch.
This is inferred from the absence of Lyman-α absorption in the spectra of quasars which have such large redshifts that the region to
the short wavelength side of Lyman-α is redshifted into the visible waveband. If there were even a very small abundance of neu-
tral hydrogen in the intergalactic medium $\Omega_{igg} \lesssim 10^{-7}$, it would
be sufficient to cause strong scattering of the continuum radiation beyond Lyman-α when it is redshifted to the Lyman-α wavelength.
Since it is difficult to understand how the intergalactic gas

Table 1

Measurements of Anisotropy of the Microwave Background
Radiation on Small Angular Scales

Observers	Wavelength λ cm	Beam-area of telescope Ω (arc min)	$\Delta T/T$
Conklin and Brace-well (1967)	2.8	300	$\leq 1.8 \times 10^{-3}$
Penzias et al. (1969)	0.35	4	$\leq 1.1 \times 10^{-2}$
Parijskij and Pya-tunina (1971)	4.0	30	$\leq 2.6 \times 10^{-4}$
Boynton and Par-tridge (1973)	0.35	4	$\leq 3.7 \times 10^{-3}$
Carpenter et al (1973)	3.56	18	$\leq 7 \times 10^{-4}$

ensity could be so low, it is assumed that it must be fully
onised, at least in the range 0 < Z < 3. It is therefore assumed
hat the intergalactic gas must have been reionised at some red-
hift in the range 100 $\gtrsim$ Z $\gtrsim$ 3. The optical depth to Thomson
cattering is therefore

$$\tau_T = \int_0^{Z_{max}} \sigma_T n_e(Z) \frac{dr}{(1+Z)} \approx 0.025 \ \Omega^{1/2} Z_{max}^{3/2} \ \text{for} \ Z_{max} \gg 1 \ (3.4)$$

he amplitude of intensity fluctuations is damped by a factor $e^{-\tau_T}$
nd thus if $\tau_T \gg 1$, the primordial temperature fluctuations will
e very strongly damped. For example, if $Z_{max} = 100$ and $\Omega = 1$, we
ind $\tau_T = 25$ which would certainly damp out the fluctuations ex-
ected even in the whirl theory.

There is no consensus about the redshift of reheating the
ntergalactic gas. Most workers who adopt conventional sources
f ionising the intergalactic gas such as the ultraviolet radia-
ion of quasars, conclude that it is difficult to account for the
igh degree of ionisation unless Ω is small, say 0.1, and the
poch of reheating was relatively recent, Z $\leq$ 5. This argument
avours the adiabatic and isothermal models since $\tau_T \ll 1$ in this
ase. However, it is impossible to exclude unknown sources of
ntense ultraviolet radiation which formed at, say, Z = 100. These
ould occur in the whirl model and quite possibly in isothermal
odels.

Therefore, only conditional statements can be made about the
xpected intensity of temperature fluctuations in the microwave
ackground radiation due to collapsing protogalaxies and proto-
lusters. On the other hand, this approach does offer one of the
ery few methods by which it is in principle possible to obtain in-
ependent evidence on the spectrum of primaeval fluctuations.

Finally, we should consider other sources of fluctuations
hich might mask those due to galaxy formation.
(i) <u>Discrete radio sources</u>. Sunyaev and I have made estimates
f the contribution of discrete sources for a wavelength of 4 cm.
e found that they should result in temperature fluctuations
$T/T \approx 3 \times 10^{-5} - 10^{-5}$ on all angular scales greater than about
0" arc. This provides a rather firm lower limit to observable
luctuations at this wavelength. However, the limits are much
etter at shorter wavelengths because of the difference in spectral
ndices between the background radiation and discrete sources.
ear the peak of the microwave background radiation sources should
nly contribute $\Delta T/T \approx 10^{-6-7}$.
ii) <u>Extensive regions of hot gas</u>. Zeldovich and Sunyaev pointed
ut that if the photons of the microwave background radiation
ropagate through a region of hot gas, the photons are scattered
o higher energies by the Compton scattering process so that a
light decrease in the intensity of the microwave background radia-
ion would be expected in such directions. The effect of these
cattering is measured by the parameter

$$y = \int \frac{kT_e}{m_e c^2} d\tau_T \ , \tag{3.5}$$

where T_e is the temperature of the gas and τ_T is the Thomson scattering optical depth. The factor $kT_e/m_e c^2$ measures the average fractional change in photon energy per scattering. The magnitude of the temperature fluctuation is $\Delta T/T = -2y$. Zeldovich and Sunyaev suggested that such an effect would be detectable in clusters of galaxies which were extended X-ray sources. Typically, for a cluster such as the Coma cluster $T_e \approx 10^8$ K and $n_e \approx 10^{-3}$ cm^{-3}. For such a cluster $\Delta T/T \sim 10^{-3} - 10^{-4}$. Such depressions have been found in the direction of a few rich Abell clusters of galaxies by Gull and Northover. If there are corresponding depressions in a cosmological distribution of such Abell clusters, they could provide another source of fluctuations in the microwave background radiation at the level $\Delta T/T \sim 10^{-4}$.

Similar fluctuations may be associated with other hot regions in the Universe. There may exist "hot regions" heated and ionised by quasars. The reheating of the intergalactic gas is presumed to be associated with discrete sources and there will be statistical fluctuations in the distribution of these sources in space.

The problem with the fluctuations produced by such hot regions is that the radio spectrum of the fluctuations is similar to that produced by collapsing proto-objects, both resulting from Compton scatterings of the microwave background radiation.

Whilst on the subject of fluctuations in the microwave background, it is useful to summarise some of the astrophysical aims of the COBE experiment described by Moorwood (this volume). It can study anisotropies on an angular scale 7° and greater (and presumably statistically limits can be set to fluctuations on smaller scales).

(i) Dipole anisotropy - 24 hour period: measurement of the absolute motion of the Earth with respect to the frame of reference defined by the microwave background radiation.

(ii) The search for "hot-spots" in the microwave background radiation. As pointed out by Zeldovich and Novikov, some highly anisotropic world models evolve into quasi-isotropic models at the present epoch. Many of these possibilities can be excluded already by the observed isotropy of the radiation. However, some classes of model would leave only a "hot-spot" which would only be detected by sensitive all-sky surveys of the background radiation. These observations would provide important clues to the very early evolution of the Universe

(iii) Limits to the primaeval fluctuation spectrum on mass scales $M = 10^{19} - 10^{20}$ M$_\odot$ (see Fig. 6).

(iv) The detection of millimetre Galactic emission.

4. ENERGY DEPOSITION IN THE INTERGALACTIC GAS

The dissipation of energy results in heating of the intergalactic gas and if the electron temperature is increased above the radiation temperature, we expect distortions of the spectrum of the microwave background radiation. The physical process which brings about the redistribution of photon energies is Compton scattering. We can distinguish two general results.

(i) In the case of <u>thermodynamic equilibrium</u>, the single parameter T_{rad} determines both the energy density of the radiation ε_{rad} and the number density of photons n_γ. The photon distribution function is the Planck distribution. It is assumed that there is always time for the physical processes which produce or destroy photons and redistribute their energies to set up this equilibrium.

(ii) If, however, ε_{rad} and n_γ are both fixed and they cannot be related by the Planck distribution but the energy may be redistributed among the available photons, a <u>Bose-Einstein distribution</u> is found in equilibrium

$$\varepsilon_\nu = \frac{8\pi h \nu^3}{c^2} \left(\exp \left(\frac{h\nu}{kT} + \mu \right) - 1 \right)^{-1} \tag{4.1}$$

This distribution is defined by the two parameters T and μ, the chemical potential.

The first case corresponds to the canonical hot model in which there is no deposition of energy at any epoch. The second corresponds to what happens if energy is deposited into the intergalactic gas at redshifts in the range $10^8 \gtrsim Z \gtrsim 10^3$. After the annihilation of the electrons and positrons at $Z \sim 10^9$, the number density of electrons in the universal plasma is too small to produce a significant number of new photons if more energy is injected. Therefore, the energy must be redistributed among the available photons by the Compton scattering process. Up till the epoch of recombination, there is always time for the Bose-Einstein distribution to be established. At very low frequencies, bremsstrahlung can provide a small additional contribution. Examples of the types of distortion expected are shown in Fig. 7. ε is the energy density of the radiation at the relevant cosmological epoch. It can be seen that enormous energy depositions are required to distort the Planck spectrum and that the strongest effects are in the Rayleigh-Jeans part of the distribution.

The types of process which could produce distortions of this type are:

(i) Dissipation of primaeval adiabatic fluctuations,

(ii) Dissipation of primaeval turbulence,

(iii) Matter-antimatter annihilation in symmetric world models,

(iv) Other more exotic possibilities involving primordial black holes, etc.

Physically, it does not matter what the origin of the energy is. It is only the total energy and the redshift at which it is deposited which are important. It can also be seen that the distortions are broad-band and mostly confined to the Rayleigh-Jeans region.

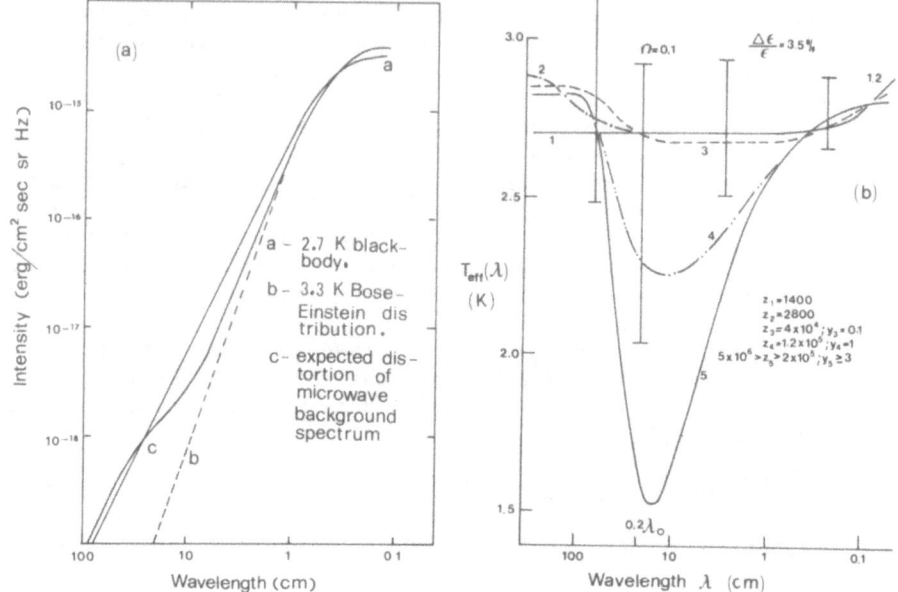

Fig. 7. (a) The form of distortion of the spectrum of the micro-wave background radiation if there is a large release of energy prior to recombination.
(b) Comparison of the predictions of some of the models of energy deposition with the observed spectrum of the microwave background radiation.

This method thus gives us the possibility of investigating the physical evolution of the Universe at redshifts greater than 10^3 but it is rather crude information and certainly not unambiguous in its interpretation. The problem is exacerbated by the fact that we have essentially no independent evidence on these processes. For example, we have no a priori evidence on the form of the spectrum of primaeval adiabatic perturbations.

Energy deposition at epochs after recombination also results in distortions due to Compton scattering but there is not time to set up an equilibrium Bose-Einstein distribution. The distortion again depends upon the parameter y. Examples of the form of these distortions are shown in Fig. 8. Existing observations enable limits of $y \lesssim 0.1$ to be set with confidence. Possible sources of such heating include:

(i) Dissipation of supersonic turbulence following the epoch of recombination,

(ii) The reheating of the intergalactic gas,

(iii) Energy dissipation at the epoch of galaxy formation.

Similar comments to those made above are in order. The predicted distortions are broad-based and are sensitive only to the

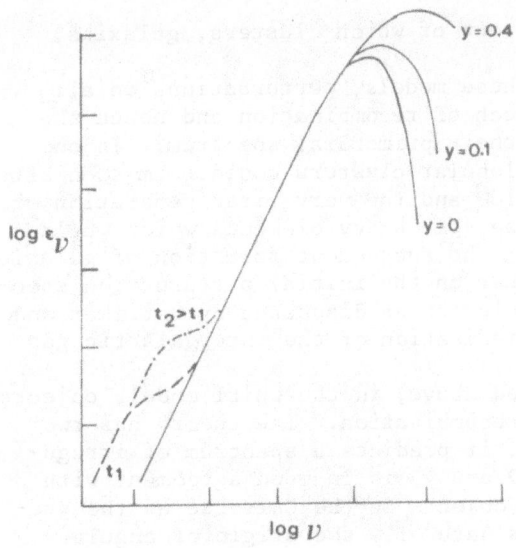

Fig. 8. Distortions of the microwave background radiation due to energy release after the epoch of recombination. The excess radiation at low frequencies is due to bremsstrahlung of the plasma with $T_e > T_{R-J}$.

total amount of energy deposited, not to its form. Again, very large depositions of energy are required to produce a significant distortion.

5. THIRD APPROXIMATION – NON-LINEAR STAGES AND YOUNG GALAXIES

The overall evolution of a perturbation has been illustrated in Fig. 4. The development of the fluctuation can be described by a linear theory until the amplitude has attained the value $\Delta\rho/\rho \sim$ ~ 1. The subsequent development is non-linear and depends sensitively upon the initial conditions. Rather than call them theories, the evolution of different types of perturbations through the non-linear stages to the formation of real objects should be called scenarios. Some of the features of these pictures are mentioned briefly below.

(i) <u>Adiabatic models.</u> All mass scales having $M \lesssim 10^{12} M_\odot$ dissipate prior to recombination. Thus the objects which form are on the scale of massive galaxies, groups and clusters with subsequent fragmentation of these large scale systems into less massive galaxies. The epoch of formation of galaxies is thus determined by the epoch at which clusters separate out and this must occur at about $Z = 10$. The non-linear development of the fluctuations has been studied in considerable detail by Zeldovich, Sunyaev, Doroshkevich and Shandarin who show that the large scale system will collapse into "pancakes" out of which clusters, galaxies, quasars, etc. may form.

(ii) <u>Isothermal Models</u>. In these models, perturbations on all mass scales survive to the epoch of recombination and hence the order of events depends upon their primordial spectrum. In one scenario discussed by Rees, globular clusters could form soon after recombination, at say $Z \sim 20\text{-}100$ and the very first generation of star formation would synthesise some heavy elements which would catalyse further generations. The subsequent formation of galaxies and clusters would depend either on the initial perturbation spectrum or upon astrophysical evolution as discussed by Ostriker and Rees. In this model, early reionisation of the intergalactic gas is feasible.

(iii) <u>Whirl Model</u>. As indicated above, in the whirl model, objects form soon after the epoch of recombination. The theory has two important attractions. First, it predicts a spectrum of irregularities which, according to Ozernoy, is in good agreement with the observed mass spectrum in objects in the Universe at the present day. Second, it explains naturally the origin of angular momentum in systems such as galaxies. In other theories, the origin of angular momentum must be ascribed to tidal interactions between galaxies.

There are many uncertaintities in these scenarios and the subject would make great advances if it were possible to observe directly the first generations of galaxies - i.e. young systems shortly after the epoch of formation. If this were possible, the problems would be set on a more secure observational foundation. It is therefore of interest to investigate the evidence on the <u>epoch at which galaxies formed</u>. Few of these arguments are unambiguous. The first two are quite strong arguments.

(i) <u>The mean sizes and densities of clusters of galaxies</u>. If we smooth out the mass in clusters of galaxies within their present volumes, we may ask at what epoch the clusters were touching and at which they have densities equal to the mean mass density in the Universe. If one adopts typical figures for a protocluster of $n = 10^{-3}$ cm^{-3}, $M = 10^{16}$ $M_{\odot}$, $R = 5$ Mpc, these epochs roughly coincide at $Z \approx 10\text{-}20$. It is therefore likely that the clusters become distinct entities at a somewhat later epoch, $Z \lesssim 10$. Notice that this argument refers only to clusters. In some models, the galaxies themselves may have formed much earlier.

(ii) <u>The cosmological evolution of quasars and radio galaxies</u>. The V/V_{max} test for quasars and the counts of radio sources indicate that the comoving space density of these objects has evolved strongly with cosmological epoch in the sense that they were roughly a thousand times more common at a redshift $Z \approx 3$ than they are at the present epoch. These changes are over and above the changes in density due to the expansion of the Universe. Models for these changes in the distribution of quasars and radio galaxies with cosmological epoch suggest that an exponential law $\exp(-t/t_0)$ can account for the observations where t_0 is the time scale over which the density decreases by a factor e. For the $\Omega = 0$ model, $t_o = 0.1/H_o$. In some models which account for the overall counts of radio sources, there is in addition a cut-off in the source distribution at $Z \sim 3-4$.

These observations indicate that the Universe went through a phase of much more violent activity when it was $\sim 1/5$ to $1/10$ of its present age in comparison with the present epoch. It is natural to associate this phase with the epoch when galaxies first formed and this would also be consistent with the formation of galaxies at a late epoch, $Z \sim 3-10$.

The other arguments are not unambiguous, but they may be strengthened by further studies.

(iii) <u>The epoch of reheating of the intergalactic gas</u>. As discussed above, if known objects such as quasars are to ionise the intergalactic gas, the heating must occur at relatively late epochs $Z \lesssim 3-4$. It is natural to associate this epoch with the epoch of formation of galaxies in general. However, as discussed above, we cannot exclude the possibility that the ionisation could have taken place at much larger redshifts as is possible in certain versions of the isothermal and whirl models for the origin of galaxies.

(iv) <u>Fluctuations in the microwave background radiation</u>. From Fig. 6, it can be seen that in the adiabatic model, if the formation of galaxies took place at a redshift $Z \gtrsim 10$, the predicted fluctuations would exceed the present limits to the temperature fluctuations in the microwave background radiation. As indicated in Section 4, however, these predictions are modified if there is early reheating of the intergalactic gas. This argument could be strengthened with an improved knowledge of the temperature history of the Universe.

(v) <u>Theory of collapse in low density Universes</u>. As shown in Section 4, fluctuations grow as $(1+Z)^{-1}$ in a dust-filled Universe until $\Omega Z \approx 1$ after which they grow very slowly. Therefore, knowing Ω, we can estimate the last redshift at which objects can attain $\Delta\rho/\rho \approx 1$. If they have not attained this amplitude by a redshift Ω^{-1}, they are unlikely to have formed bound systems by the present epoch. There is some evidence that Ω is small, ~ 0.1, and therefore the last redshift at which objects can grow to $\Delta\rho/\rho = 1$ is $Z = 10$. Since there are likely to be more small amplitude fluctuations than large ones, the bulk of objects in the Universe will attain $\Delta\rho/\rho = 1$ at this epoch. Therefore, the epoch of formation, corresponding to $\Delta\rho/\rho \gg 1$, would occur at $Z \approx 5$. The problems with this argument are that the value of Ω is still uncertain and also that there is nothing to prevent objects forming at a larger redshift.

None of these arguments forces us to conclude that the epoch of galaxy formation occurred at redshifts $Z < 10$, but they are sufficiently encouraging that it is very much worthwhile searching for young galaxies in this redshift range. As emphasised above, such a discovery would put many aspects of these problems on a firm observational basis.

It is therefore worthwhile speculating what the observable
properties of young galaxies might be.

(1) Optical luminosity. The intrinsic optical luminosity of a
young galaxy is expected to be significantly greater than normal
galaxies at the present epoch. First, there is much more gas
present and the rate of star formation must have been greater.
Second, because of the uniformity of the metal abundances of old
disc stars in our Galaxy, it may be necessary to synthesise most
of the metals in the early stages of evolution of the galaxy through
the rapid formation and evolution of massive stars. This hypothesis is not essential because there exist models in which star
formation takes place preferentially in regions of enhanced metal
content. However, these arguments suggest that young galaxies
may have been $\sim$10-1000 times brighter than our own galaxy which
would make them as bright as quasars and hence potentially observable to redshifts Z $\approx$ 10.

(2) Angular sizes. If we suppose that the bulk of star formation
takes place within a region of size 10 kpc, then for the world
model with Ω = 0.1, the predicted angular size of the young
galaxy is θ < 1" arc throughout the redshift range 3 < Z < 10.
Therefore young galaxies will appear to be star-like. Previous
estimates have suggested larger angular sizes because larger
physical sizes were assumed, the Ω = 1 world model used and a
Hubble constant of 100 kms^{-1} Mpc^{-1} rather than 50 kms^{-1} Mpc^{-1}
adopted.

(3) The optical/infrared spectrum. As a crude approximation, the
young galaxy may be envisaged as a giant HII region.

(i) If the stars alone dominate the optical spectrum and there
is little gas present, we would expect a spectrum which extends
into the far ultraviolet because of the presence of large numbers
of young massive stars. Quite possibly the total luminosity is
$\sim$10$^{46-47}$ erg s^{-1}, mostly in the far ultraviolet as indicated by
the solid line in Fig. 9.

(ii) If there is a substantial quantity of gas present, all the
Lyman-continuum radiation will be absorbed and the emission spectrum of the young galaxy will resemble that of an HII region with
prominent hydrogen recombination lines and, if some element formation has already taken place, forbidden lines of oxygen, nitrogen
etc. The continuum spectrum will cut-off abruptly at 912 Å.

(iii) If there is a substantial quantity of dust present, much of
the ultraviolet and optical continuum may be absorbed by the dust
and reradiated at far infrared wavelengths, $\lambda \sim$ 10-100 μm, as in
the case for those HII regions in which there is evidence of recent
star formation in our own Galaxy.

In all three cases, the observed spectrum and colours of the
young galaxy depend upon the redshift at which the galaxy formed.
If the redshift of formation is Z $\approx$ 3-10, the spectra are all
shifted by a factor of 4-11 to longer wavelengths. Thus, in case
(i) the observed spectrum would not be too different from normal
galaxy colours. In case (ii), a somewhat more likely model in my

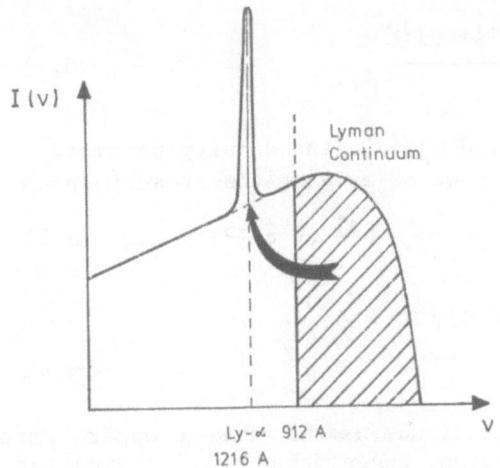

Fig. 9. Illustrating the expected emission spectrum of a young galaxy containing large amounts of partially ionised gas. The radiation in the Lyman continuum is reradiated as hydrogen recombination radiation, principally in the Lyman-α line.

view, the spectrum of the galaxy will be very red and the Balmer emission lines redshifted to infrared wavelengths. In case (iii), the young galaxy will be a very powerful emitter at submillimetre wavelengths, λ_{max} 100–1000 μm. Thus, the infrared waveband may well prove to be the most important waveband for the discovery of young galaxies.

(4) The background radiation due to young galaxies. It is of interest to enquire whether the background radiation due to young galaxies is detectable. Various estimates of the optical background radiation from a uniform distribution of normal galaxies are in reasonable agreement and suggest that it should amount to about one half of the present observational upper limits quoted by Roach and Smith. Therefore, if galaxies were brighter in the past, their integrated emission may be detectable. The most direct approach to this question is to work in terms of the energy density of optical radiation produced by young galaxies.

The first step in the synthesis of heavy elements in stars is the conversion of hydrogen to helium in the cores of stars. For each helium atom synthesised, $0.007 \, m_p c^2$ of nuclear binding energy is released per nucleon and this energy is eventually degraded into starlight which escapes through the surface of the star. This process is the essential first step in the synthesis of heavy elements and more binding energy is liberated in this step than in subsequent reactions.

Suppose that a percentage x% of the matter density in the Universe is converted into heavy elements at redshift Z_f. Then, the energy density in starlight $\epsilon_r(Z)$ at redshift Z_f is expected to be

$$\varepsilon_r(Z) = 0.007 \frac{\Omega_{gal} \, \rho_{crit} \, c^2 (1+Z_f)^3 x}{100} , \tag{5.1}$$

where $\rho_{crit} = 5 \times 10^{-30}$ g cm^{-3} and Ω_{gal} is the density parameter for galaxies. Since $\varepsilon_r \propto (1+Z)^4$, we observe at the present epoch

$$\varepsilon_r = 0.007 \frac{\Omega_{gal} \, \rho_{crit} \, c^2 x}{100(1+Z_f)} = \frac{0.2 \times \Omega_{gal}}{(1+Z_f)} \text{ eV cm}^{-3} \tag{5.2}$$

For example, if $x = 3\%$, $\Omega_{gal} = 0.03$, $Z_f = 5$, we find

$$\varepsilon_{rad} = 3 \times 10^{-3} \text{ eV cm}^{-3} \tag{5.3}$$

which is of the same order of magnitude as the present upper limits to the optical background radiation, except that this is expected to be redshifted into the infrared waveband. Harwit (this volume) mentioned the recent estimate of the optical background radiation by Matilla using a high-latitude dark cloud as an occulting screen. We could achieve this high value if, for example $Z_f = 2$ and $\Omega_{gal} = 0.1$. This estimate is confirmed by more detailed calculations by Tinsley.

(5) Comparison with the properties of quasars. It will be noticed that many of the predicted properties of young galaxies are similar to those of quasars, a point made as long ago as 1965 by Field. The young galaxies differ from quasars in the following respects:

(i) they are of finite angular size, $\theta \approx 1-0.1''$ arc,

(ii) they should not exhibit strong optical variability. Variations due to individual supernovae should not amount to more than about $\Delta m \approx 0.01$ compared with the variability of quasars which can amount to several magnitudes,

(iii) the continuum radiation should not be polarised,

(iv) they are very red or infrared objects having intrinsic luminosities $L \sim 10^{46-47}$ erg s^{-1}. They might even be intense submillimetre sources if there are substantial quantities of dust present.

These objects might be discovered in surveys designed to discover quasars. Indeed, some of the quasars already catalogued may in fact be young galaxies in disguise. Infrared surveys to detect young galaxies would seem to be one of the most exciting prospects for extragalactic infrared astronomy.

PART II - USEFUL TOOLS FOR OBSERVATIONAL COSMOLOGY

6. SOME USEFUL RESULTS

One of the basic problems in cosmology is defining distances in an

unambiguous manner. I will introduce only two which I find useful.
Fig. 10 shows a simple space time diagram for an expanding Universe in which an observer in the local standard of rest with
respect to the microwave background radiation has space coordinate
r = 0 = constant. This observer observes objects along his past
light cone and thus observes galaxies (1), (2) and (3) as they
were at earlier epochs t_1, t_2 and t_3. It is most convenient to
define distances at a particular epoch, but it can be seen from the
diagram that this is not directly possible from observation. If
we define distances at, say, the present epoch, we have to project
the positions of galaxies (1), (2) and (3) forward along their

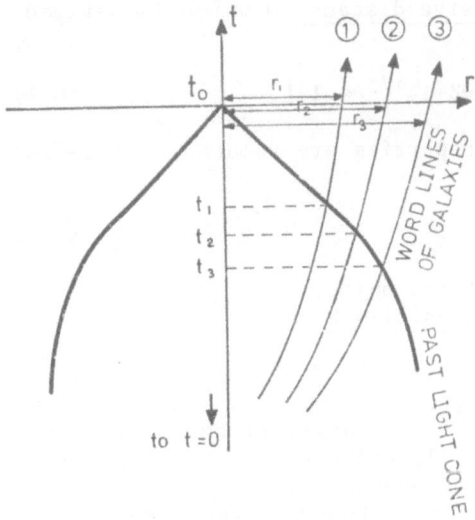

Fig. 10. A simple space-time diagram. The observer is at rest at
the origin r = 0 and cosmic time increases up the diagram. Observations are made at the present epoch t_0 and the observer sees
those galaxies which lie along his past light cone. The world
lines of galaxies (1), (2) and (3) which partake in the expansion
of the Universe are shown. Thus galaxy (1) is observed as it
was when it emitted its light at cosmic time t_1 (at redshift Z_1),
galaxy (2) at t_2 (at redshift Z_2), etc. At large distances, the
light cones are "bent round" because of the gravitational influence of the matter density upon light paths.
The diagram illustrates the definition of comoving radial distance coordinate, r. The world line of the galaxy is projected
to the present epoch t_0 adopting a suitable world model and the
radial distance of the galaxy at t_0 is defined to be the comoving
radial distance coordinate. Thus the comoving distance coordinates of galaxies (1), (2) and (3) are r_1, r_2 and r_3.

world lines to the present epoch. Obviously this requires a
knowledge of the dynamics of the world model, i.e. a knowledge
of Ω. Distance defined in this way is called <u>comoving coordinate</u>
<u>distance</u> r and the values r_1, r_2 and r_3 for galaxies (1), (2)
and (3) are indicated on the diagram. Plainly r is a function of
redshift and the world model, $r(\Omega,Z)$.

r is very useful in labelling the positions of objects in
the Universe but it is not useful for relating intrinsic properties of objects such as their physical sizes and intrinsic luminosities to observables such as angular diameters and flux densities. This is because these properties are related by the laws
of light propagation in non-Euclidean space. I find it most
convenient to introduce an <u>effective distance</u> D which is defined
by

$$D = \frac{2c}{(1+Z)H_0\Omega^2} \left[\Omega Z + (\Omega-2)\{(\Omega Z+1)^{1/2} - 1\}\right] \tag{6.1}$$

Then observables and intrinsic properties are related as follows:
(1) <u>Angular diameters θ</u>

$$\theta = \frac{\ell(1+Z)}{D} \tag{6.2}$$

where ℓ is the proper diameter of the object.
(2) <u>Observed monochromatic flux densities S</u>

$$S(\nu_o) = \frac{L(\nu_1)}{4\pi D^2(1+Z)} \tag{6.3}$$

where $\nu_1 = \nu_o(1+Z)$. $S(\nu_o)$ is the flux density of the source at
frequency ν_o (measured in W m^{-2} Hz^{-1}) and $L(\nu_1)$ is the monochromatic luminosity of the source at frequency ν_1 (measured in W
Hz^{-1}). Luminosity distance D_L is often found in the literature;
it is related to our D by $D_L = D(1 + Z)$.
(3) <u>Number of objects in the interval of comoving coordinate dis-</u>
<u>tance dr</u>

$$dN(r) = 4\pi n_0 D^2 dr \tag{6.4}$$

where n_0 is the space density of objects <u>at the present epoch</u>. This
calculation assumes that objects are distributed uniformly throughout the world model and that the total number of objects is conserved in the expansion: i.e. $N(Z) = N_0(1+Z)^3$, where $N(Z)$ is the
proper space density of objects at redshift Z.

$$dr = \frac{cdZ}{H_0(1+Z)(\Omega Z+1)^{1/2}} . \tag{6.5}$$

If the comoving density of objects changes with cosmological epoch,
i.e. $n(Z) = f(Z)N_0(1+Z)^3$, where $f(Z) \neq 1$, the number of objects in
the interval dr is modified by this factor $f(Z)$.

As an example of the use of these formulae, let us work out
an expression for the background radiation at frequency ν_o from a

uniform cosmological distribution of sources of the same luminosities $L(\nu_o)$ and the same spectrum; their local space density
is n_o.

The flux density of a source at redshift Z is

$$S(\nu_o) = \frac{L(\nu_1)}{4\pi D^2(1+Z)} = \frac{L(\nu_o(1+Z))}{4\pi D^2(1+Z)} \qquad (6.6)$$

The number of such sources per steradian in the interval Z to
Z+dZ is

$$dN(r) = n_o D^2 dr \qquad (6.7)$$

and therefore their contribution to the background intensity is

$$dI_\nu = \frac{n_o D^2 dr \; L(\nu_o(1+Z))}{4\pi D^2(1+Z)} \qquad (6.8)$$

Therefore the total background is

$$I_\nu = \int_o^\infty \frac{n_o \; L(\nu_o(1+Z)) \; cdZ}{H_o(1+Z)^2(\Omega Z+1)^{1/2}} \qquad (6.9)$$

If the comoving space density of sources has evolved with epoch,
an evolution function $f(Z)$ may be included in the integrand.

7. EXERCISES AND COMMENTS ON THE USE OF THESE TOOLS

The above relations are all that is needed to work out most of
the useful relations in observational cosmology. The classical
cosmological tests can be derived from them by considering how the
observed properties of standard objects change with redshift
as a function of Ω. If one could be certain that one had selected precisely the same type of object at different redshifts,
the variations of their observed properties with redshift would
enable Ω to be found. I refer to these tests as the classical
cosmological tests.

The following exercises will reveal a number of interesting
features of these tests:
(a) Derive the redshift-angular diameter relation for cosmological
models with $\Omega = 1$ and $\Omega = 0$ and show that these relations have a
minimum angular diameter at $Z = 1.25$ and $Z = \infty$, respectively.
(b) Derive an expression for the counts of infrared galaxies
$\Delta N(S)$ for a uniform population source of the same intrinsic luminosity $L(\nu_o)$ and identical spectra. In this calculation it is
best to evaluate the differential counts which describe the
numbers of galaxies observed in the flux density interval ΔS
at flux density S. Show that $\Delta N(S) \propto S^{-5/2} \Delta S$ at very small distances; this source count is known as the Euclidean prediction.
Compare the predictions of the world model having $\Omega = 1$ with the
Euclidean prediction.

(c) Derive the predicted redshift-flux density relation for infrared galaxies such as NGC 253 and M82 at 100 μm and 10 μm.
(d) How large a change in the intrinsic luminosity or spectrum of infrared galaxies at Z = 1 is necessary in order to make discrimination of world models having $\Omega = 0$ and $\Omega = 1$ difficult.
(e) Evaluate the background radiation expected from:
   (i) A universal population of infrared spiral galaxies which were described by Rieke (this volume).
   (ii) The background due to quasars assuming an evolution function $f(t) \propto e^{-10t/t_0}$ in an $\Omega = 0$ world model; $t_0 = H_0^{-1}$. (N.B. dt = dr/c(1+Z)).

   Ideally, in performing these classical cosmological tests, one should have good physical reasons for understanding why the objects used in the tests should have standard properties. Unfortunately, this is not the case in any of the classical cosmological tests which I am aware of up till the present time. It is only if one has a physical understanding of these standard properties that it may be possible to make corrections for differing physical conditions at earlier cosmological epochs.

   Two final comments may be useful.

(i) Interpretation of data on the background radiation. In general the background radiation may be written

$$I_\nu = \frac{f(\Omega)}{4\pi} \frac{c}{H_0} \sum_L n(L)L \qquad (7.1)$$

where c/H_0 is the cosmological distance, L is intrinsic luminosity, n(L) is the space density of objects of luminosity L and $f(\Omega)$ is a constant of order unity. Thus the background radiation gives crude information about the product $\Sigma n(L)L$. It is generally most useful to compare an observed background with the expectations from known classes of sources and if they agree within order of magnitude, that is as good as one can normally do.

   It has often been suggested that additional information can be derived from fluctuations in the intensity of the background radiation. The information to be obtained is limited and the theory of such analyses has been described in detail by Scheuer. If fluctuations of intensity ΔI are observed, one obtains information about the counts of objects contributing to the background. Roughly, the surface density of such sources must be N $\sim$ 1 per beam area of the telescope used and the typical flux density of these sources is $\Delta I\Omega$ where Ω is the beam width of the telescope. A more detailed analysis can provide information on the shape of the source count down to a flux density corresponding to one source per beam area.

(ii) Cosmological evolution. Many papers on observational cosmology discuss the effects of changes on the intrinsic properties of objects with cosmological epoch upon the classical cosmological tests. However, the direct observational evidence for such changes is limited. The most dramatic changes are those associated with
   a) the counts of radio sources,

b) the V/V_{max} test for quasars,
c) the counts for optically selected quasars.
I do not wish to go into the details of these analyses, but will
only make the point that all three results apply only to very
special sets of object in the Universe. Quasars are among the
most luminous objects in the Universe and they are very rare. The
observational evidence indicates that their comoving space density
has evolved with cosmological epoch roughly as

$$f(t) \propto e^{-10t/t_0} \; ; \; t_0 = H_0^{-1} \tag{7.2}$$

This law can also account for the radio source counts, but <u>only if
it is applied to the most powerful radio sources</u> which have space
densities similar to those of quasars. The reason for this restriction is that the anomalies of the radio source counts occur
over a relatively small range of flux densities and if all classes
of radio source evolved according to the above law, the source
counts would not converge at low flux densities, contrary to
observation.

To summarise the results, there is convincing evidence for the
strong cosmological evolution of quasars and the most powerful
radio galaxies. However, there is no evidence for the strong
cosmological evolution of weak radio galaxies, Seyfert galaxies
and normal galaxies. Indeed they must <u>not</u> evolve strongly.

The only way infrared astronomers can find out if such evolutionary changes should be included in their cosmological calculations is for them to make unbiased surveys of the high lati-
tude infrared sky. They will then perform their own source counts
and V/V_{max} tests. I wish them the very best of luck in these
programmes and hope the results will be as exciting as those already obtained in the radio and optical wavebands.

REFERENCES

The material on which the above lectures are based is derived
from a very wide range of sources. Rather than attempt to give
a complete bibliography, the following list of review articles
and books should serve as a source for more detailed presentations
of the material:

Bondi, H., 1961, <u>Cosmology, Second Edition</u>, Cambridge University
Press.
Longair, M.S., 1971, Observational Cosmology, Rep. Prog. Phys.,
<u>34</u>, 1125.
Longair, M.S. and Sunyaev, R.A., 1977, <u>Matter and Radiation in
the Universe</u>, Cambridge University Press, in press.
Peebles, R.J.E., 1971, <u>Physical Cosmology</u>, Princeton University
Press.

Rees, M., Ruffini, R. and Wheeler, J.A., 1974, Black Holes, Gravitational Waves and Cosmology – An Introduction to Current Research, Gordon and Breach Publishers.
Schatzman, E. – editor, 1973, Cosmology – Cargese Lectures on Theoretical Physics – Vol. 6, Gordon and Breach Publishers.
Sciama, D.W., 1971, Modern Cosmology, Cambridge University Press.
Weinberg, S., 1972, Gravitation and Cosmology – Principles and Applications of the Theory of Relativity, John Wiley and Sons, Publishers.
Zel'dovich, Ya.B. and Novikov, I.D., 1977, The Structure and Evolution of the Universe, University of Chicago Press, in press.

REFERENCES TO TABLE 1

Boynton, P.E. and Partridge, R.B., 1973, Astrophys. J. 181, 243.
Carpenter, R.L., Gulkis, S. and Sato, T., 1973, Astrophys. J. (Letters) 182, L61.
Conklin, E.K. and Bracewell, R.N., 1967, Nature 217, 777.
Pariiskii, Yu.N. and Pyatunina, T.B., 1971, Sov. Astron-AJ. 14, 1067.
Penzias, A.A., Schraml, J. and Wilson R.W., 1969, Astrophys. J. (Letters) 157, L49.

OBSERVATIONAL TECHNIQUES IN INFRARED ASTRONOMY

P. Lena

Observatoire de Paris
Meudon, France

1. INTRODUCTION

Infrared photons from celestial sources are very hard to detect because of the surrounding radiation emitted, by virtue of Planck's law, by everything which has a temperature different from absolute zero. The infrared astronomer is usually contending with an average radiance of everything (the instrument, the detector, the sky, the telescope itself), which is roughly from 10^4 to 10^6 times brighter than the source he is hunting for. Another problem is that infrared photons are of very low energy. A 1 μm photon has an energy of ~ 1 eV and a 100 μm photon has an energy of ~ 0.01 eV. Therefore, even the best detectors which are now available are unable to give a signal unless they receive a large number of photons per second. In the visible there are photon counting cameras which can count individual photons. Yet in the infrared we need many more photons: e.g. at a few μm we need 10^{-16} Watts on the detector, that is to say, about 1500 photons/sec to have a detectable signal. At 100 μm, the situation is worse: we need at least 10^{-14} Watts on the detector, and because these photons have lower energy, about 5×10^6 photons/sec are required. Fortunately many sources are bright enough in the infrared so that in a reasonable observing time we can still obtain a signal. We can summarize the above by saying that below the level of about 10^3 to 10^5 photons per second, for a given source, the sky is just unknown.

The first lecture will deal with the *earth's atmosphere*; despite the fact that this has nothing to do with astrophysics. It is so important to gather infrared information about astrophysical objects that one needs to know exactly what the properties of the atmospheric screen are. The second lecture will deal with *imaging*, the various ways to obtain images in the infrared and information

231

G. Setti and G. G. Fazio (eds.), Infrared Astronomy, 231-269.

about the spatial structure of the source. The third lecture will deal with *detecting* those scarce, expensive photons.

2. BASICS OF THE EARTH'S ATMOSPHERE

The earth's atmosphere is really the limiting factor for infrared astronomy on the ground or higher in an aircraft, or still higher in a balloon. The earth's atmosphere absorbs infrared radiation by pure rotation, or vibration rotation, transitions of its various molecules. An easy number to remember is that short of 25 μm, maybe 35 μm in certain dry sites, and beyond 350 μm, one can observe from the ground. Between these values the atmosphere is optically thick and what one sees from the ground is solely the emission of the atmosphere. So one has to go above at least the troposphere with an aircraft or even higher with a balloon. When it is optically thin, the atmosphere has a *radiance* due to its thermal emission, and this very crudely can be written

$$\text{Emission} \atop \text{(sky radiance)} \quad = \quad \tau_a(\lambda) B_\lambda(T_a), \qquad (2.1)$$

where $B_\lambda(T_a)$ is the Planck function at some average atmospheric temperature, $\tau_a(\lambda)$ is the optical depth. This emission is important only when the Planck function has a significant value, which happens for wavelengths $\lambda(\mu m) \geq 1000/T_a$. The temperature T_a is about 300 K, so beyond 2 or 3 microns, the emission is not negligible. The absorption of the atmosphere occurs everywhere and it is simply given by:

$$I_\lambda/I_0 = e^{-\tau_a(\lambda)} . \qquad (2.2)$$

Another factor to consider when discussing the detection limits imposed by the atmosphere is that of *scattering* i.e., the Rayleigh and Mie scattering of sunlight by the constituents of the atmosphere. IR astronomy has at least one advantage and that is the potential for daytime observing. Since the sky is thousands of times brighter than the sources to be looked at, it is immaterial whether it be day or night, and one may as well use the large telescopes during the daylight hours as well as night.

To further understand why this is so, consider Fig. 1 wherein we have plotted the sky brightness (magnitudes) versus the wavelength. The curve represents a kind of average emission (in the sense that it has no spectral features) of the Planck function, as expressed by eq. (2.1), with an average emissivity of $\sim$ 1%. This shows that the atmosphere is becoming extremely bright at 5 and 10 μm. The measurements have been made with a 6 arc sec aperture. The figure also shows the contribution due to scattering of sunlight by the various particles in the atmosphere. We see that beyond 3 μm scattering is really much smaller than the thermal emis-

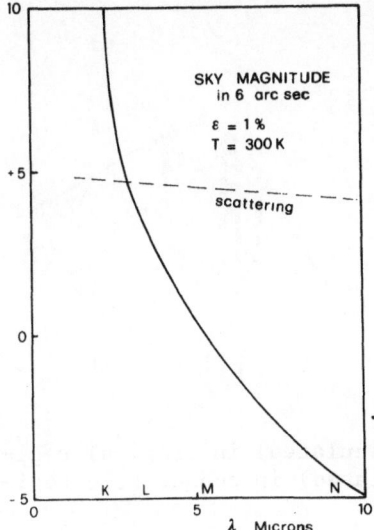

Fig. 1. Near infrared sky radiance.

sion of the sky. So there is no point in making any distinction
between day and night, because the crucial consideration is the
thermal emission of the atmosphere. The only difficulty in daytime is pointing the telescope in an absolute sense without guide
stars, but with modern techniques this is not a real problem.

Let us consider the thermal emission and see what problems it
creates. First of all it decreases the contrast. The source has
a contrast of only 10^{-3}, or maybe 10^{-5}, with the atmosphere. Secondly, it floods any detecting system with a number of photons
which have nothing to do with the source. These background photons
will undergo fluctuations and therefore will create a signal called
background noise . This noise is present even when using the most
sensitive detectors and persists whether the source is there or not.
Therefore it represents a sort of fundamental limit and the higher
the thermal emission of the atmosphere, the more numerous these
photons, and therefore the higher the noise. Finally, by virtue of
Kirchoff's law, which says that when the emission is high, absorption is also high, this thermal emission corresponds to a reduced
transmission which attenuates the very source you want to measure.
If the emission had a steady state value there would be only these
problems. But there is a fourth problem which arises from the fact
that this thermal emission has *fluctuations*.

Let us consider the *thermal emission*, without considering the
fluctuations for the moment. Of course, if you know the atmosphere,
its constituents, their molecular and atomic characteristics, and
vertical distribution, you can predict what the emission of the atmosphere would be and from that get an accurate prediction of what

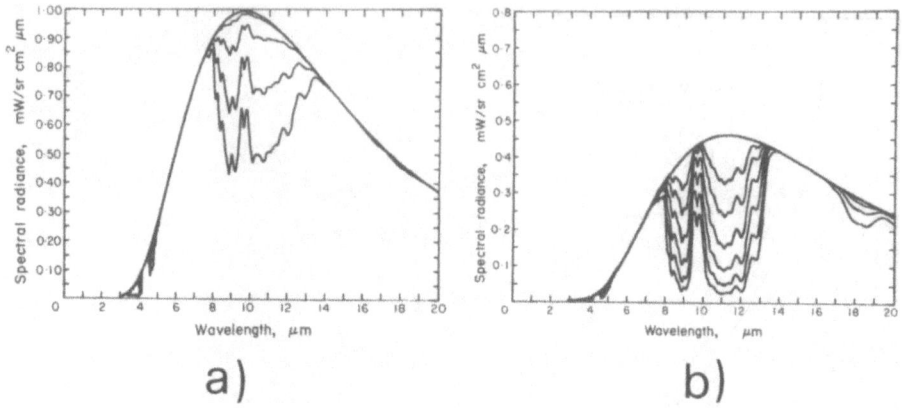

Fig. 2.   a) Sky background radiance (calculated) in tropical regions.   b) Sky background radiance (calculated) in subantartic regions.   (Tam and Corriveau, 1976).

this sky background, or sky radiance, would be. Fig. 2 gives an example of what is called a very good window for astronomical work, the 10 μm window. The spectral radiance of the atmosphere is plotted versus wavelength. It resembles very closely a blackbody, which means that everywhere out of the 8 to 12 μm band, it is optically thick; but in the window it deviates from a blackbody. In the case of Fig. 2 a) the departure from the blackbody curve is a factor of 2, which means that the emissivity is only 0.5 and therefore the transmission is 50%. A better case would be obtained in different atmospheric conditions (Fig. 2 b)): with a 5% emission, one has a 95% transmission. As a check Fig. 3 shows the difference between the calculated and the experimental values. The fit is reasonable.

Let us now consider the sky background *photon noise*. Here one introduces a very useful quantity called the Noise Equivalent Power (NEP). The meaning of this quantity is that if one has that much

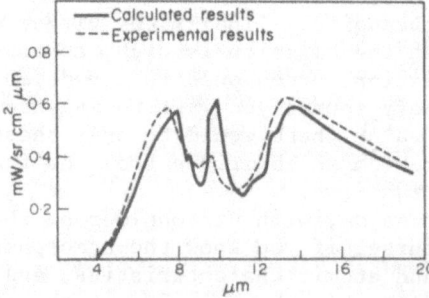

Fig. 3.   Spectral sky radiance. (Tam and Corriveau, 1976).

Table 1

The Sky Background Photon Noise Evaluated for a 3.6 m diameter telescope with a 6 arc sec field of view and an assumed emissivity of 1%.

Band	L	M	N
NEP(W Hz$^{-1/2}$)	8×10^{-16}	10^{-15}	4×10^{-14}

power on a detector due to the source, then because of the noise created by the photon fluctuations of the sky background, one gets a signal-to-noise ratio of unity in 1 second. Typical NEP values are summarized in Table 1. We see that the brightness of the sky, flooding the detector with photons, is already limiting, and that at 10 μm it is limiting in a severe way. And this is very optimistic, because we took ∿ 1% sensitivity. We'll come back to these sensitivity problems in our discussion of detectors.

Another example of the importance of the effect of atmospheric emission is the case of spectral lines emission. Fig. 4 shows an example of how necessary it is to predict, with high accuracy, the emission of the atmosphere. It shows how a 3°K blackbody emission due to the cosmic background (dashed curve) would look compared to the atmospheric emission. To deduce the flux, which really is the astrophysical quantity of interest, we need first to have minimal noise, that is to say, a good detector and low background radiation, and then to know exactly the predicted emission of the atmosphere to be able to subtract all the atmospheric lines.

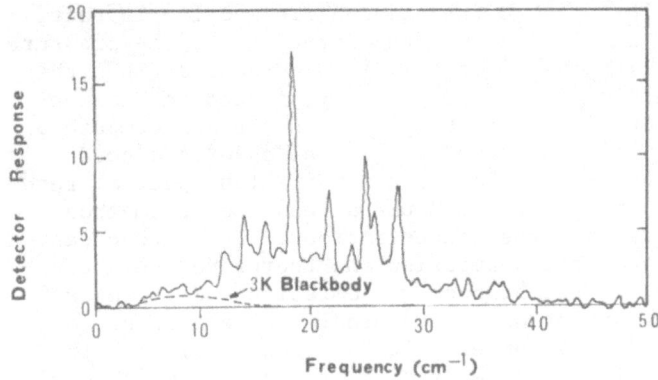

Fig. 4. Observed instrumental response to the night sky compared with a 3 K blackbody curve. (Adapted from Richards et al., 1975).

What is true for the cosmic background also applies to a number
of other problems in spectroscopic work such as line emission from
HII regions and line emission from molecules in the far infrared.
Of course one could use a satellite to get above the atmosphere
and eliminate its interference with detection, but we don't expect
to have any infrared satellite, or telescope in orbit, before
three or four years.

In principle, there is no difficulty in having an exact prediction of the atmospheric behaviour at any wavelength because one
has a thermal equilibrium situation. In the earth's atmosphere,
collisions are dominant, everything is thermalized, and it is easy
to measure the temperature and the pressure. So, at a given frequency, the intensity at an altitude z is

$$I_\nu(z) = \int_z^\infty B_\nu(z') e^{-\tau(z,z')} \kappa_\nu(z') dz', \tag{2.3}$$

where $B_\nu(z')$ is the Planck function, multiplied by the absorption
coefficient $\kappa_\nu(z')$, and attenuated by the optical depth $\tau(z,z')$
between the heights z, and z'. A horizontal stratification is assumed, which is reasonable in a stratified model of the atmosphere
in thermal equilibrium. In this equation are hidden the variables
concerning molecular constituents (H_2O, CO_2, O_3, N_2O, CO, O_2),
which are a function of altitude, pressure and temperature. The
line profile appears in the absorption coefficient, but it is well
known from the theory of line broadening dominated by collisions.
The absorption coefficient is the sum of the coefficients of each
constituent i times its concentration in the atmosphere: $\kappa_\nu =$
$\Sigma_i \kappa_i(\nu) N_i$. How is it possible to predict the exact transmission
at each frequency? Let us take the molecular model, which gives
the absorption coefficient, and the atmospheric model. In fact,
instead of using the altitude z, one can directly use the pressure
as a variable in the integral in eq.(2.3). Thus one defines the
temperature as a function of the pressure, $T(p)$, and the concentration of a given constituent as $N_i(p)$. Then one can compute a
certain intensity of emission. Now one can also, with a good
spectrometer, measure the emission with a very high spectral resolution, and then compare the two and use a least square method
(Marten, 1977) to determine the concentrations. It is then easy
to compute the atmospheric transmission and therefore the attenuation for any source, or the atmospheric emission and therefore
whatever will be in front of an astronomical source. To be con-
vinced that the method does converge let us look at Fig.5. The
solid line gives an initial concentration of a constituent from
which one computes the emission, assumed thereafter to be the
measurement. Then take a trial distribution N°, compute the emission, apply the least square method and change N° to N'....., to
minimize the quantity $\Sigma |I_\nu'^2 - I_\nu^2| N_i(p)$. N^8 is the final distri-
bution obtained, and is absolutely identical to the one one start-

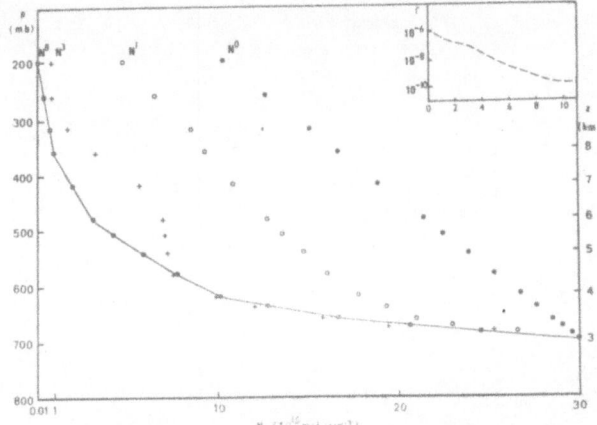

Fig. 5. Test of the Marten method. __ Profile used for simulating the emission ("true"); $N°$, initial trial profile to compute the e-mission; N^8, final profile to fit computed emission and "true" emission.

ed with. This means that the inversion of the emission profile of a line, or a series of lines, in the atmosphere provides an extremely good knowledge of the atmosphere itself. This is a very classical problem. The method works well because of thermal equilibrium and a good knowledge of absorption coefficients. Fig. 6 shows the difference between computed and "true" emission. Fig. 7 is what a spectrum looks like when it is observed through the atmosphere: the region is around 80 μm and the observation is made from the C141 aircraft with a resolution of 0.02 cm^{-1} (unapodized). Although the spectrometer was looking at the Orion nebula, one sees

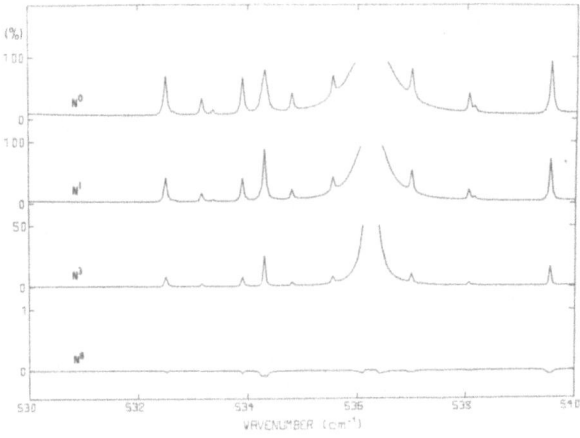

Fig. 6. Difference between computed and "true" emission.

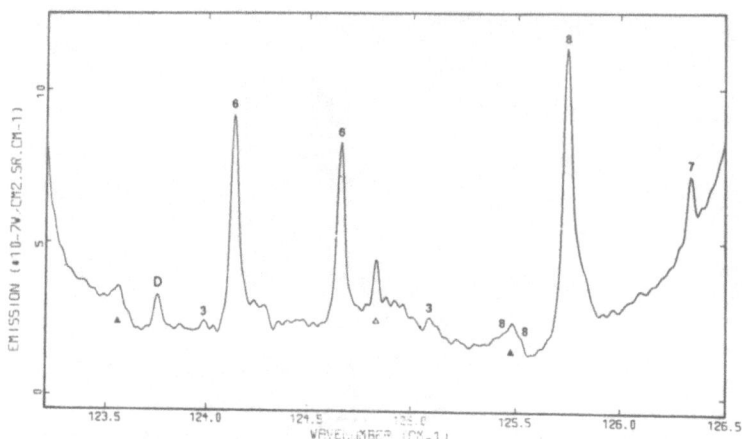

Fig. 7. C141 measured spectrum. $\Delta\sigma = 0.02$ cm^{-1}, Altitude = 12 Km.

primarily atmospheric lines. If you want to know anything about Orion nebula, you have to first remove them.

Comparison with a synthetic spectrum will provide the free spectral ranges where Orion nebula emission may be observed. A synthetic spectrum allows one to compute the synthetic *transmission* spectrum (see Fig. 8 for another wavelength range around 20 μm). There are certain wavelengths which are completely blocked, and nowhere the transmission reaches 1, implying simply that there are numerous lines with overlapping wings and, even at a place without lines, there is still some absorption. At the C141 altitude, which is about 13.5 Km, the situation is better than from the ground; the lines are narrower because pressure is lower and

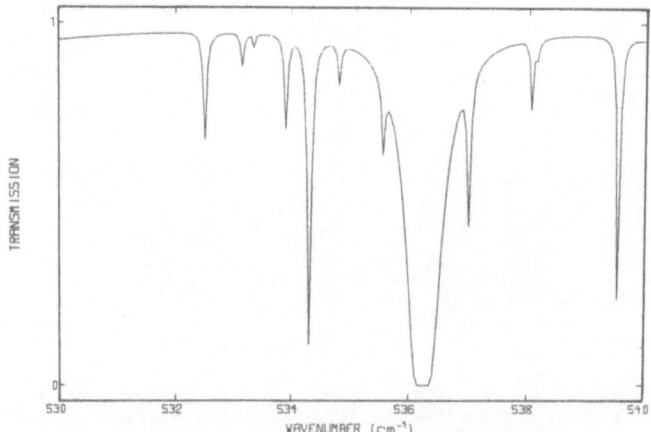

Fig. 8. Synthetic spectrum (3 Km altitude).

in between there are really clean regions. So, if an astronomical
source has a line at the right place, one really has a chance to
measure it. Of course, at the higher and higher altitudes attained by balloons and satellites, the lines become narrower and of
lower intensity.

All of what we have said up to now concerns mainly the lower
atmosphere where CO_2 and H_2O are the principal emitters. There is
also line emission much higher in the atmosphere due to atoms. It
is worthwhile to say a few words about these lines because one
must contend with them even at satellite altitudes. It will not
be easy to get rid of them. Let us consider atomic oxygen, which
in fact begins to form around 80 Km and above, by photodissociation of molecular oxygen. This atom has a triple ground state
which gives two lines at 63 and 147 μm. Assuming reasonable numbers at this altitude for the concentration of oxygen atoms ($2x10^{11}$
to 10^{12} at. cm^{-3}), the temperature ($300°$ K) and the scale height
(35 Km), one can compute the width of the oxygen lines. One finds
that they are extremely narrow, i.e. collisions are no longer important. But the medium is also very optically thick. The computed strengths of the atomic lines are shown in Fig. 9. The amount
of radiation is not very small, in fact, and the fluctuations of
those photons are *not* completely negligible. Even at satellite
altitudes it could contribute significantly to the background. On
the other hand, those very narrow lines may provide a very good
tool to calibrate spectrometers, once they are well known.

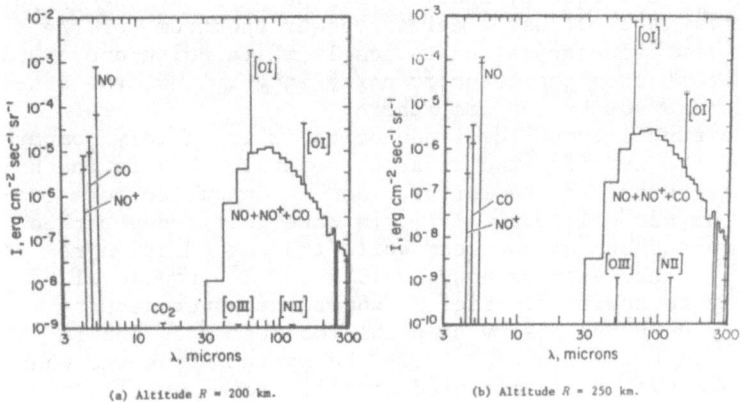

(a) Altitude R = 200 km.   (b) Altitude R = 250 km.

Fig. 9. The infrared emission from the upper atmosphere of the
earth at zenithal line of sight is plotted as a function of wavelength. A column density for the average kinetic gas temperature
T = 1000 K was used. For the NO^+, CO and NO vibration-rotation
bands, the intensity on the sunlit side of the earth is given by
the upper tick mark and the intensity on the dark side of the earth
is given by the lower tick mark. The NO, NO^+, and CO rotation
lines have been summed into 10 μm wide bands; for these bands the
units are then erg cm^{-2} sec^{-1} sr^{-1} 10 $μm^{-1}$.

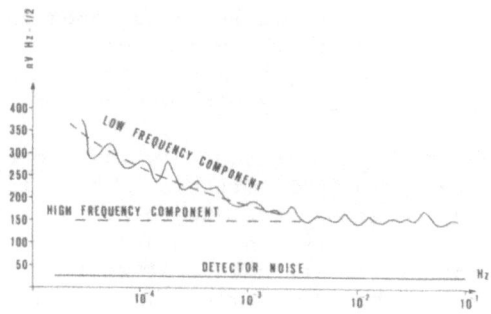

Fig. 10.   Atmospheric noise in the 8-14 μm window measured at sea
level in Firenze.   (From Blanco et al., 1976.)

2.1 Sky Noise

Let us consider now another phenomenon which is *sky noise*. Up to
now we have discussed the steady state emission of the atmosphere.
Unfortunately, a telescope, with a certain beam and a certain field
of view, looking at an atmosphere which is moving all the time,
will collect not only the photons coming from the source, but also
a fluctuating power due to the atmospheric emission.
    Fig. 10 shows a frequency analysis of the noise at the output
of a photopolarimeter. It has a certain power spectrum, of the
1/"f" type, at low frequencies and typically white noise above 0.01
Hz. Since, in this case the detector noise is very low, the sensitivity limits are set by the atmosphere.
    One way to cancel those low frequencies is to rapidly compare
the emission of the sky from two adjacent points. This is the basic observing technique in the infrared and it is called beam compensation or beam cancellation. If it is done fast enough, then
the source is in and out of the beam while the sky emission remains the same. Cancellations down to 10^{-3} to 10^{-4} the level of
the emission may be achieved. Fig. 11 shows the power density
versus frequency when one observes in the two beam mode and in the
one beam mode. In this figure one sees the system noise and what
is left as a residual in the measured signal. Beam cancellation
will essentially suppress slow drifts of the atmospheric emission.
But all this is done at a limited speed, because the detectors will
not follow very fast chopping and, also, mechanically it is difficult to chop very fast. Some atmospheric noise power will remain
at a reasonable chopping frequency, say 20 Hz.
    Now I will give you an example concerning airborne observa-
tions. The experimental conditions are as follows.
    An on board telescope looks at the sky with a certain field of
view, namely the angle θ, defined by a certain diaphragm while the
aircraft flies through the sky containing water and other constitu-

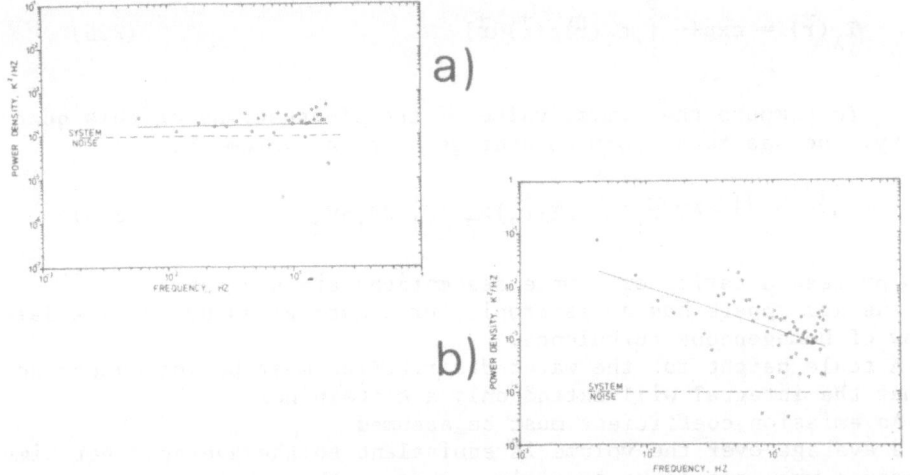

Fig. 11. a) A representative power density spectrum of atmospheric fluctuations while operating in the two beam mode. The vertical axis is scaled in terms of the power spectral density of the
stochastic process representing the fluctuations in antenna temperature produced by the atmosphere. The system noise is that not
arising from the atmosphere. b) A representative power density
spectrum of atmospheric fluctuations while operating in the one
beam mode. The units are the same as in Fig. 11 a). (From
Sollner, 1977).

ents with fluctuating temperatures and parameters; the telescope
sweeps through that ambient at a speed of ∿ 1000 Km/h and observes
this weak emission noise. Note that the sweeping exists for
ground based observations too, but in a reverse sense in that the
observatory is fixed and the wind is sweeping above and creating
the same differential effect. This noise is a function of the
wavelength and of the angle θ; it is larger at short wavelengths
and for larger diaphragms.
To try to understand these facts, let us write down the measured
flux as an integral over a certain volume, i.e. the volume that
the telescope is looking through, of the Planck function $B(\lambda, T)$
multiplied by the emissivity $\varepsilon(T,p)$, the water density $\rho(\vec{r})$, and
the transmission $T_\lambda(\vec{r})$ between this mass of water and the observer,
all this being done assuming spherical symmetry for the emission:

$$S_\lambda = \iiint_V \frac{B(\lambda,T)\varepsilon(T,p)\rho(\vec{r})T_\lambda(\vec{r})d\vec{r}}{4\pi r^2} \qquad (2.4)$$

The transmission $T_\lambda(\vec{r})$ is a function of the emissivity and of the
water density:

$$T_\lambda(\vec{r}) = \exp\{- \int_0^r \epsilon_\lambda(\vec{r})\rho(\vec{r})d\vec{r}\} \qquad (2.5)$$

To compute the r.m.s. value of the fluctuations of this quantity, one has to perform an average over the volume V:

$$<S_\lambda^2> \propto \iint <\Delta T(\vec{r}_1) \cdot \Delta T(\vec{r}_2)>_{\vec{\tau}} \ldots dV_1 dV_2 \qquad (2.6)$$

To proceed a series of simple assumptions are necessary:
- The atmosphere has an isotropic turbulence given by the standard law of homogeneous turbulence.
- A scale height for the water distribution must be introduced so that the integral will extend only a certain height.
- An emission coefficient must be assumed.
The average over the volume is equivalent to the average over time, because when one looks at a given instant through the atmosphere, and then again some time later, the volumes concerned are completely independent statistically.
All these assumptions lead to a value for the r.m.s. fluctuations given by

$$< \Delta S^2>^{\frac{1}{2}} = \kappa_\lambda' \theta^2 \qquad (2.7)$$

The term κ_λ' is wavelength dependent because of the Planck function and of the emissivity. Comparisons have been made of the predicted value of fluctuations using the above formula and the measured values. The agreement is rather good within a factor of 2.

2.2 Infrared Seeing

This section concerns the structure of the image, or of the wavefront, which reaches the telescope. The inhomogeneities in the atmosphere not only emit power and fluctuating power, but they also introduce phase differences between the various points. Considering the seeing as an assembly of physical phenomena which produce the final intensity distribution in an image, i.e., the quality of an image, it is interesting to investigate the physical causes of those degradations. But be careful, for this does not have anything to do with the instrument itself such as diffraction, aberration, geometrical distortion, bad mirrors, etc. Everything connected with the telescope is supposed to be perfect and we consider only what happens in the atmosphere. Consider a point source such as a point-like star, for instance. Its plane wave goes through the very inhomogeneous atmosphere and then starts to be distorted before it reaches the entrance pupil of the instrument. The wavefront will be distorted in phase and also in amplitude. This problem can be treated in a very general way given all the statistical moments of the distribution of the index of refraction in the med-

ium. Here, we take a simple case, a transparent medium, so that
the index of refraction is purely real. We study the correlation
in the pupil plane between the electric field and average over the
pupil. This autocorrelation of the pupil is, as well known, the
modulation transfer function.

Application of the theory of the influence of a turbulent homogeneous layer (Kolmogorov spectrum) through which the wave propagates gives:

$$<\psi(\vec{r}) \cdot \psi^*(\vec{\rho}+\vec{r})>_r = \exp\{-1.45 \frac{4\pi^2}{\lambda^2} \Delta h \; C_N^2 \rho^{5/3}\}, \qquad (2.8)$$

where ρ is the radial distance in the pupil plane, λ the wavelength, and Δh the height of the turbulent layer.

C_N measures the r.m.s. dispersion of the index of refraction, and
relates to the same quantity for the temperature of the air by

$$C_N = 8 \times 10^{-5} \frac{P}{T^2} C_T \quad \text{with} \quad <|\Delta T(\vec{r}+\vec{\rho})-\Delta T(\vec{r})|^2> = C_T^2 \rho^{2/3}. \qquad (2.9)$$

ΔT is the local departure from the average temperature T (Kelvin),
p being the pressure (mb), at the point $\vec{r}$ (or $\vec{r}+\vec{\rho}$) in the volume
where the index fluctuates.

   The correlation function for the wave ψ over the pupil may
therefore be written:

$$<\psi(\vec{r}) \cdot \psi^*(\vec{r}+\vec{\rho})> = \exp\left(- \frac{1}{2\Delta a^2} \; \frac{\rho}{\lambda}^{5/3}\right) \qquad (2.10)$$

in a quasi-gaussian shape of width Δa. Since this function describes the MTF (modulation transfer function) of the "distorted"
pupil, Δa is the width (in spatial frequency units) of this function. The important variable for the structure of the image is
not the geometric dimension ρ, but ρ/λ.

It is interesting to see if the width
of this correlation function is really wavelength dependent and why. If
a telescope is used at a given wavelength such that the pupil stays
within a coherence area, then from
the telescope's point of view, a
plane and a distorted wavefront are
indistinguishable. The distortion
is small enough that it remains coherent over the whole aperture, and
the image is given by the ideal image,
namely the diffraction limit. On the

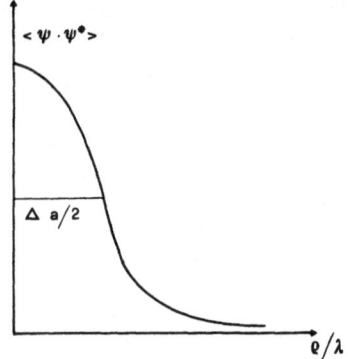

other hand, if the pupil is much larger than this coherence area,
then the wavefront will be very incoherent in phase from one point
to another, and the image will be completely distorted. Is this
effect wavelength dependent? To answer this question it is neces-

sary to see what there is in the factor Δa derived from eq.(2.8), remembering that the variable is ρ/λ. Since Δh and C_N are independent of λ, and depend only on the atmosphere, we see that the wavelength dependence of Δa is very low, only $\lambda^{1/6}$. Between 0.5 and 10 μm there is a ratio of 20 in wavelength, but there is only a gain of $20^{1/6} \simeq 1.65$ in width. So the seeing image spread, excluding diffraction, is not very strongly wavelength dependent for a given telescope size as long as the assumptions made here are valid.

There is still another effect: the wavefront, although it is coherent, may be distorted at a certain angle when large turbulent structures and wind effects are involved. Let us consider the situation drawn schematically in the figure. The first aperture will give an image in position (1) and the second aperture an image in position (2). This will happen if large space cells in the atmosphere exert a sort of bending or prism effect on the wavefront. Thus there will be no overlap of the images given by the two apertures despite the fact that there is coherence on both apertures. This is called the wave tilt effect.

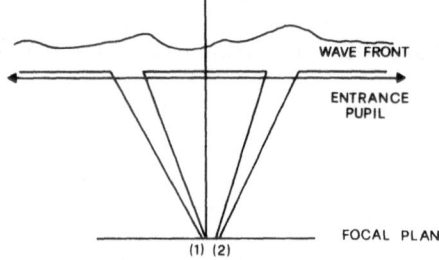

Within this distorted image of typical angular width λ/Δa, an instantaneous exposure can identify smaller details called *speckles*, which have a size distribution pattern down to the diffraction limit of the full aperture. We shall study them later. The point to note here is that the instantaneous intensity distribution in the image will vary with time: from one image to the next the correlation is almost lost.

We saw that the wavefront is changing from point to point, it is the phase effect; now it is interesting to study what happens to the wavefront at a given point versus time. Is there a time correlation? The time correlation can be written as:

$$< \psi(\vec{r},t)\psi^*(\vec{r},t+\tau)>_t \propto \exp(-\tau^2/\tau_c^2) + \text{constant}. \qquad (2.11)$$

The constant simply expresses the fact that no matter how long you wait, the image is always roughly at the same place, so the correlation remains. But, yet, the fine structure has moved and this is why there is a second term in the correlation function, corresponding to a certain correlation time τ_c, the time during which the image can be considered as frozen. This time is characteristic of the motions in the atmosphere. Of course these motions have a certain cut-off frequency. The velocity of sound, for instance, sets an upper limit. The characteristic time τ_c is such that in an image observed for a time shorter than τ_c the speckles are present, but for a time longer than τ_c everything is washed out and the fine structure disappears. τ_c is of the order of 20 to 50 milliseconds.

2.3 Conclusion

Let us summarize the conclusions of this incursion into the atmosphere and its effects on the observations. With a good knowledge of the properties of the absorbers, and because of thermal equilibrium, it is possible to predict and understand emission and absorption in a steady state condition. From this knowledge one can decide, for a given observation, which altitude is required and eventually which observing platform. Then the sky noise phenomenon, or fluctuations of atmospheric emission, is in many cases the limiting factor, more than the quality of the detector. This is why it is important to understand sky noise and, as far as possible, to cancel it. There are different ways to do that: either by adjusting the beam throw of the telescope, or by selecting an appropriately high frequency.

And then there is another phenomenon, completely independent of the emission of the atmosphere, which exists even for a purely transparent medium: this is the distortion of the wavefront by inhomogeneities in the atmosphere, both space and time dependent. This will influence the structure of the image, and will affect the spatial characteristics of the instrument, that is to say its resolving power and therefore the image quality.

3. BASICS OF DETECTORS

The objective is to convert infrared photons energy into electrical signal: current or voltage. *Signal* photons are usually much less numerous than *background* photons created by thermal surroundings; hence a minute fluctuation of the background can easily be much larger than the signal. These fluctuations are indeed larger for periods of time of seconds or minutes, therefore one usually needs to use beam cancellation at a fast rate, and the electrical signal on the detector is an AC signal. Additional reasons to modulate at or above 10 Hz reside in the better noise characteristics of the preamplifiers when the frequency increases. Because the surrounding background decreases rapidly with temperature, there is every reason to cool the detector environment, and the detector itself, independent of its own physical characteristics.

The fluctuations on the detector are set by the background incoming photons. Assume N photoelectric events per second, due to background. The fluctuation of the current will be (per 1 Hz bandpass)

$$< \Delta i^2 >^{\frac{1}{2}} = i/N^{\frac{1}{2}} \qquad (3.1)$$

where i is the current corresponding to N. This is multiplied by $B^{1/2}$ for a bandpass of B(Hz), assuming Poisson statistics for the photons. When the background fluctuations dominate, then one speaks of background limited operation of the detector.

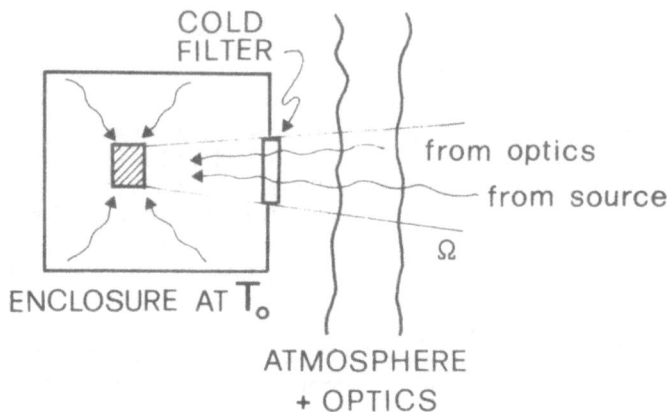

Fig. 12. The sources of photon noise.

The *Noise Equivalent Power* (N.E.P.) of the detector, in a particular configuration, corresponds to the power reaching the detector and giving a signal-to-noise ratio unity for a 1 Hz bandpass.
This power is *not* the power referred to the source: many intervening elements (optics, atmosphere...) are then to be taken into account, if one wants to determine the Noise Equivalent Flux outside
the Earth's atmosphere of a given system, namely the flux which
will give a signal-to-noise ratio of unity.
    Let us now consider how to improve the NEP in a background
limited case. Fig. 12 shows a typical configuration: a detector
enclosed in a box at a temperature as low as possible, with an aperture for the outside flux. The beam from the source has to go
through a more or less opaque medium (atmosphere, window) at 200-
300 K, hence emitting background photons. For simplicity, we will
assume that the detector is a quantum detector, converting individual photons. The detector sees the source photons (N_S), intermediate temperature photons (N_T), and low temperature photons (N_O).
Usually, but in space eventually, $N_S \ll N_O, N_T$; N_T sets the fluctuations, or ultimately N_O.
In the quantum limit, $h\nu \gtrsim kT$, the fluctuation per Hz is:

$$N' = \left\{ \int_0^\infty \eta(\nu)\varepsilon(\nu)t_\nu \frac{B_\nu(T)}{h\nu} (S_\nu\Omega)d\nu \right\}^{\frac{1}{2}} \quad , \qquad (3.2)$$

where $\eta(\nu)$ is the quantum efficiency, $\varepsilon(\nu)$ the background emissivity, T the background temperature, S_ν the detector cross section,
which may be frequency dependent if diffraction effects are important, Ω the solid angle and t_ν is the transmission of the cold filter placed in front of the detector at low temperature.
To improve this background limited case, one may:
- suppress the intervening medium, and window, by going to space.
- decrease $\varepsilon(\nu)$ by proper design of the optics.

- decrease T by cooling the optics, but here the atmospheric condensation is a potential problem.
- decrease the throughput $S_\nu \Omega$, but there is obviously a limit given by the source area one wants to study.
- decrease the integration interval by limiting the spectral range where warm radiation enters the detector cavity with a cold (T_O) filter keeping only the useful spectral interval.
If the N' fluctuation is made negligible, then N_O fluctuations may take over, since the solid angle ($\sim 4\pi$) is much larger. At this point, further cooling of the detector is necessary. It is also the case in which intrinsic noise sources in the detector (e.g. Johnson noise) will become significant.

The order of magnitude of achievable limits can be summarized as follows:
- in the near infrared ($\lambda < 5$ μm), the background emission is negligible, if the throughput is small, and the dominant term may be signal fluctuations.
- at longer wavelengths, a very coarse estimate of the lower limits would result from the following. Assume a diffraction limited detector ($S\Omega = \lambda^2$) receiving the signal photons under $\Omega = \pi$. The area is then 3×10^{-9} m^2 at $\lambda = 100$ μm. $N_O \approx \sigma T_O^4 2S/h\nu_O$, where ν_O is the frequency of the Planck function maximum, assuming a uniform value $\eta(\nu) = 1$. One gets $N_O^{1/2} h\nu \sim 10^{-16}$ W $Hz^{-1/2}$. Real cases will lie between 10^{-16} and 10^{-14} W $Hz^{-1/2}$.

A more accurate theory will include both thermal and quantum fluctuations (see, e.g., Harvey, 1970) when dealing with quantum detectors, and thermodynamic fluctuations in detectors' lattice, when dealing with thermal detectors (see, e.g. Coron, 1976).

Without yet being too specific about a detector, let us define the most important parameters for detectors:

Quantum Efficiency: this factor η is the ratio of photoelectrons produced per incident photon ($\eta \lesssim 1$) for a quantum detector. For a thermal detector the equivalent term would be the absorptivity of the detector.

Area: the maximum throughput a detector may receive is πS (one side) to $4\pi S$ (if placed in an integrating cavity). For a given telescope of area A, the solid angle on the sky is defined by $\omega \geq \pi S/A$.

Responsivity: the ratio of electrical signal (current or voltage) to incident power (optical responsivity), or to truly absorbed power (electrical responsivity).

Response Time: we have seen that AC response is needed. The time constant of the detector will be set either by its own electrical capacity, or by the lifetime of the generated charges (photo-conductors) or by the coupling of a thermal detector to the thermostat. Time constants may vary from 10^{-1} to 10^{-9} s depending on the particular detector. An heterodyne spectroscopy experiment may require a nanosecond response, while slow scanning with a thermal bolometer would accept a 10 Hz frequency cut-off.

Wavelength Response: this is set by the surface and/or the bulk

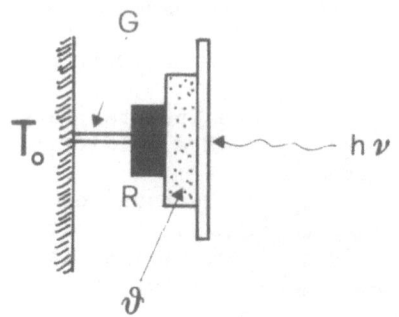

Fig. 13.   The Bolometer: schematic principle.

properties of the material, on one hand, and by its coupling to
the radiation field, on the other; here diffraction plays an important role, as do detector size effects. As an example, photo-
conductors have high frequency pass characteristics, determined by
the gap, and detect only photons with $h\nu \gtrsim E_g$ (E_g = gap energy).
    Operating Temperature: one sets $T_{det} \leq T_0$, where T_0 corresponds
to the thermal excitation of transitions in the detector with energy comparable to photons energy: $T_0 \sim 2800/\lambda_{ph}$. Lower values of
T_{det} may improve responsivity of the detector and/or reduce background noise.

3.1 The Bolometer

Since most of the astronomical work has been essentially carried
out with two types of detectors, *bolometers* and *photoconductors*,
we will deal with these as typical examples, and then we will discuss briefly recent developments in *array detectors* with special
read-out techniques.
    The bolometer, a thermal detector developed by Low, by Coron
and by Richards among others, absorbs photons on an absorbing surface A (Fig. 13), whose absorption can eventually be matched to the
wavelength of interest. A thermal capacitance θ is then heated by
conduction, and a thermometer R measures this temperature change by
the variation of its electrical resistance. Heat is evacuated to
a thermostat T_0 through a thermal conductance G. The time constant
of the system is given by $\sim \theta/G$, and the steady state temperature
of the detector, when submitted to chopped incident radiation, remains slightly above T_0. The sensor is usually made up of a doped
semi-conductor (Ge or Si), the resistance of which may be written:

$$R \propto R_0(T_b)^A, \text{ with } A \sim 6. \tag{3.3}$$

Here the noise sources are:
- fluctuation in photon exchange between bolometer and thermostat:

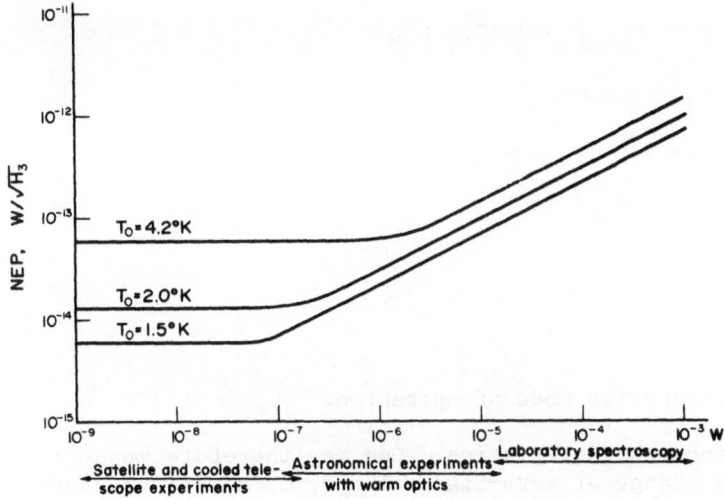

Fig. 14. Ultimate possible NEP for a bolometer as a function of
the background received at different bath temperatures (the photon
noise is not accounted for). This is given for the case A = 6 and
the horizontal limits are those reached so far with the three-part
bolometer (from Coron, 1976).

$$(NEP)^2 \propto (T_b - T_o). \qquad (3.4)$$

- Johnson noise of the resistance:

$$(NEP)_J^2 \propto \frac{4kTR}{\mathcal{R}^2} , \text{ where } \mathcal{R} \text{ is the responsivity.} \qquad (3.5)$$

To optimize such a detector, one ideally decouples the absorptive,
thermal and electrical properties, and one selects the optimum value for the thermal conductance G: in the presence of a high background, to keep (T_b-T_o) small G must be large, while for low background the time constant will set G. The size is essentially determined by the wavelength of operation and the desired throughput.
Fig. 14 gives achieved performance versus size.

3.2 The Photoconductor

The photoproduction of free charges in a semiconductor is the basic effect observed. It can be used in two different modes:
i) photoconductive mode (Fig. 15), in which the free charges produce a current being driven by an electric field imposed on the
bulk material. ii) photovoltaic mode (Fig. 16), in which the liberated carriers are separated in the electrical field internally
generated in a diode junction. The measureable effect is then a

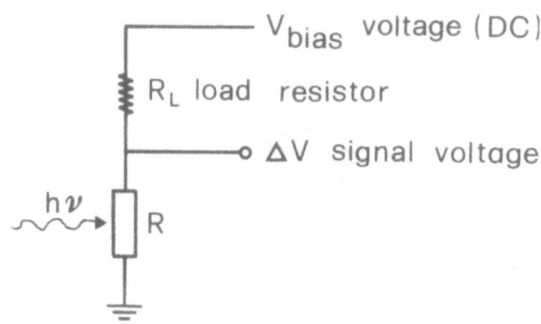

Fig. 15. Photoconductive mode of operation.

change in the diode characteristics. One may therefore measure
either a current change at zero-bias voltage, or a voltage change
at zero-current. AC operation will be limited by the carrier lifetime. The photoconductive effect uses: i) <u>intrinsic</u> photoconduc-
tivity, namely transition in pure semi-conductor material between
conduction and valence bands (if E_g is the gap energy, the photons
are only detected for $\nu > E_g/h$, and the detector temperature must
be below E_g/k). Typical cases cover Si (gap at 1.2 µm), Ge (1.8
µm), InSb (5.6 µm), $Hg_{1-x} Cd_x$ Te (8-15 µm); ii) <u>Extrinsic</u> photo-
conductivity, wherein dopant impurity levels are introduced in a
pure material. Assuming a group IV matrix (Ge, Si), dopants from
groups I,II,III, such as Cu, Au, Zn, Hg, Ga, and B will produce
acceptor levels of p-types, dopants from groups V, VI will produce
donor levels, such as St, P, and As. The response is set by the
relative change in the number of charge carriers. To reduce them,
one must cool the detector to avoid thermal excitation of these
dopant levels. The amount of background photons must also be decreased for the same reason, independent of noise considerations.
Because of thermally excited residual impurities, and even in the

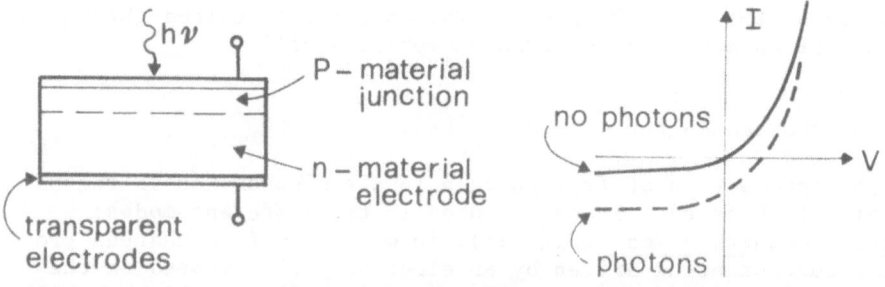

Fig. 16. Photovoltaic mode of operation.

absence of incoming photons, the resistance remains finite, but may be very high ($10^{11}-10^{12}\Omega$). These high resistance values create matching difficulties with preamplifiers and are a source of microphonics.

The doping is a compromise in that it has to be large enough for the detector to offer an optical depth close to unity to incoming photons, but small enough to keep the resistance high, and hence the responsivity. Yet the dopant photoconduction must be larger than the conduction due to thermally ionized impurities. In Ge:Ga for instance, which is a very good extrinsic photoconductor below 120 μm, dopant concentration is 10^{14} cm^{-3}, and impurities are all in the $10^{12}-10^{13}$ cm^{-3} range.

The *photoconductive gain*, g, is an important parameter in specifying a detector. It measures the ratio between the number of carriers contributing to the current at a given time, to the number of absorbed photons per second. If the free path of the photogenerated charges in the material is much smaller than the detector dimension, it is clear that the effective current will be very small, corresponding to g << 1. g is given by the ratio of the lifetime of a carrier to its transit time, and may be close to or much less than unity, depending on the number of available traps in the material. Of course, increasing the carrier lifetime also increases the time constant of the detector.

For a given bias voltage, one may write

$$\Delta v_{signal} = \frac{R_L R}{(R_L+R)^2} V_{bias} \frac{\Delta R}{R} \quad \text{with} \quad \frac{\Delta R}{R} = \frac{\Delta n}{n} \quad , \qquad (3.6)$$

$\Delta n/n$ being the relative change in carrier number density n. V_{bias} is maximum with respect to impact ionization ($V_b \lesssim 20^V$). We have: $\Delta n = \eta Q_s \tau A$, where η is the quantum efficiency, Q_s the photon flux, τ the carrier lifetime, A the detector area. n is fixed by dopant concentration, thermally excited carriers (if any) and background photons. The objective is to maximize $\Delta n/n$, in order to bring Δv above V_p, preamplifier noise for a given input impedance R.

Other causes of noise are: signal photon noise, V_s, background photon noise, V_B, Johnson noise of detector and load resistance, V_J, generation-recombination noise of carriers, V_{GR}. (For a more detailed description, see various papers in Hudson and Hudson, 1975). The best current photoconductors seem to be:

InSb	(intrinsic)	for	$\lambda < 5.6$ μm	(77 K)
Si:As	(extrinsic)	for	$\lambda < 23$ μm	(4 K)
Ge:Ga	(extrinsic)	for	$\lambda < 120$ μm	($\lesssim 4$ K)

with typical performances in the range of $10^{-16}-10^{-14}$ W Hz$^{-1/2}$.

3.3 Mosaic Detectors

The simplest way to best use the incoming photons is to collect

every *pixel* (or resolution element) of the image on an individual
detector. The resulting mosaic is then a one or two dimensional
array of detectors of the above type. This has been slowly becoming available in the past year with moderate size arrays.

A still newer development is the use of integrated circuitry,
which gathers not only the photosensitive material but also the
storage and read out capabilities on a single semi-conductor chip.
These systems are called Charge Coupled Devices (CCD) or Charge
Integrating Devices (CID).

Fig. 17 describes the geometry and physical arrangement of the
CCD. The basic principle is to create behind and in electrical
contact with every photoconductive element a potential well at a
MOS junction (Metal Oxide Semiconductor). The photoproduced charges will therefore be locked in this well and accumulate, at least
for times $\tau < \tau_d$, where τ_d is the diffusion time ($\tau_d \sim 1$ min to
several hours). A readout logic will empty the wells in an orderly
sequence and deliver a current measuring the charges. Silicon
technology allows miniaturization of such arrays and achieves up to
64 x 64 elements with a good filling factor.

Infrared CCD's have now been developed with InSb. Two types
of devices are under consideration: a) monolithic, in which the
whole electronic circuitry is made up of the photo-conductive material itself and b) hybrid, wherein the photoconductive material
is deposited on silicon (see Philips et al., 1976).

4. BASICS OF IMAGING

We have already seen how important it is to have good information
on the spatial structure of sources and how crucial it is for the
understanding of the physical phenomena. This is one rationale
behind imaging; the other one comes from the high cost of the detection of infrared photons. It is really worthwhile to make every

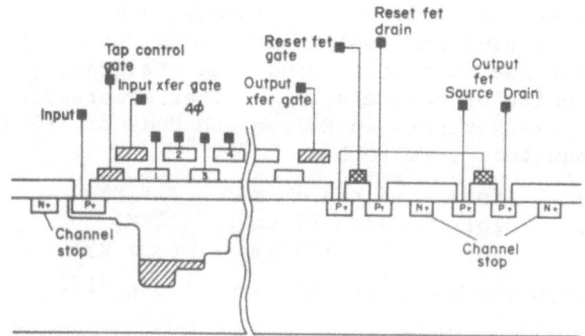

Fig. 17. Structure of two level, overlapping gate CCD. Note surface potential profile for charge injection and confinement into
first cell (from Steckl, 1976).

effort to improve the capability of collecting them and to use them
efficiently.

This lecture is organized into three different topics: i) a
rather simplified discussion of what I call the standard pupil,
namely a circular telescope of a given diameter. For people un-
familiar with the rather specialized language of image processes
(spatial frequency, image filtering, etc.), this first discussion
provides a little bit more knowledge. ii) a discussion of the
methods of image mapping which tries to answer the question: "What
is the most efficient way to do mapping?" iii) a discussion of
the problems associated with seeing effects, which were mentioned
in the first lecture. Here, we examine some existing ways to overcome the seeing limitations and to reach the diffraction limit of
the largest telescopes we can build. Also, a few words will be
said about aperture synthesis in the infrared.

4.1 The Standard Pupil

We have already seen that a perfect wavefront, a plane wave given
by a point-like source, is usually distorted when it reaches the
entrance pupil of an instrument and it has a phase which is position dependent on the pupil. The pupil is described by a transmission function $P(\vec{R})$, where $\vec{R}$ is the vector on the pupil plane
(Fig. 18). In the case of a distorted wave, it is equivalent to
say that we are dealing with a plane wave and a complex pupil of
the form $P(\vec{R}) = \exp\{i\phi(R)\}$ inside and equal zero outside, where
$\phi(\vec{R})$ contains the inhomogeneities of the wave phase. This phase
factor can also be a function of time. In the focal plane of the
instrument, or in the image plane, we have a certain intensity

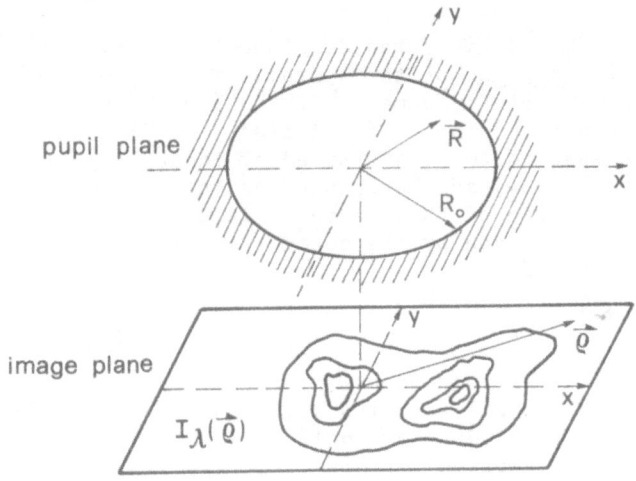

Fig. 18. The standard pupil.

distribution $I_\lambda(\vec{\rho})$ which is usually a function of wavelength and of position $\vec{\rho}$. The intensity distribution of the object, $I_0(\vec{\rho}) = I_0(x,y)$, is of course a function of the same variables. In the following, we shall use angular variables in arc min or arc sec units. They are related to the linear variables in the focal plane by the focal distance.

This information about the spatial frequency content of the object, or of the image, is given by taking the two dimensional Fourier transform of these two functions:

$$I_0(\vec{\rho}) \overset{F.T.}{\Longleftrightarrow} \tilde{I}_0(\vec{w}) \qquad \text{object}$$

$$I_\lambda(\vec{\rho}) \overset{F.T.}{\Longleftrightarrow} \tilde{I}_\lambda(\vec{w}) \qquad \text{image,}$$

the Fourier transform being defined by

$$\tilde{I}_0(\vec{w}) = \iint I_0(\vec{\rho}) e^{2i\pi\vec{\rho}\cdot\vec{w}} \, d\vec{\rho}, \qquad (4.1)$$

where $\vec{w}$ is the conjugate vector of $\vec{\rho}$; its coordinates (u,v) are the conjugates of (x,y). So, $\vec{w}$ is an angular spatial frequency.

The basic relation of imaging is that the image is the convolution product in two dimensions of the object with the instrumental response of the system. It is expressed by the equation

$$I(\vec{\rho}) = I_0(\vec{\rho}) * T(\vec{\rho}). \qquad (4.2)$$

Or, if we are in the conjugate Fourier plane (u,v) which is very familiar to radioastronomers, this relation becomes a simple multiplication product between complex functions:

$$\tilde{I}(\vec{w}) = \tilde{I}_0(\vec{w}) \cdot \tilde{T}(\vec{w}) \quad , \qquad (4.3)$$

where $\tilde{T}(\vec{w})$ is called the modulation transfer function (MTF) and represents the frequency response of the system. This result holds only if there is aplanatism, namely, if the linearity of adding intensities is respected in the focal plane. And this is a condition which can be assumed to be satisfied at least for a few arc second field of view. Now, another basic result, that we will use all the time in the course of this lecture, is that the modulation transfer function is given by the autocorrelation product of the pupil $P(\vec{R}/\lambda)$ by itself

$$\tilde{T}(\vec{w}) = P(\vec{R}/\lambda) \otimes P^*(\vec{R}/\lambda + \vec{w}). \qquad (4.4)$$

The autocorrelation product in two dimensions is simply defined as the overlapping area of the pupil with itself, being displaced by an amount $\vec{w}$. (The complex conjugate P^* applies to the case of a complex pupil with phase distortion). This is a very simple way of building the frequency response of the system. Because the pupil always has a finite support in the pupil plane (X,Y), the au-

tocorrelation product always reaches zero for any value $\vec{w}$ larger
than a certain cut-off frequency w_c. This cut-off value corresponds to displacement of the pupil with respect to itself such
that it does not overlap anymore. Thus, we can state the very
general conclusion that any physical pupil is a low-pass, or in
some cases a band-pass spatial filter in two dimensions. The image, in the sense of spatial frequency content, is just the object
multiplied by this low pass filter. Then we lose everything in
the object beyond the cut-off frequency w_c. Moreover, we somewhat affect the whole spectrum below w_c, and if the filtering is
time-dependent, we shall affect it in a time-dependent fashion.
     Finally, since there is a cut-off frequency in this function,
then it is a band limited one, and the Shannon sampling theorem
tells us that such a function needs only to be sampled at discrete
points separated by intervals of width $1/2w_c$. Therefore, we define in the image a resolution element or a pixel. There is no
point in making measurements closer than $1/2w_c$, because it would
only provide redundant information.
Let us apply the concepts we have just introduced to the standard
circular pupil. Since we have circular symmetry, the MTF can be
computed as a function of the modulus w. This function monotonically decreases from 1 at the zero frequency, to 0 at the cut-off
frequency D/λ. (D is the diameter of the telescope.) The curve
is shown in Fig. 19, with and without central obscuration. If we
are dealing with a point source, then the image is the Fourier
transform of this curve. The result, as expected, is the Airy
function. Its first zero occurs for an angular radius $\alpha = 1.22\lambda/D$
(Fig. 20) which is the size of the Airy disk. In Table 2 characteristic figures of cut-off frequencies and of widths of the Airy
function computed from 1 µm to 1000 µm are listed for a typical
telescope diameter of 1 m. What is really meaningful is the cut-
off frequency and not the width of the Airy function. The latter
gives us a sort of Rayleigh criterion for the resolving power.

Table 2

Cut-off Frequencies and Widths of the Airy Function for a 1 m
Diameter Telescope.

λ (µm)	1	10	100	1000
w_c(arc sec)$^{-1}$	5	0.5	0.05	0.005
α(arc sec)	0.24	2.4	24	240

     As we have seen in the first lecture, for ground based work,
which is limited to the 1-30 µm spectral range, the resolution
quoted here, and a fortiori the theoretical resolution of a big-

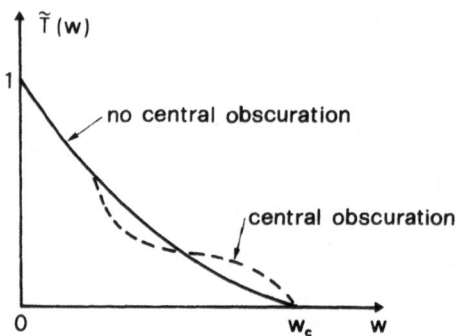

Fig. 19. MTF of a circular pupil.

ger telescope, is hardly attained, because the image is always blurred by the coherence effect to 1-4 arc sec, depending on the seeing. On the other hand, in space, at longer wavelengths, the seeing disappears, but the diffraction limit gets worse (24 arc sec at 100 μm to 4 arc min at 1 mm). To restate this, let us say that there is no simple way to go below a few arcsec resolution, just by using the standard pupil. Now, one may question whether diffraction is a real absolute limit. It is an absolute limit in the sense that this low-pass filter cuts off anything beyond w_c. But let us have a closer look at this problem.

If we invert equation (4.3), we see that in the Fourier plane (u,v) the object is given by the image divided by the transfer function:

$$\tilde{I}_o(\vec{w}) = \frac{\tilde{I}(\vec{w})}{\tilde{T}(\vec{w})} \quad . \tag{4.5}$$

So, if we know the transfer function, and in theory we do, then we can recompute the object from the image for any value of $\vec{w}$ which

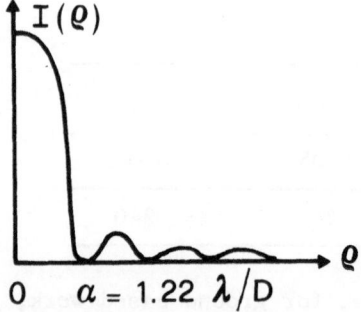

Fig. 20. The Airy function.

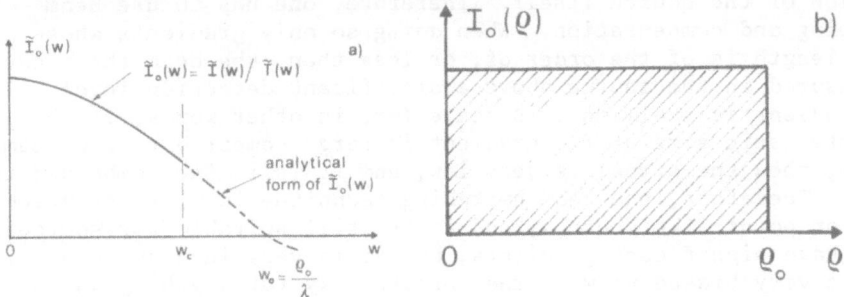

Fig. 21. a) The object transform versus spatial frequency, assuming there is no noise; b) The object as a uniform disk of radius ρ_o.

gives $\tilde{T}(\vec{w}) \neq 0$, that is to say for $|\vec{w}| < w_c$. Infact, it would be possible except for the noise. Because $\tilde{I}(\vec{w})$ is always determined with some noise, when $T(\vec{w})$ becomes very small, near w_c, then the ratio in eq. (4.5) becomes erratic. So, the smaller the noise, the closer one can approach w_c. Let us assume for a moment that there is no noise, so that we can restore $I_0(w)$ to w_c. This is shown as the solid line in Fig. 21. Now let us suppose (although we don't know it) that the object is just a uniform disk of radius ρ_o, as in Fig. 21 b). Then, if we could completely recover it up to an infinite frequency, $\tilde{I}_0(w)$ would appear like the Fourier transform of this particular object. In this case, it would be the Bessel function of the first kind divided by w. Therefore, with the assumption of uniform brightness, we can match the measured part of $I_0(w)$ with this analytical form and this enables us to determine the value w_0 corresponding to the first zero of $\tilde{I}_0(w)$.

This method is called super resolution and it is simply the idea that if we know something a priori about the source, then we know a priori the form of its spatial frequency spectrum. Unfortunately, even to gain a factor 2 or 3 in resolution, one needs very high signal-to-noise ratios (S/N > 100 ÷ 1000). So, though in theory one can do a lot of things, in practice this method is limited.
Another idea would be to use the fact that the object is always positive ($I_0(\rho) \geq 0$), which places additional constraints upon $I_0(w)$.

To conclude, the diffraction does not constitute an absolute limit, if a priori information on the object is known or can be acquired.

4.2 Mapping

In the first lecture we have seen that in most cases the thermal

emission from the atmosphere is always much more intense than the
emission of the source itself. Therefore, one has to use beam
switching and compensation. When doing so only gradients whose
scale length is of the order of, or less than, the beam throw can
be measured in the source above a significant detection level. If
the gradient is small on this scale (or, in other words, if the
characteristic size of the gradient is large compared to the beam
throw), then the signal is very low, and it is in fact embedded in
noise. Therefore, the beam switching technique is very satisfactory for point-like sources, and it is still suitable for sources
which have significant gradients, but it is very inadequate and it
gives a very biased view of the infrared sky for anything which
has very small gradients within the beam size. Usually, one works
with a beam throw of the order of the beam diameter. The beam diameter is taken so as to cover the central part of the Airy disk
($\phi \sim 2.5\lambda/D$). Therefore, gradients in the infrared sky which are
on the degree scale, or even larger, never show up. Besides, the
atmosphere itself also has the same type of large scale gradients,
so detection is further complicated. The consequence is that the
large scale diffuse emission of our Galaxy, or even of other galaxies or clusters, is rather ignored compared to studies of point
sources or clouds. Undoubtedly, this biased view will change in
the future and very important new information can be expected from
the study of the diffuse emission.

With beam switching in action, the normal way to do mapping
is to make raster scans like the one in Fig. 22. Actually, most
of the maps presented in the other lectures have been achieved by
this method. But this is a very poor method, because it does not
use photons efficiently. At a given instant, all the photons
which are not inside the beams are just thrown away. Fortunately,
there are other methods available which can deal with this problem. One method is to use somewhat of an equivalent to the photographic plate in the visible; namely, instead of one single detector, one-, or even better a two- dimensional array of detectors, each detector having the size of one pixel in the image.
More details will be given later as to the availability of such
arrays. In fact, they are difficult to get, they don't work at
all wavelengths and they are rather expensive. Another method has
been developed by several groups, e.g. the Berkeley group: It uses
up conversion of the photons, which is a transformation of every

beam throw

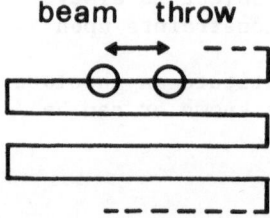

Fig. 22. Raster scanning is the common mapping method.

infrared photon to a visible photon which can be detected with
conventional detectors like photographic plate, television camera,
photo-counting tube and so on. If this system works, then it will
also deal with the problem of mapping efficiency.
The third method is multiplexing of the image. The principle is
the following: We superimpose on an image given by the instrument
a mask with transparent and opaque areas, each area being of the
size of one pixel. All the light issued from the image and encod-
ed by this mask is concentrated on one single detector. Then, the
energy received by the detector is the cumulative energy of light
in every pixel multiplied by the corresponding transmission of the
mask. If we define P lines and Q columns in the image with indices i,j for each pixel, and if we repeat the encoding with P·Q
different masks (labelled k,l), then we have the following set of
relations:

$$n_{k,l} = \sum_{i=1,P} \sum_{j=1,Q} a_{ij}(k,l) \cdot x_{ij} \qquad (4.6)$$

with x_{ij} = intensity in the (i,j) pixel of the image, $a_{ij}(k,l)$ =
transmission of mask (k,l) in the (i,j) pixel, $n_{k,l}$ = signal received on the detector when using the (k,l) encoding mask. This
set of equations can be represented by a tensorial relation

$$N = A \cdot X \qquad (4.7)$$

where X is the intensity distribution matrix of the image, A is
the encoding tensor, and N is the signal matrix. Now if one can
choose the set of masks such that the A tensor is inversible, one
can simply restore the intensity distribution of the image from
the set of P·Q successive measurements:

$$X = B \cdot N , \qquad (4.8)$$

where $B = A^{-1}$. Various sets of functions can be used in principle
to construct the inversible tensor A; a trivial one could simply
be the Fourier base of functions. So, one mask would be a sinu-
soidally transparent object with P lines, or Q columns, and a
given spatial frequency, another mask would have another spatial
frequency, and so on. But then, one needs a mask with a continuously variable transparency, which is difficult to realize. It is
much easier to make a binary mask, coded in either black or transparent (0 or 1 transmission). Fortunately, there exists a set of
orthogonal functions on this basis, which are called the pseudo-
noise functions (they won't be described here). Another interesting feature of these masks is that each mask in the series can
simply be deduced from the previous one merely by shifting one
line, instead of changing the whole mask.
What is the advantage of multiplexing? Let us suppose we have a
total observing time T. If we do a raster scan on the image, which

contains PQ = n^2 pixels (for simplicity let us take P = Q = n), we spend T/n^2 time per pixel. If we perform multiplexing half of the photons reach the detector at the same time. So, if we neglect the very small fraction of time during which we have to switch the masks, we spend T/2 time per element. This means that we have gained a factor $n^2/2$ in observing time. We can even gain a factor n^2, if the black areas of the masks are made reflecting, so that another detector can simultaneously receive the complementary part of the intensity reaching the first detector. As far as the signal-to-noise ratio is concerned, this method leads to an improvement by a factor $n/\sqrt{2}$ (or n). If n is large, this may constitute a very important advantage provided that, in the course of the operation, we have not increased the basic noise of the system. However, while on the one hand we receive the full area of the image on one detector, on the other hand we also increase the number of background photons coming from the telescope and the atmosphere. The fluctuation of these photons is also increased by the same factor n, so that the advantage can be completely destroyed.

The conclusion is that multiplex advantage holds only when one is system-noise limited and not background-noise, or image-noise limited. This is exactly the same situation that one has in Fourier spectroscopy and there it is known as the Felgett advantage of noise. This multiplex method has also fringe benefits, such as giving the opportunity of restoring the true shape of the image because there is no wriggling of scanning rasters.

4.3 Across the Seeing Limitation

One way or another, we have obtained the image, but this image is seeing limited. Is it possible to overcome this seeing limitation to reach the diffraction limit? It is really worthwhile to try to reach this limit, because at this point we know what we have to do to improve the resolution (larger instruments, superresolution).

Let us consider an ideal wavefront, a plane wave, and its distortion by the atmosphere. For distances of the order of ρ_0, the coherence length that we introduced in chapter 1, the distortion is such that the phase remains constant. This phase cell is of the order of a few centimeters. For an exposure time less than the correlation time τ_c, which characterizes the atmospheric turbulence, the pupil gives us an instantaneous image, in the sense that everything is as if the atmosphere were frozen. What does this image look like? It looks like a spot, more or less irregular, whose typical angular size is given by λ/ρ_0, as the diffraction figure given by a pupil of diameter ρ_0 (Fig. 23). But within the image, there are still high frequency features up to the cut-off frequency D/λ. If the wave were perfectly planar, the image would simply consist of the Airy figure of typical size λ/D. In the case of wave distortion, the spatial frequency transfer of the

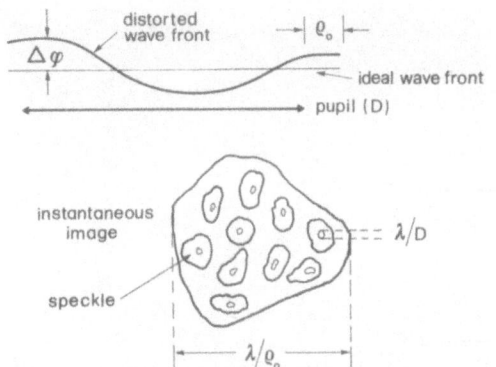

Fig. 23. Crossing the seeing limit.

complex pupil is essentially correct up to frequency ρ_0/λ, and it
is weaker beyond this value, although it is not completely zero up
to the cut-off frequency $w_c = D/\lambda$. So, there is still some information on the high frequency content of the image which can eventually be gathered. To do so, one can investigate two approaches:
the first one using the given fact of coherence over ρ_0 to achieve
interferometry, the second one restoring whatever information is
left over in the high frequency cells of the instantaneous image.
In the following analysis of the two methods, we shall deal with a
point source, or a Dirac peak, whose spatial frequency spectrum is
a constant. Because of linearity of convolution and because of
aplanatism, the results will apply to any distribution of intensity
$I(\vec{\rho})$.

First approach: Michelson Interferometry. This method, developed in the beginning of the century, has been recently revived
first in the visible and now in the infrared. Here, one uses the
fact that the wavefront is not distorted on a scale length ρ_0. So,
the pupil is composed of two holes of diameter ρ_0 separated by a
distance d (Fig. 24 a). Between these apertures, there is a phase
difference, $\Delta\phi$, which will undoubtedly vary with time (Fig. 24 b).
In real cases, we have already seen that one wavefront can be til-

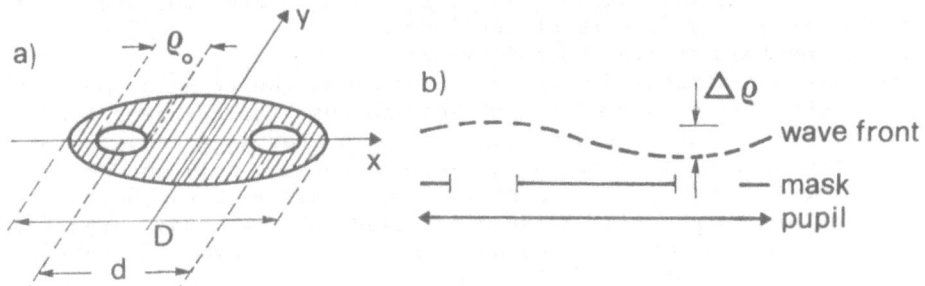

Fig. 24. Michelson Interferometry.

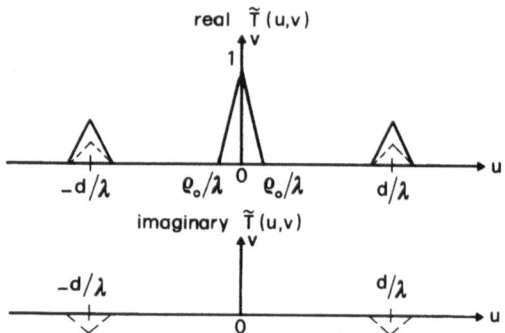

Fig. 25.   Modulation transfer function of the pupil in the Michelson case.

ted with respect to the other, so that there is no overlap of the
two images in the focal plane.   But, let us assume that there is
actually overlapping, the condition being that the two wavefronts
remain parallel although they are affected by some phase difference.
What is the modulation transfer function of this pupil?   It is giv-
en by the autocorrelation product of the pupil function.   It is a
very simple computation and the result is given in Fig. 25.   The
solid lines represent the MTF in the ideal case of a plane wave,
the dashed lines showing the real case with phase distortion.   For
the latter, the non-zero frequency features of the real part are
reduced by half, giving rise to an antisymmetric imaginary part.
This simply reflects the phase displacement $\Delta\phi$ between the two apertures of the complex pupil, that will introduce a corresponding
displacement of the fringes in the image plane.   Then, the transmission of the pupil is 1 for the D.C. part of the object (zero
frequency) and is 1/2 for the A.C. part (non-zero frequency) around
$u = d/\lambda$.   Here, the MTF is no longer a low-pass filter, but a bandpass filter with two bands of width $2\rho_0/\lambda$, centered at $\pm d/\lambda$.   If
one varies d from the minimum value to the maximum value allowed by
the telescope size, and changes the orientation of the holes, one
can then explore the Fourier plane, as shown in Fig. 26, and determine the frequency content of the image.
So, the conclusions are the following:
a) Despite the effects of seeing, which moves the fringes, one can
restore the true values of the object, because the transfer function is 1/2 at every frequency up to w_c.   b) For a given value of d
and a given orientation of the holes, one obtains only one point
in the (u,v) plane.   c) Obviously, one does not make the best possible use of the photons, because one uses only two small areas of
the telescope.   So most of the photons are thrown away in the ratio
$N = (D/\rho_0)^2$ (typically 10^2-10^3).   d) Finally, the fringes may or
may not be constant, depending on the seeing; they move all the
time with a time dependent seeing.

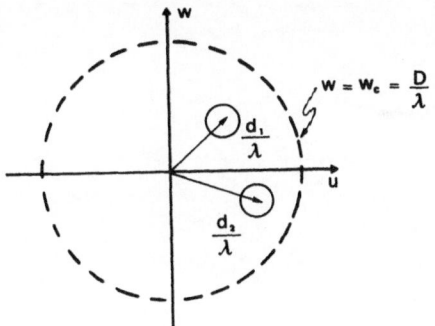

Fig. 26.   Exploration of the Fourier plane of the image.

The method of using Michelson interferometry in the infrared (5–12 μm) and some results obtained on the sizes of a number of stars are illustrated in Figs. 27 to 30.

Second approach: Speckle interferometry. The instantaneous image, represented in Fig. 23, shows small cells within the large spot produced by turbulence effects. The cells which contain the high frequency information in the range $w_0 = \rho_0/\lambda < w < w_c = D/\lambda$ are called "speckles". This means that the MTF of the complete pupil is non-zero in this domain, but we know that this function is highly time dependent. With exposure times greater than τ_c, the image is totally blurred. A simple derivation of $\tilde{T}_t(\vec{w})$ leads to

$$\tilde{T}_t(\vec{w}) = \tilde{T}_0(\vec{w})^{1/2} e^{-\delta\phi_t(\vec{w})} N^{-1/2} \qquad (4.9)$$

where $N = (D/\rho_0)^2$ is an attenuating factor proportional to the number of speckles, $\phi_t(\vec{w})$ is the random phase of the complex pupil, and $\tilde{T}_0(\vec{w})$ is the unperturbed MTF. The poorer the seeing, the smaller ρ_0, the larger the N, and the poorer the transmission of higher frequencies. If we do a simple average of $\tilde{T}_t(\vec{w})$ over time, because of random phase $\phi_t(\vec{w})$, the result will be zero; $< \tilde{T}_t(\vec{w}) >_t = 0$. This simply reflects the loss of information when making long exposure images. But, if we take the square modulus of $\tilde{T}_t(\vec{w})$, the phase term disappears and the average over time becomes

$$< |\tilde{T}_t(\vec{w})|^2 >_t = |\tilde{T}_0(\vec{w})| \; N^{-1} \qquad (4.10)$$

which is the unperturbed MTF attenuated by $1/N$ for every frequency, at least between w_0 and w_c (the formula above is only valid in this range). See also Fig. 31.

So, with a point-like source, one can restore the true profile of the MTF just by scaling by an arbitrary factor the curve measured between w_0 and w_c. Fig. 32 shows such a restoration made at 2.2 μm on Arcturus, with the 4m telescope at Kitt Peak. The restoration of $\tilde{I}_0(w)$ obeys the following equation, which is derived from

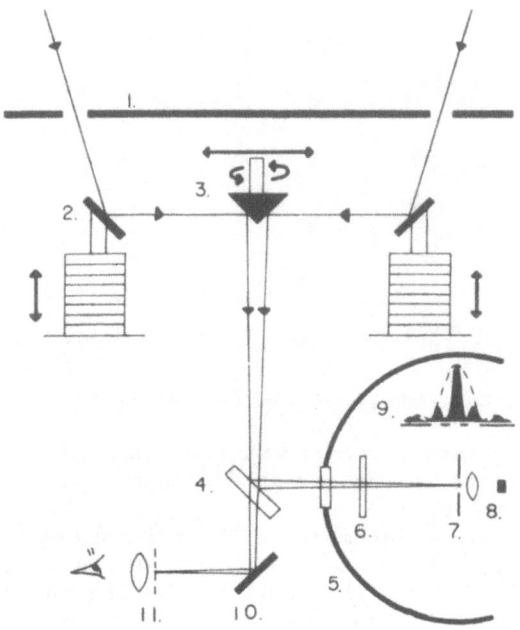

Fig. 27. Schematic diagram of the Michelson instrument. (1) mask; (2) plane mirror on piezoelectric stack; (3) roof mirror (4) dichroic (5) 77 K dewar; (6) filters; (7) slotted mask; (8) field lens and InSb detector; (9) interference fringes on slotted mask; (10) plane mirror; (11) visual focus, reticle, and eyepiece. (From McCarthy and Low, 1975).

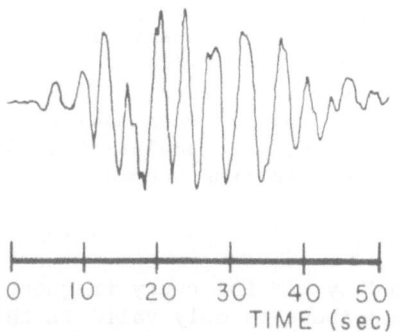

Fig. 28. 5 μm interferogram of α-BOO obtained on 155 cm telescope during a slow scan (∿ 0".25 s^{-1}) in right ascension with the fringe chopper operating and the two apertures aligned in the east–west direction. Visual seeing was 1". Aperture diameter and separation were 20 and 102 cm. (From McCarthy and Low, 1975).

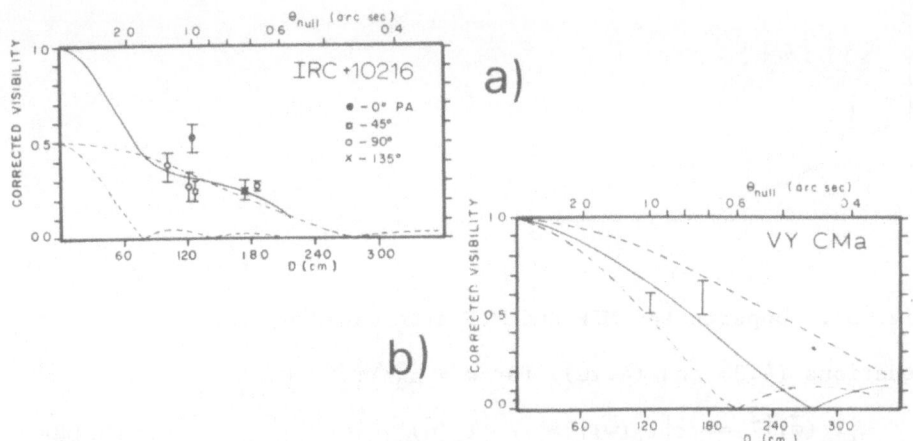

Fig. 29. Fringe visibility results as a function of aperture separation (D). a) IRC+10216. Symbols refer to different position
angles of the line joining the centers of the two apertures. The
solid curve represents a theoretical two-component model in which
the individual components are represented by the dashed curves and
have angular diameters (θ_{null} = 1.22λ/D) of 0".44 and 1".6. b) VY
CMa. The solid and dashed curves are theoretical, single-source
fits to the data and correspond to an angular diameter of 0".45 ±
0".15.

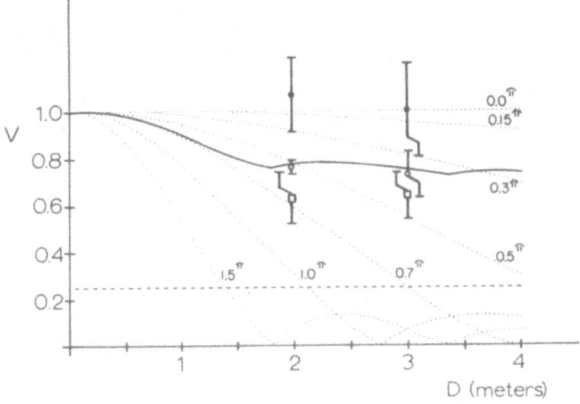

Fig. 30. Visibility V as a function of baseline D, for αOri at
three wavelengths: 8.3 μm (filled circles), 10.2 μm (open squares),
11.1 μm (open circles). Dotted lines are visibility functions at
11.1 μm for various diameter uniform circular disks as indicated.
The dashed line represents the flux emitted by the 3500 K stellar
core of diameter 0".05. The solid line is a possible fit to the
11.1 μm data indicating that most of the flux is emitted by a shell
≳ 1".5 with the remaining 75% unresolved. (From McCarthy et al.,
1977).

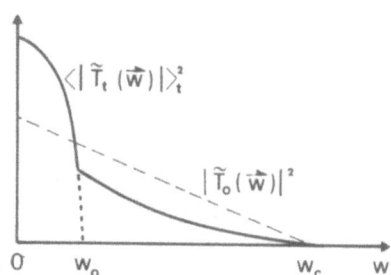

Fig. 31.   Unperturbed MTF and MTF attenuated by $1/N$.

equations (4.3) and (4.10), for $w > \rho_0/\lambda$:

$$|\tilde{I}_o(\vec{w})|^2 = N<|\tilde{I}_t(\vec{w})|^2>_t \,/\, |\tilde{T}_o(\vec{w})| \qquad (4.11)$$

Knowing $|\tilde{T}_o(\vec{w})|$, one can restore the object $|\tilde{I}_o(\vec{w})|$ by scaling the measured $<|\tilde{I}_t(\vec{w})|^2>_t$ up to $w = w_c$.

The comparison of the two methods is given in Table 3. Putting together all these factors, one could build up a merit factor, which would be the equivalent for spatial multiplexing of the merit factor defined in spectroscopy.

For the sake of completeness I shall just mention a third

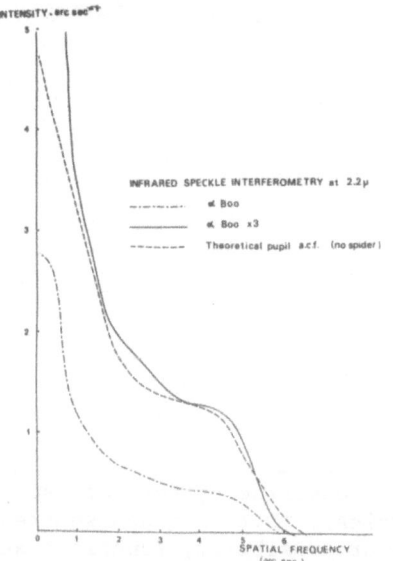

Fig. 32.   Infrared Speckle interferometry at 2.2 µm on Arcturus. The theoretical pupil autocorrelation function of the KPNO 4 m telescope has been drawn and vertically scaled for a best fit (unpublished results from the author).

Table 3

Michelson Interferometry	Speckle Interferometry
Divide up the observing time in $N^{1/2}$ to obtain the spectrum.	All spectrum obtained.
Attenuation is 1/2 and is known.	Attenuation is $1/(\tilde{T}_0(\vec{w}))^{1/2}$.
Only a fraction $(\rho_0/D)^2$ of the photons are used.	All photons are used.
Background photons, if generating noise, can be reduced with cold masks.	Background radiation is fixed and corresponds to full aperture.
Fringes are difficult to track.	Works with any seeing.

method. It consists of measuring angular sizes of sources which are occulted by the moon. But in this case, the telescope only takes part as a collecting surface and no longer as a filtering pupil, the resolution being set by the diffraction on the lunar edge. Here, the limitation would be in the signal-to-noise ratio, because the sweep produced by the apparent movement of the moon is fast (880 m/s), and because of the thermal emission from the moon's edge and its fluctuation due to improper tracking.

To conclude with imaging processes, let me repeat that we have seen that either in space with existing telescopes, or on the ground, with these methods, one can reach the diffraction limit of a given aperture. To go beyond that limit, one really needs to build either larger telescopes, or do aperture synthesis. Instead of putting a mask on the pupil, one can use two separate telescopes and then recombine the two beams in order to obtain the fringes. This is the classical method used in radioastronomy. It has difficulties in the infrared, because the wavelength is much smaller and the stability of the baseline has to be accordingly better. Also, the seeing which is worse in the infrared than in the radio range makes the enterprise more difficult. But it has been tried, and in some cases it has been successful. At least, in space, where the environment can probably be controlled much better and where there is no seeing effect, it could become a very useful tool.

Let us conclude by taking a brief look at what telescopes are available for infrared observations. The different families of infrared instruments are summarized in Table 4.

Acknowledgements. These lecture notes were kindly prepared by Marie de Muizon and Regis Courtin.

Table 4

Different Classes of Telescopes Available for Observations in the
Infrared

Type	Diameter	Examples
Ground-based	1-2 m	several 60 inches instruments Pic du Midi telescope (2 m)
	3.6-6 m	ESO/Chile, CFHT/Mauna Kea (3.6 m) UKIRT/Mauna Kea (3.8 m) Mayall telescope/Kitt Peak (4 m) Hale telescope/Mt. Palomar (5 m) Zelentchouk telescope/USSR (6 m) Multimirror telescope/Arizona (4.5 m)
Airborne	32 cm	IRS-Meudon telescope on Caravelle
	91 cm	G. Kuiper Observatory on C141-NASA ARC
Balloon-borne	60 cm	University College London
	1 m	Harvard-Smithsonian/University of Arizona
Space Instruments	60 cm	IRAS (1981)
	2.4 m	Space Telescope (> 1983)

REFERENCES

Atmosphere

Baluteau, J.P., Marten, A., Bussoletti, E., Anderegg, M., Beckman,
J.E., Moorwood, A.F.M., Coron, N., 1977, Infr. Phys. 17, 283.
Blanco, A., Bussoletti, E., Melchiorri, B., Melchiorri, F., Natale,
V., 1976, Infr. Phys. 16, 569.
Marten, A., IRS Internal Report, Meudon, 1977, "Determination de
la Transmission atmospherique à partir de l'e'mim'on".
Richards, P.L., Woody, D.P., Mather, J.C., Nishioka, N.S., 1975,
Phys. Rev. Letters 34, 1036.
Righini, G., and Simon, M., 1976, Infr. Phys. 16, 543.
Simpson, J.P., "Infrared Emission From the Atmosphere Above 200
Km", NASA TN D-8138.
Sollner, G., 1977, Astron. and Astrophys. 55, 361.
Tam, W., Corriveau, R., 1976, Infr. Phys. 16, 129.
Wijnbergen, J., Lena, P., Celnikier, L., 1977, Infr. Phys., in
press.
Wolfe, W.L., ed., 1965, Handbook of Military Infrared Technology,

(U.S. Government Printing Office).

Detectors

Coron, N., 1976, Infr. Phys. 16, 411.
Harvey, P., Coherent Light, (J. Wiley and Sons, London and New York, 1970).
Hudson, R.D. and Hudson, J., editors, Infrared Detectors, (J. Wiley and Sons, London and New York, 1975).
Scorso, J.B., Thom, R. and Philips, J., "Focal Plane Mosaic Technology Using Intrinsic Detector Materials", NASA-JPL Symposium on CCD Technology, 1976.
Steckl, A.J., 1976, Infr. Phys. 16, 65.

Images

Aime, C., 1977, Thèse de Doctorat, Université de Nice.
Born, M. and Wolf, E., Principles of Optics, 4th. ed., (Pergamon Press, London and New York, 1970).
de Batz, B., Bensammas, S., Delavaud, J., Gay, J. and Journet, A., 1977, Infr. Phys. 17, 305.
McCarthy, D.W. and Low, F.J., 1975, Astrophys. J. 202, L37.
McCarthy, D.W., Low, F.J. and Howell, R., 1977, Astrophys. J. 214, L85.
Progress in Optics, Vol. XVI, 1977.

INFRARED ASTRONOMICAL SPECTROSCOPY

Martin Harwit

Cornell University, Ithaca, New York, U.S.A. and
MPI für Radioastronomie, Bonn, West Germany

1. OBSERVATIONAL METHODS

Spectroscopic instruments are designed to divide radiation into
its component wavelengths (or frequencies) and to provide the
observer with a display of this information at high signal-to-noise
ratio (SNR).

This simplest kind of spectral work involves the use of a
succession of relatively broadband filters to observe an astronomical source. This kind of observation is often termed spectrophotometry. In general the resolving power is very low; $R \equiv \lambda/\delta\lambda$,
the wavelength of the observation divided by the resolved wavelength
band, normally has a value of 10 or less in spectrophotometry.
Much of the earliest infrared spectrophotometric work was done by
H.L. Johnson and his coworkers in the early 1960's. In addition
to the three normal filters of visual astronomy, U, B, and V, they
used five additional filters to conduct observations out to 4 μ
wavelengths. The spectral resolution in these observations was
roughly 3 to 5.

In the late 1960's spectrophotometric work was extended out
to about 20 microns. The University of Minnesota group in particular studied a wide variety of different cool stars by these means,
and Woolf and Ney discovered excess radiation in the 10 micron
region of the spectra of cool stars. This is the feature now
generally attributed to silicate components in interstellar and
circumstellar dust.

Spectroscopy is the term generally used for work at higher
resolution than spectrophotometry provides. Usually the resolving
power in spectroscopy is $R \geq 10$. A variety of different instruments
can be used for this purpose, and these will first be briefly described before their relative merits are discussed.

G. Setti and G. G. Fazio (eds.), Infrared Astronomy, 271-283.

1.1 Fabry-Perot Interferometers and Interference Filters

In Fig. 1 we present a schematic diagram of a Fabry-Perot interferometer. It consists of two finely ground plane transparent
plates separated by a gap D. The faces bounding the gap are reflection coated to transmit a small fraction of the radiation, $\underline{t}$,
to absorb as small a fraction of the radiation as possible, $\underline{a}$, and
to reflect a comparably large fraction of the radiation, $\underline{r}$: $\underline{r}$ = 1 -
- $(\underline{t+a})$.

    When radiation is incident on the plates at an angle θ con-
structive interference between multiply reflected rays occurs provided the path lengths in the gap between the plates differ by an
integral number of wavelengths. If the refractive index of the
medium in the gap between the mirrored surfaces is $\underline{n}$, this condition occurs whenever the wavelength λ_m has a value satisfying

$$2 \, Dn \cos \theta = m \, \lambda_m \qquad , \quad m = 1, 2, \ldots \qquad (1.1)$$

Here m is the order of the transmitted spectrum - the number of
wavelengths by which successive interfering rays are out of step.
Since the relation (1.1) is obeyed by a whole series of different
wavelengths λ, it is customary to use two or more Fabry-Perot interferometers in series. They are arranged so that all of them transmit one of the wavelengths, say λ_j, but that all other wavelengths

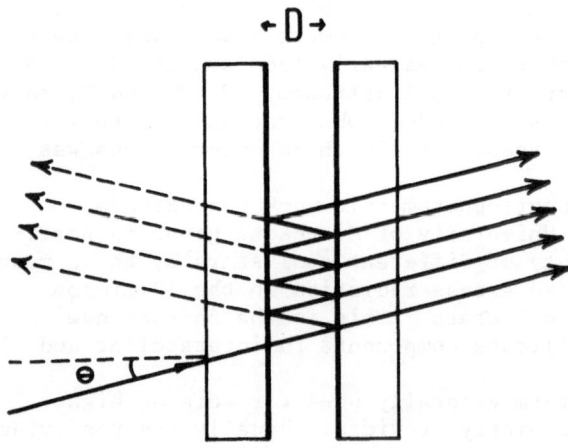

Fig. 1. Fabry-Perot Interferometer.

transmitted by one set of plates interfere negatively and therefore are rejected by other plate pairs. The wavelength transmitted
by a Fabry-Perot interferometer, or by a series of such devices
can be varied either by changing the refractive index of the medium
between plates - generally done by varying the gas pressure in the
gap - or by changing the plate separation D, or else by changing
the angle of incidence θ.

An interference filter essentially is a non-variable Fabry-
Perot interferometer or sequence of such interferometers, often
manufactured by depositing films of different thickness and indices
of refraction onto a single substrate.

A wedge filter - often called a circular variable filter -
is an interference filter in which the thickness of the deposited
layers varies across the face of the filter in a regulated way.
D therefore varies from one position on the filter to the next,
and λ_m therefore varies in proportion. Such filters tend to have
a resolving power of ∿70, while Fabry-Perot interferometers can
have resolving powers of the order of several hundred thousand.
For the determination of strong absorption features, such as the
3 μ feature that Cohen observed in the Rosette Nebula - a feature
attributed to interstellar ice grains - a wedge filter provides
adequate resolving power. Similarly the Ne^+ 12.8 μ emission line
seen in a variety of planetary nebulae can often be detected with
such a filter.

For higher resolution work a wedge filter often does not suf-
fice. A Fabry-Perot interferometer, however, can still provide
adequate resolving power in many such cases. One particularly clear
demonstration of the usefulness of these instruments is the detection of the isotope ^{17}O in a carbon monoxide absorption line
of the star IRC + 10216.

1.2 Grating Spectrometer

The optical configuration of a grating spectrometer is schematically
shown in Fig. 2. The wavelengths for which rays diffracted off
successive adjacent grating facets constructively interfere are
given by the grating equation

$$\sin \alpha + \sin \beta = m \lambda_m/a \quad , \quad m = 1, 2, \ldots \ldots \tag{1.2}$$

Here a denotes the distance between adjacent grating rulings, and
as before, m refers to the order of the spectrum. The angles of
incidence and diffraction, respectively α and β are shown in the
figure. Radiation of different orders usually is subtracted out
by means of a blocking filter, so that only a given wavelength
λ_j in the jth order is passed.

Aitken and Jones have used grating instruments extensively
in the 10 μ region, while the Cornell University group has used a
variety of liquid helium cooled grating spectrometers at wavelengths between 16 and 120 μ in observations carried out from

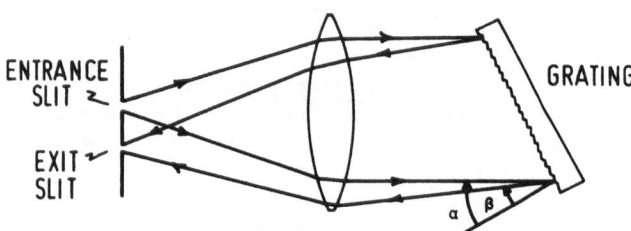

COLLIMATING
AND
FOCUSING
OPTICS

ENTRANCE
SLIT

EXIT
SLIT

GRATING

α β

Fig. 2. Grating Spectrometer.

aircraft. Grating spectrometers can be constructed to have resolving powers as high as several hundred thousand.

1.3 Michelson Interferometers

A Michelson interferometric spectrometer is shown schematically in
Fig. 3. As the movable mirror translates along the direction shown,
the path lengths in the two arms of the instrument can be made to
differ by increasing amounts. Each time the separation in path
lengths becomes an integral number of wavelengths λ, constructive
interference occurs for that wavelength. This type of instrument
can be made to have resolving powers about an order of magnitude
larger than Fabry-Perot or grating instruments. The initial discovery of molecular hydrogen in the Kleinmann-Low Nebula of the
Orion region was made by means of such a spectrometer.

1.4 Resolving Power and Energy Throughput

In order to provide a spectrum having the highest signal-to-noise
ratio a spectrometer often has to have a high energy throughput,
particularly if the source that is viewed is extended. The
throughput T is the product of the area A of the instrument's
limiting aperture, and the solid angle Ω over which radiation can
be accepted:

$$T = A\Omega \tag{1.3}$$

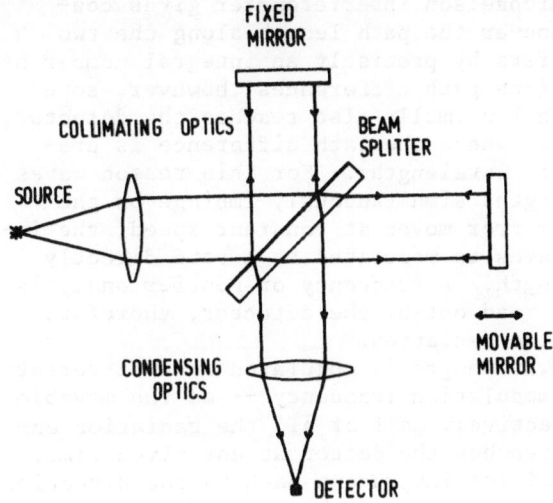

Fig. 3. Michelson Interferometer.

For a Michelson interferometer or a Fabry-Perot instrument, the diffraction limited resolving power is given by

$$R = 2\pi/\Omega \tag{1.4}$$

and we see that resolving power can be increased only at a loss of throughput. A useful figure of merit therefore is the product of resolving power and throughput. For an interferometer this is

$$M = RA\Omega = 2\pi A \tag{1.5}$$

For a grating instrument the limiting aperture can be taken to be the grating area, and the solid angle of acceptance then becomes the angle subtended by the slits as seen from the grating. That angle, in the diffraction limit, can have a value $\sim 1/R$ along the direction of the slit width and approximately $(8/R)^{1/2}$ along the length of the slit. The figure of merit M then becomes

$$M = A\sqrt{8/R} \tag{1.6}$$

With multislit grating instruments, this figure of merit can be increased to a value similar to that of interferometers. However, in infrared astronomy few sources have proved to be extended enough to strain instrumental throughput.

1.5 Multiplexing

We had noted that the Michaelson interferometer gives constructive interference whenever the path length along the two
arms of the instrument differs by precisely an integral number of
wavelengths. For intermediate path differences, however, some
radiation at any wavelength λ normally also reaches the detector,
except for those differences where the path difference is precisely an odd number of half wavelengths. For this reason waves
at widely differing wavelengths simultaneously impinge on the
detector. If the movable mirror moves at constant speed, the intensity of the different waves is modulated at a rate directly
proportional to the wavelength. A frequency or Fourier analysis
of the signal continuously read out by the detector, therefore
can provide a spectrum of the radiation.

In this technique each wavelength is modulated in a different
fashion -- at a different modulation frequency -- as the movable
mirror translates, and effectively half of all the radiation entering the interferometer reaches the dector at any given time.
The other half passes out of the instrument back in the direction
of the source. If the limiting noise of the observation is detector noise, a SNR advantage is provided by the multiplex technique, since any given wavelength can be viewed by the detector
up to N/2 times longer than it would if each wavelength were
scanned in succession, as in the simplest types of grating instruments or Fabry-Perot interferometers. SNR improvements
quite generally increase as the square root of the time available
for the observation.

Length of time, however, is not the only factor that has to
be considered in judging the SNR advantage to be gained through
multiplexing. Some multiplexing schemes are very efficient while
others are poor. Tai and Harwit have compared the SNR improvements obtained by different classes of instruments all of which
are capable of receiving N different wavelengths at one time.
For a Michaelson interferometer the SNR improvement is $(N/8)^{1/2}$
while for a multislit grating instrument it can be $(N/4)^{1/2}$ if a
particular arrangement of open and closed slits is used. In this
mode the instrument is said to be a Hadamard transform instrument.
The Hadamard transform is a digital transform involving digits
1 and 0, and data reduction in this type of instrument is achieved
in a manner analogous to that of the Fourier transform (Michaelson
interferometer) instruments.

1.6 Heterodyne Spectrometry

In heterodyne spectrometry an astromical signal is beaten
against a standard wavelength produced in a laboratory laser.
Since the laser radiation has great spectral purity an extremely
high spectral resolving power can be attained; in practice this
has given resolving powers of $R{\sim}10^7$ in observations of wing velocities in the upper atmosphere of Venus, at 10μ.

At the moment such instruments can only be operated in the

immediate vicinity of strong laser and maser lines, and the spectral coverage is therefore very narrow, and limited to quite specific regions. With the expected introduction of tunable lasers
of sufficiently high power during the coming years this technique
should, however, provide infrared astronomical spectroscopy with
an increasingly important tool.

2.   INFRARED SPECTRA

Electrons, Ions, Atoms and Molecules are able to radiate energy
by a variety of different mechanisms. Free electrons and ions
radiate when they are accelerated for example in collisions with
other charged particles -- when they emit free-free or bremsstrahlung radiation. They also emit at high energies in a
strong magnectic field, where their spiral motions produces synchrotron radiation. Both of these radiation mechanisms produce
a continuum spectrum. Synchrotron emission however, is not a
dominant source of radiation in the infrared part of the spectrum,
as far as we can determine. Among sources observed thus far it
may dominate the spectrum only of the Crab Nebula pulsar. Free-
free emission on the other hand can be important in the infrared
spectra of hot stars, compact H II regions and envelopes of X-
ray stars.

Another mechanism of continuum radiation in atomic systems is
the free-bound process, in which a free electron recombines with
an atom or ion. The upper energy state of such an atomic system
lies in the continuum, while the lower state has a fixed energy.
Much of the visible and near infrared radiation we receive from
the sun and other cool main sequence stars has been emitted in a
free-bound transition of an electron about to become attached to
a hydrogen atom to produce an H^- ion.

Atoms, ions and molecules can also give rise to spectral line
emission by a variety of physical processes. The least energetic
of these transitions are the hyperfine transitions which involve
the flip of an electron spin with respect to the direction of
the nuclear spin orientation. Such transitions occur primarily
in the radio spectrum, since the transition energies are so low.
The 21 cm hydrogen line observed extensively in radio astronomy
is such a hyperfine transition.

Somewhat more energetic than the hyperfine transition are the
rotational transitions of small molecules. Simple diatomic molecules, for example, can be characterized by a rotational quantum
number J which reflects a rotational angular momentum hJ.
The corresponding rotational energy of the molecule is

$$\varepsilon_{rot.} = \hbar^2 J(J + 1)/2I \qquad (2.1)$$

where $\hbar$ denotes Planck's constant h divided by 2π, and I is the
molecule's moment of inertia. If rotational transitions

$J \longrightarrow J - 1$ are allowed, the transition energy simply becomes the energy difference between these two rotational states

$$\delta\varepsilon_{rot.} = \hbar^2 J / I \tag{2.2}$$

Transitions of this type are permitted by quantum mechanical selection rules, when the molecule has a dipole moment. If it only has a quadrupole moment - like H_2, the most abundant cosmic molecule - then transitions $J \longrightarrow J - 1$ are not permitted and the molecule can only undergo an angular momentum change $J \longrightarrow J - 2$. Moreover, the transition probability also is some eight orders of magnitude lower in this case.

We can estimate $\delta\varepsilon_{rot.}$ rather easily for a diatomic molecule if we remember that typical interatomic distances are of the order of 10^{-8} cm, and that a hydrogen atom has a mass of 1.6×10^{-24} g. The moment of inertia of the molecule therefore becomes roughly 10^{-40} m cm^{2}g, where m is the mass of the lighter atom, measured in atomic mass units. h^2/I is therefore $\sim 10^{-14}$ m^{-1} ergs. Since the transition frequency is given by the relation

$$\delta\varepsilon_{rot.} / h = \nu_{rot.} = \hbar J / (2\pi I) \tag{2.3}$$

we see that typical rotational transitions for low values of the quantum number J will occur in the frequency range

$$\nu_{rot.} \sim 10^{12} J/m \sim 10^{12}/m \quad Hz \quad . \tag{2.4}$$

For carbon, m = 12, and we therefore find the rotational transition for interstellar CO molecules in the millimeter wevelength range: $\lambda = c/\nu$ and 10^{11} Hz corresponds to a wavelength of 3 mm. Similarly hydrogen-containing diatomic molecules-may be expected to have rotational transitions at frequencies some ten times higher, in the far infrared, at wavelengths around 100μ.

Fine structure transitions tend to have energies roughly comparable to rotational transitions. For hydrogen-like ions, ordinary electronic transitions which we will discuss below have energies that are of the order of $(Z^2\alpha^2)^{-1}$ larger than fine structure transitions, where Z is the charge of the atomic core and α is the fine structure constant, $\alpha \sim 1/137$. The fine structure transitions involve only a change in the orbital angular momentum of the ion or atom. Because of the very small mass of the electron which makes the bulk of the contribution to the orbital angular momentum, the usual angular momentum change $\hbar$ requires only a very small energy adjustment of the atom, and typical transition energies are of the order of $mc^2 Z^4 \alpha^4 / n^4 \sim 10^{-14} Z^4 / n^4$. For $Z/n \sim 1$ this implies transition energies of the order of 10^{-14} erg, or transition wavelengths $\lambda \sim (hc/\delta\varepsilon)$ in the 10^{-2} cm or 100μ range. The first discovered far infrared fine structure transition observed astronomically is a doubly ionized oxygen fine structure line at about 88μ.

Molecules can also undergo vibrational transitions. Typically these are found in the near infrared spectral range. The inter-atomic distance in molecules tend to be somewhat larger than the sizes of typical atoms. Since molecules, just like atoms, are kept intact largely by coulomb forces, this larger size implies that the molecules dissociate rather more readily than do atoms. As a consequence, for example, we do not expect to find molecules in interstellar space where ultraviolet photons from stars or H II regions could lead to dissociation. For molecular hydrogen H_2, this energy amounts to 4.5 eV; while the ionization energy for a hydrogen atom is 13.6 eV. During vibrational transi-tions the interatomic distance also changes appreciably less in molecules than does the atomic radius in electronic transitions. This leads to vibrational transition energies that typically are measured in tenths of electron volts, while electronic transitions in atoms or molecules involve several electron volts when only the lowest energy levels come into play.[†]

In summary then, electronic transitions occur largely in the visible part of the spectrum which corresponds to transition energies of the order of electron volts. Vibrational transitions generally are found in the wavelength domain beyond 1 micron. Fine structure and rotational transitions populate the region beyond about 10 microns, and can reach out into the microwave range.

The processes we have described here combine to yield the spec-tra we observe in stars, ionized hydrogen (H II) regions, or cool interstellar clouds. Spectral lines are not generally sharp, but involve a finite transition probability even at frequencies quite far removed from the line center. As a result, radiation passing through a thick slab of material may be absorbed in the wings of spectral lines, and the slab appears opaque. Such a slab, when fully opaque will exibit a blackbody emission spectrum if it has a well defined temperature throughout.

Interstellar dust grains, in this sense exhibit properties part way between molecular lines and bulk material. Their emission spectra approximate blackbody characteristics, but the small size of the grains makes them inefficient radiators at long wavelengths in the infrared, and in addition, the material properties of the grains lead to some broad characteristic spectral features. Interstellar dust grains, as well as dust seen in comet tails for example have a broad spectral feature around 10 microns. This may be providing us with evidence that at least some cosmic grains consist of silicates. Laboratory studies show that most silicates have a broad spectral feature at 10 microns: For hot emitting material such a feature should appear in emission, while for cold absorbing matter it would produce an absorption

---

† Electronic transition energies typically are of order $mc^2 Z^2 \alpha^2 / n^2$ which for $Z/n \sim 1$ corresponds to tens of electron volts for low lying states.

band.   In interstellar clouds different regions do exibit this
'silicate feature' absorption (or emission). When both absorption and emission take place in the same extended cosmic cloud,
the interpretation of observations becomes very difficult.   In
comets the feature is found only in emission. An interstellar
feature at 3μ is believed to be providing evidence for H_2O ice.

For furthur study, the reader is referred to the following
reference works, some of which are quite general, others more
specifically oriented examples representative of current infrared
astronomical work.

Acknowledgements

The author is grateful to the Alexander von Humboldt Stiftung
of the Federal Republic of Germany for a travel grant to the
NATO Advanced Study Institute on Infrared Astronomy, and in particular for a U.S. Senior Scientist Award that enabled him to
spend a year at the Max-Planck-Institute for Radioastronomy in
Bonn. He thanks Prof. Peter Mezger, director of the Institute
for Radioastronomy, for his hospitality during that year.

REFERENCES TO CHAPTER 1

General References

James, J.F., and Sternberg, R.S., 1969, The Design of Optical
Spectrometers, (London: Chapman and Hall)
A slim volume that emphasizes the most important theorectical
and practical considerations in the design of spectrometers.
Mertz, L., 1965, Transformations in Optics, (New York: John Wiley
and Sons)
This book is imaginative, brief, enjoyable and full of insights
Stewart, J.E., 1970, Infrared Spectroscopy, (New York: Marcel
Dekker)
This is an excellent comprehensive book on infrared spectral
techniques.
Jacquinot, J., 1954, Optical Soc. Amer. 44, 761.
This classical paper provided the first systematic comparison of
the energy throughput of different classes of spectrometers.
Nelson, E.D., and Fredman, J;, 1971, J. Optical Soc. Amer. 60.
1664. Another classical paper which derives the optimum multiplex
advantage permissible in spectrometry.
Tai, M.H., and Harwit, Martin., 1976, Appl. Opt. 15, 2664
Treffers, R., 1977, Appl. Opt., to be published
The last two papers deal with the coding efficiencies of different
classes of multiplexing spectronometers.

Specific References

Spectrophotometry
Woolf, N.J., and Ney, E.P., 1969, Ap. J. 155, L 181
This article first noted the 10μ excess found in the spectra
of many cool stars and attributed it to emission by silicate
grains.
Ney, E.P., 1974, Ap. J., 189, L 141
Here the author obtained spectra of different parts of comet
Kohoutek near perihelion. One of the interesting features of
cometary spectra is the close resemblance of dust emission fea-
tures seen in comet dust tails to those observed in circumstellar
dust shells.
Wedge Filters
Cohen, Martin., 1976, Ap. J., 203, 169
This article decribes the observation of an ice grain absorption
feature at 3μ in the spectrum of the Rosette Nebula.
Gillett, F.C., Forrest, W.J., Merrill, K.M., Capps, R.W., and
Soifer, B.T., 1975, Ap. J., 200, 609
In this paper spectra of a variety of compact H II regions are
compared. A broad absorption feature centered at 9.7μ is attri-
buted to silicate dust.
Fabry-Perot Interferometer
Rank, D.M., Geballe, T.R., and Wollman, E.R., 1974, Ap. J. 187,
L111
These authors detected the isotope ^{17}O in a CO absorption line
seen at 4.67μ in the atmosphere of the carbon star IRC +10216
Grating Spectrometers
Aitken, D., and Jones, B., 1975, MNRAS 172, 141
An 8 - 13 μ spectrum of η Car is one of the brighest sources in
the sky, possibly a nova or oddly behaved supernova that reached
high visual brightness in the middle of the nineteenth century.
Decker, J.A.Jr., and Harwit, M., 1968, Appl. Opt. 8, 2252
This discribes the operation of a Hadamard transform spectro-
meter.
Harwit, M., and Decker Jr., J.A., 1974, Progress in Optics XII,
(Amsterdam: North Holland) ed. E. Wolf, p. 102
This is a survey of the capabilities of multislit grating spec-
trometers.
Ward, D.B., Dennison, B., Gull, G.E., and Harwit, M., 1976, Ap. J.
205, L75
This paper provides a composite spectrum of the Orion Nebula
from 16 to 120μ.
Interferometric Spectrometry -- Michelson Interferometers
Gautier, T.N. III, Fink, U., Treffers, R.R., and Larson, H.P.,
1976, Ap.J. 207, L 129.
This paper describes the discovery of vibration-rotational infra-
red transitions of molecular hydrogen in the 2μ spectral range.
The existence of these lines had been predicted some decades earlier but had never been detectable before.

Larson, H.P., and Fink, U., 1975, Appl. Opt. 14, 2085.
This article shows the powerful uses to which a Michelson interferometer can be put in the spectral range between the visible
region and 5.6μ. Both ground based and airplane observations are
described and illustrated with spectra of planets and their moons.
Vanasse, G.A., and Sakai, H., 1967, Progress in Optics Vol. VI,
(Amsterdam: North Holland) ed. E. Wolf, p. 261. This review article presents interferometric spectrometry and particularly Michelson interferometry in concise terms. It provides useful references to earlier work.

Heterodyne Spectroscopy
Betz, A.L., McLaren, E.C., Sutton, E.C., and Johnson, M.A., 1977,
"Infrared Heterodyne Spectroscopy of CO_2 in the Atmosphere of Mars"
to be published. This paper gives a summary both of the instru-
mentation used and results obtained in observations of Mars.
Betz, A.L., Johnson, M.A., McLaren, E.C., Sutton, E.C., 1976,
Ap. J. 208, L 141. This paper discusses heterodyne observations
of wind velocities in the atmosphere of Venus.

REFERENCES TO CHAPTER 2

General References

Radiative Processes Important in Astronomy
Harwit, M., 1973,"Astrophysical Concepts", (New York: John Wiley
and Sons) Chap. 6 and 7.
Herzberg, G., 19 ,"Spectra of Diatomic Molecules", (Holland:
Van Nostrand)
Herzberg, G., 19 , "Infrared and Raman Spectra of Polyatomic
Molecules", (Holland: Van Nostrand)

Fine Structure and High Rydberg Transitions
Petrosian, V., 1970, Ap. J. 159, 833.
Simpson, J., 1975, Astron. and Astrophys. 39, 43.
These two references show how the fine structure line strengths
are related to specific properties of emitting H II regions, and
predict line strengths for transitions in some of the brighter
known sources. Petrosian's paper also relates the strength of
high lying Rydberg transitions in the infrared to the observed
Hβ line strengths in H II regions.

Radiative Transfer (Dust)
Schmidt-Burgk, J., and Scholz, M., 1976, Astron. and Astrophys.
51 , 209.
This paper demonstrates that interpretation of dust cloud spectra
leads to wide ranging ambiguities, unless high spectral and spatial resolution observations are employed. A quantitative relation
between spectral resolving power and ambiguity in the models is
provided in some simple situations that are readily analyzed. The
paper is an important warning that shows how difficult it is to
draw conclusions about physical conditions in dust clouds on the
basis of infrared observations alone.

Specific References

Free-Free and Free-Bound Transitions
Gilman, R., 1974, Ap. J. 188, 87.
This article shows the spectrum of the supergiant variable star
VX SGR, and interprets it in terms of free-free and free-bound
emission in the star's outer layers as well as blackbody stellar
radiation and radiation by a circumstellar dust cloud.
Panagia, N., and Felli, M., 1975, Astron. + Astrophys. 39, 1.
This paper discusses free-free emission in early stars.
Molecules in Stars
Hall, D.N.B., and Ridgeway, S., 1977, Symposium at Liege, Belgium,
to be published.
This article reports on a variety of molecules observed between
1.2 and 14μ in variable and late stars. HF, CO, HCN, H_2, C_2H_2,
HCL molecules are observed (these include the first polyatomic
hydrocarbons detected in stars), and isotopes ^{12}C, ^{13}C, ^{17}O, ^{18}O,
^{35}Cl and ^{37}Cl are detected in some of these molecules.
Molecules in Interstellar Space
Gautier III, T.N., Fink, U., Treffers, R.R., and Larson, H.P.,
1976, Ap. J. 207, L129
The authors reported the first observations of H_2 in a dark cloud
in interstellar space.
Field, G.B., Somerville, W.B., and Dressler, K., 1966, Ann. Rev.
of Astron. and Astrophys. 4, 207.
This paper gives a detailed review of molecular hydrogen transitions to be expected in interstellar space, and includes a useful
energy level diagram.
Fine Structure Transition Observations
Baluteau, J.P., Bussoletti, E., Anderegg, M., Moorwood A.F.M.,
and Coron, N., 1976, Ap. J. 210, L45.
In addition to (O III), these authors also detected (S III) in
the Orion Nebula.
Ward, D.B., Dennison, B., Gull, G., and Harwit, M., 1975, Ap. J.
202, L31.
This paper reports the first observation of a far infrared line,
(O III) at 88μ, seen in an astronomical source, M 17.

INFRARED ASTRONOMY FROM SPACE: A REVIEW OF FUTURE POSSIBILITIES

A.F.M. Moorwood

Astronomy Division, Space Science Department,
European Space Agency, Noordwijk, Holland

1. INTRODUCTION

After more than ten years spent exploiting the possibilities for
groundbased, aircraft, balloon and rocket observations, the logical
step of going into space is now set to occur in the early 1980's.
These will not strictly be the first infrared astronomical observations to be conducted from outside the earth's atmosphere as
there has been some military interest in this area for a number
of years. A 2.7 μm equatorial sky survey is currently in progress
using US Air Force satellites, for example (Heinseimer et al., 1977),
and results from the Celestial Mapping Programme satellite launched
several years ago are now being made available. The results obtained from these programmes, however, do not invalidate in any
way the more astronomically sophisticated and ambitious plans now
being proposed by the infrared astronomical community. Rather,
the technical expertise gained through these missions has laid much
of the present foundation on which infrared space astronomy is
based. Both cryogenic and detector technology have been stimulated
and to a considerable extent proved. The Air Force Geophysics
Laboratory (formerly AFCRL) rocket sky survey at 4 μm, 11 μm, 20 μm
and 27 μm, for example, has demonstrated the considerable sensitivity advantages possible using low background detectors in space
(Price and Walker, 1977) and has increased the motivation for long
duration survey satellite flights.
　　This rather new confidence in the technical feasibility of
infrared space missions coupled with the increasing difficulty of
improving available instruments has already provoked a large
number of ideas within the infrared community. Some diversity
arises through the somewhat conflicting requirements of survey instruments and those intended for detailed studies of the extensive

G. Setti and G. G. Fazio (eds.), Infrared Astronomy, 285–316.

variety of infrared objects already known. Differences in detection and observational techniques over the three decades in wavelength between 1 μm and 1 mm also lead to different experimental
requirements and emphasize that a comprehensive programme of
infrared space astronomy will eventually demand a range of complementary facilities. Various space agencies have already
sponsored studies of many of the proposed facilities and brief
summaries of them are given here.

The first project to have been approved and funded however
is the IRAS satellite which will become the first major infrared
space project when it is launched in 1981. Appropriately, its
prime scientific aim is to conduct an all sky survey at mid and
far infrared wavelengths. This will yield the first complete
census of infrared objects whose prototypes are already known
and may additionally reveal completely new classes of infrared
objects. Whatever the results, however, they will clearly increase
rather than decrease the demand for other space instruments capable of high spectral and spatial resolution investigations of individual sources.

2. WHY SPACE EXPERIMENTS?

Infrared astronomy has so far been a relatively low cost and
low manpower activity. Considerable ingenuity has been displayed,
however, in opening up the entire 1 μm to 1 mm spectral region
and demonstrating its importance despite the severe difficulties
caused by atmospheric emission and absorption.

It has always been recognized, however, that the ultimate observing conditions would be found in space and that many observations would have to wait until telescopes could be operated above
the atmosphere. No groundbased surveys have been successfully
carried out beyond 2 μm, for example, and all observations from the
ground have been very selective through their restriction to a
number of atmospheric "windows" in the 1 μm – 2 μm region. At
very high and dry sites it has been possible to make some observations at 34 μm and 350 μm, but these windows are of very poor
quality. For most far infrared work it has been necessary to go
higher on aircraft, balloons and rockets. The total amount of
observation time which has been achieved by these techniques is
relatively small, however, and the largest airborne and balloon
telescopes flying are only around 1 m in diameter, compared with
those of up to 5 m used on the ground. In spatial resolution
consequently they face the double penalty of longer wavelengths
and smaller apertures relative to the near and mid infrared.
Typically the beam sizes used in the far infrared are 1 arc min.
or more compared with those of a few arc secs. used in groundbased
work. This is already posing difficulties for the interpretation
of data on extended and structurally complex sources.

Most of the observations made until fairly recently have also

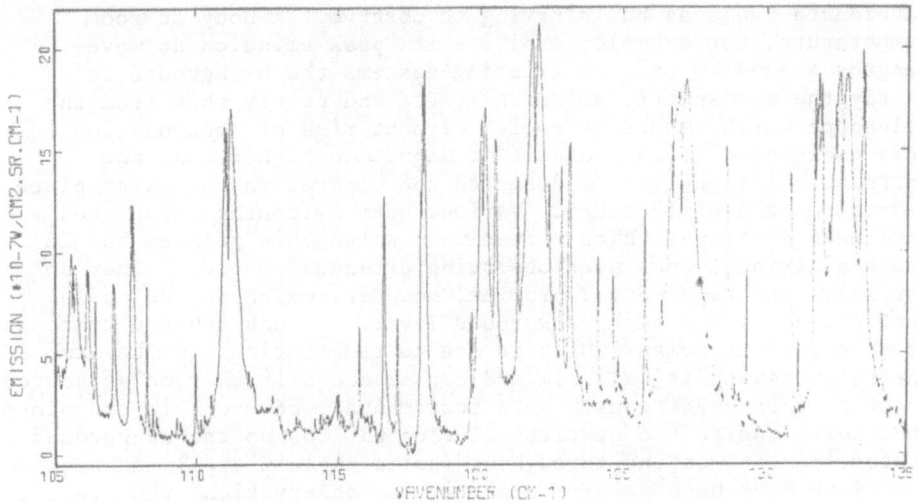

Fig. 1. Stratospheric emission spectrum obtained at 0.02 cm^{-1} resolution on the Gerard P. Kuiper Airborne Observatory (Baluteau et al., 1977).

utilized relatively broadband photometric systems. For spectroscopic work, however, the precise details of the atmospheric absorption spectrum become of crucial importance. This absorption arises as a result of transitions in a wide range of molecular species, notably H_2O, CO_2, O_3, O_2, CH_4. The actual transmission of the atmosphere is consequently a rapid function of frequency and may vary between 0 and 100% within very narrow spectral intervals. This is true even at aircraft altitudes where many of the lines are still saturated as can be seen in Fig. 1, which is an emission spectrum of the atmpsphere around 80 μm obtained on the Gerard P. Kuiper Airborne Observatory. (Baluteau et al., 1977). Searching for astrophysical emission lines in the infrared can be quite a problem, therefore, because the uncertainty in their rest frequencies coupled with their Doppler shift makes it impossible in many cases to predict whether or not they will be coincident with atmospheric features. For continuum spectroscopy, the problem is that of accurately removing the atmospheric structure from the recorded spectrum. This is already the limiting factor in the accuracy of spectral observations of the brighter sources.

Clearly the frustrating effect of these limitations increases as better data is sought to answer specific astrophysical questions. The most compelling argument in favour of space observations has also not yet been mentioned. That is the ultimate sensitivity which can be achieved when the radiation background viewed by the detector is reduced. The infrared astronomer can be compared with a visible astronomer working during the day in the sense that everything in front of his detector is radiating in exactly the same

wavelength range as he is trying to observe. A body at room
temperature, for example, exhibits its peak emission at wavelengths around 10 μm. In existing systems the background is
partly the atmospheric emission itself and partly that from the
telescope which cannot be cooled without risk of condensation.
This background is many orders of magnitude higher than the
astronomical signal to be detected and creates in the first place
a problem of dynamic range. Various beam switching techniques
have been developed which effectively solve this problem but do
cause additional ones when observing extended sources. They do
not solve the fundamental problem, however, which is the noise
contributed by the high background level. A much debated component
of this is "sky noise" which is due to fluctuations in the atmospheric emission, is variable and can exceed all other noise sources
under certain conditions. More fundamental, however, is the photon
shot noise due to the statistical fluctuations on the background
photon flux. Once the throughput (solid angle x area) and spectral
bandwidth have been set for a particular observation, the overall
sensitivity is determined by either this photon noise or by the
intrinsic detector noise. Both of these contributions can be expressed as a noise equivalent power (N.E.P. W $Hz^{-1/2}$) i.e. the
signal power equal to the r.m.s. noise in a 1 Hz electrical bandwidth. Typically the background limited N.E.P. is around 10^{-14}
W $Hz^{-1/2}$ in existing systems. In space, however, the atmospheric
contribution disappears and the background can be further reduced
by cooling the telescope.

The effect on the photon noise is illustrated on Fig. 2 for
the useful reference condition of a diffraction limited beam and
a spectral band $\Delta\lambda = 0.1\lambda$. (When working at the diffraction limit
the telescope throughput is proportional only to λ^2, i.e. it is
independent of telescope size. The photon noise for other conditions can also be scaled from these curves. It varies as
$(\varepsilon/0.04)^{1/2}$ $(\Delta\lambda/0.1\lambda)^{1/2}$ $(\Omega A/3.7\lambda^2)^{1/2}$ where ε is the telescope
emissivity, ΩA is the telescope throughput and λ is the wavelength
in cm.) These curves show that the photon noise limit of the telescope can be reduced by several orders of magnitude by cooling the
mirrors to very low temperatures. The fundamental sensitivity
limits in space, therefore, are determined either by the intrinsic
noise of the detectors themselves or, ultimately, by another
background which now becomes important, the zodiacal light from
dust within the solar system. This depends on observing direction,
and the curve shown on the figure represents a mean value for the
noise contributed by this background. Inspection of the absolute
scale, however, shows that this limit is roughly four orders of
magnitude lower than that currently faced. The critical factor
in assessing the sensitivity advantage of space observations,
therefore, is the intrinsic performance of detectors under low
background conditions. This is an area where rapid progress is
being made at the present time. State of the art photoconductive
detectors (Si/As, Si:P, Ge:Ga, etc.), however, are already capable

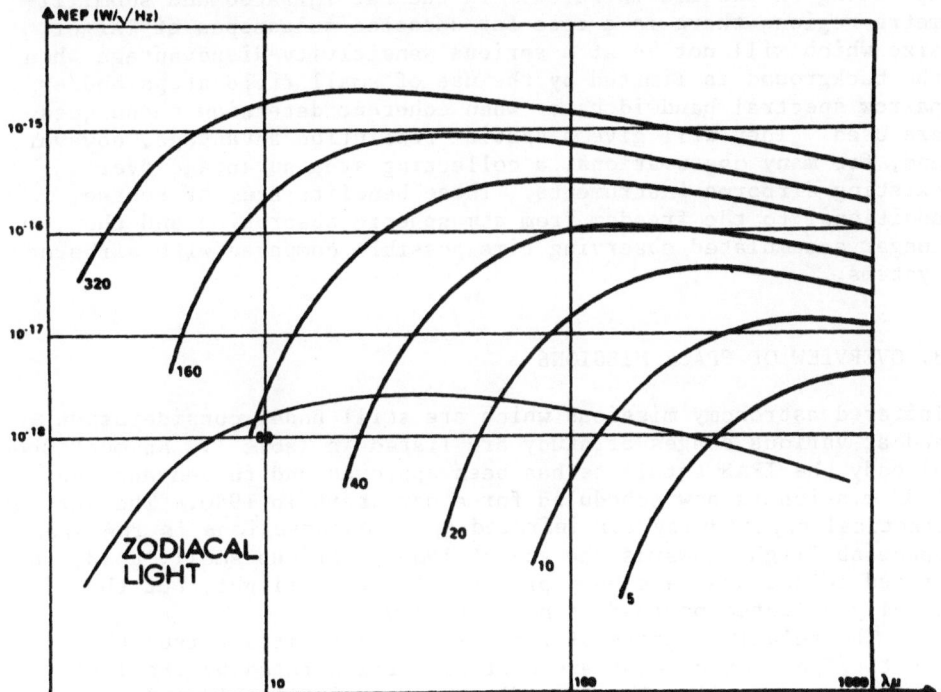

Fig. 2. Background noise equivalent power for a diffraction limited telescope as a function of wavelength and for various mirror temperatures. Emissivity of the telescope is taken as 0.04 and the spectral bandwidth is given by $\Delta\lambda/\lambda = 0.1$. Also shown is a mean curve for the expected noise contribution from the zodiacal light.

of yielding N.E.P.'s approaching 10^{-17} W Hz$^{-1/2}$ and these detectors will be available out to wavelengths beyond 100 μm even for the early 1980's missions. Throughout the mid infrared, therefore, the sensitivity advantage expected of cold space telescopes is in the two to three orders of magnitude range. N.E.P.'s at longer wavelengths are not likely to improve as rapidly, but figures between 10^{-15} and 10^{-16} W Hz$^{-1/2}$ can be anticipated from various types of bolometer. An alternative to cooled telescopes in the far infrared and submillimetre regions, however, is to maximise telescope diameter and hence spatial resolution at the expense of some sensitivity. For many narrow band, and particularly for spectroscopic applications utilizing coherent detection techniques, this does not even compromise the sensitivity, in fact, and the large uncooled telescope becomes the optimum facility.

In summary it is to be expected that available detectors will lead to sensitivity improvements throughout the infrared once the atmospheric background is eliminated. This is likely to amount to some three orders of magnitude for a cooled survey type instrument

operating in the mid infrared. In the far infrared and submillimetre regions there is a case for uncooled telescopes of larger
size which will not be at a serious sensitivity disadvantage when
the background is limited by the use of small field stops and/or
narrow spectral bandwidths or when coherent detection techniques
are used. They will give a spatial resolution advantage, however,
and, for many observations, a collecting area advantage over
existing airborne instruments. These benefits are, of course,
additional to the freedom from atmospheric absorption and the
longer accumulated observing time possible compared with airborne
systems.

3. OVERVIEW OF SPACE MISSIONS

Infrared astronomy missions which are still under consideration
and at various stages of study are listed in Table 1. As mentioned
already the IRAS satallite has been approved and funded and the
COBE mission is now scheduled for a new start in 1980. The earliest
practical opportunity for infrared space observations is the second
Spacelab flight towards the end of 1980. Various small, cold, infrared telescopes have been proposed for this flight, but the
finally selected payload is not yet known.
      The relative merits of free flying satellites versus the
Shuttle/Spacelab concept are partially illustrated by the list
of missions itself. The free flyers carry small cold telescopes,

TABLE 1

Overview of Proposed Space Missions

Name	Origin	Telescope	Launcher	Launch
Free Flyers				
IRAS	NL–UK–NASA	0.6 m ($\sim$15 K)	TD	1981
COBE	NASA	HORN ($\sim$ 3 K)	TD ?	?
IRSAT	ESA	0.3 m (· 2 K)	SCOUT	--
ST	NASA	2.4 m (294 K)	SHUTTLE	1983
Spacelab				
SIRTF	NASA	1.16 m ($\sim$15 K)	SHUTTLE	>1984
LIRTS	ESA	2.8 m ($\sim$280 K)	"	"
EAR's	JPL/NASA	>10 m (ambient	"	?
GIRL	D	0.4 m (10 K)	"	1982
OTHER COLD TELESCOPES}	--	<0.4 m	"	1980

can operate continously on orbit for about one year (set by their
cryogenic lifetime) and are primarily designed for all sky surveys.

The major Spacelab telescopes are larger, operate for up to
30 days on any one flight and can be flown repeatedly with different complements of focal plane instruments. They are aimed primarily at detailed studies of sources already known and those
anticipated from the surveys. Somewhat in between these two
categories is the Space Telescope which is not a dedicated infrared facility but may carry an infrared instrument within its focal
plane complement. It is a free flying system but the infrared
lifetime will be limited by the cryogenic supply and the time
allocated to each of the focal plane instruments. A decision on
the instrument complement is not expected for several months.

It is interesting to note that all of the missions utilize
helium for cryogenic cooling of the detectors and in some cases
the telescope as well. They divide between the use of supercritical
helium gas and superfluid helium liquid. In the supercritical
systems the gas is maintained at around 5 K and 5 atmospheres
although lower temperatures which may be required locally (e.g. for
bolometer detectors) can be achieved by incorporating Joule-
Thompson expansion nozzles. Superfluid helium on the other hand
only exists below about 2 K and the pressure must be maintained
around a few torr by venting gas to space. Because the gas and
liquid do not separate out under zero-g, however, it is necessary
to incorporate a porous plug in the vent line to prevent catastro-
phic liquid loss. Development and testing of this item has been
one of the most important technical factors in establishing the
feasibility of infrared space experiments.

4. REVIEW OF SPECIFIC MISSIONS

The following sections aim to summarize the main instrumental
features and the principal scientific objectives of the various
missions already listed in Table 1. As no attempt is made here
to enter into detailed aspects, it is only fair to the many people
involved to state that such details do exist and that a great deal
of effort has gone into the various studies from which this information has been distilled. Directing the reader to these more detailed sources of information is rather difficult, however, as
most of it exists in the form of contractor, consultant and, in
some cases, partially completed reports which are only produced in
limited numbers. I have not referenced information in the text,
therefore, but have listed some references at the end. It should
also be borne in mind that whether approved or not these missions
are all still being subjected to study and none of the concepts
presented can yet be considered as "frozen".

4.1 IRAS (Infrared Astronomy Satellite)

The sun-synchronous orbital configuration is illustrated on
Fig. 4.   Orbital parameters (altitude and inclination) have been
chosen such that the orbital plane is always perpendicular to the
Sun vector i.e. it precesses at a rate of 1° per day.   The solar
panels always face the Sun and the satellite rolls such that its
viewing direction is always away from the Earth.   A circle on the
sky is thus scanned at a rate of 3.6 arc min/sec during each orbit
and the optical axis moves over the whole sky in a period of 6
months.   In practice the satellite can also be tilted out of its
orbital plane and the survey conducted in the more efficient manner
described below.

The Survey.   The photometric bands and detectors of the main
survey instrument are summarized in Table 3, together with their
fields of view and the expected overall sensitivities.   Additionally
there will be a number ($\sim$4) of superconducting tin bolometers
operating between about 120 μm and 300 μm.   Fig. 5 illustrates the
number of detectors foreseen and their possible arrangement
in the focal plane.   As the only signal modulation is that produced by the satellite orbital motion moving the source across the
detector, the survey will be most sensitive to sources of small
angular extent.

The sensitivities quoted correspond to detectable bolometric
luminosities of about 1 L_0 at 1 kpc and 10^8 L_Θ at 10 Mpc for objects
cooler than 1000 K.   Most known types of galactic infrared sources
should be observable, therefore, to the limits of our Galaxy.   These
include protostars (e.g. BN), late type stars (e.g. α Ori, IRC +
+ 10216), HII regions (e.g. M42, M17, W51), planetary nebulae
(e.g. NGC 7027) and molecular clouds (e.g. Sgr B2).   The more luminous sources will be observable in other galaxies within the local
group.   The sample of known classes of extragalactic objects will
be substantially increased, particularly in the far infrared, and
the most luminous members will be seen out to cosmological distances
(e.g. $\sim$v/c = 0.5 for the quasar 3C273).   Even without the unexpected
discoveries which one hopes will be made with the satellite, IRAS
will certainly detect a large number of sources during its mission.

Table 3

IRAS Survey

BAND	DETECTOR	DIMENSION (arc min)	E F D (Jy, s/n=10)	MAG.
8 - 15 μm	Si : As	1.2 x 2.4	0.04	7.5
15 - 30 μm	Si : Sb	1.2 x 2.4	0.08	5.4
30 - 60 μm	Ge : Be	1.2 x 2.4	0.8	0.4
60 - 120 μm	Ge : Ga	1.8 x 4.8	3	- 2.3

   Project Description. This joint project between the Netherlands, United Kingdom and the United States has the primary aim
of achieving an all sky survey at mid and far infrared wavelengths.
Some general features of the mission are summarized in Table 2.
The Netherlands is responsible for providing the main body of the
satellite and for some focal plane instruments, NASA for the cryo-
genically cooled telescope, the array of survey detectors and the
launch, and the United Kingdom for the flight operations via a
ground station in England and for additional long wavelength detectors. An on-board computer, similar to that used on the Dutch
ANS satellite, will be used to sequence the satellite operations
over a 12 hr. period. Commands will be loaded during a ground
station pass when data recorded on tape during the preceding 12
hrs. will also be dumped to ground. This will comprise a large
number of data bits which will be subjected to quick-look analysis
at the ground station before full data processing is carried out
in the United States.
   The general satellite configuration is shown in Fig. 3. The
telescope is mounted centrally in a toroidal cryogenic tank of
superfluid helium and the various detector packages are located at
the Cassegrain focus in the space behind the primary mirror. A
large external baffle protects both the detectors and the cryogenic
system from excessive radiation loads from the Sun. Power is provided by the solar panels and the satellite is stabilized in three
axes.

Table 2

IRAS Summary

PRIME SCIENTIFIC MISSION:	ALL SKY SURVEY
TELESCOPE	: 0.6 m RITCHEY CHRETIEN, COOLED ~ 15 K
POINTING	: 3 AXIS STABILIZED. ACCURACY 30", STABILITY 10" 60° OFFSET CAPABILITY FROM SOLAR VECTOR
MODES	: SCAN (SURVEY) RASTER (EXTENDED SOURCES) POINT (INTEGRATIONS)
DATA	: $\sim 5 \cdot 10^8$ BITS/DAY
OPERATIONS	: ON BOARD COMPUTER COMMAND LOAD + DATA DUMP EVERY 12 HRS. DURING GROUND STATION PASS
SATELLITE MASS	: ~ 900 KG
ORBIT	: 900 KM, 99° SUN SYNCHRONOUS
LAUNCH	: 1981
LIFE	: 1 YEAR

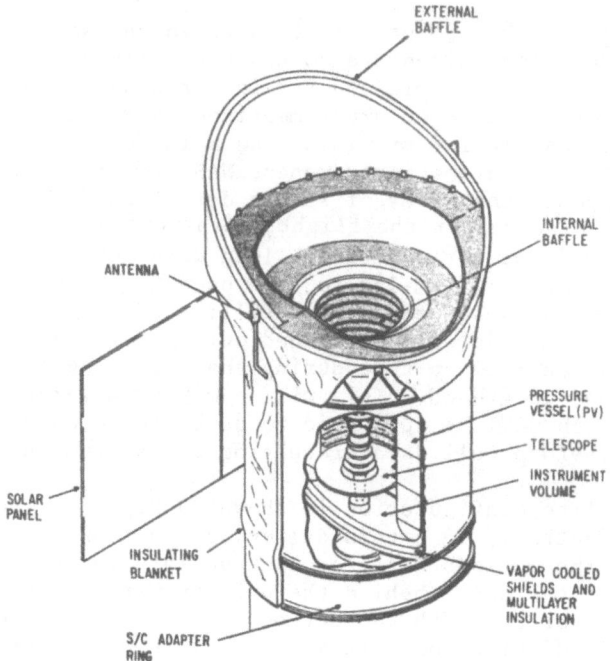

Fig. 3. View of a possible configuration for the IRAS satellite.

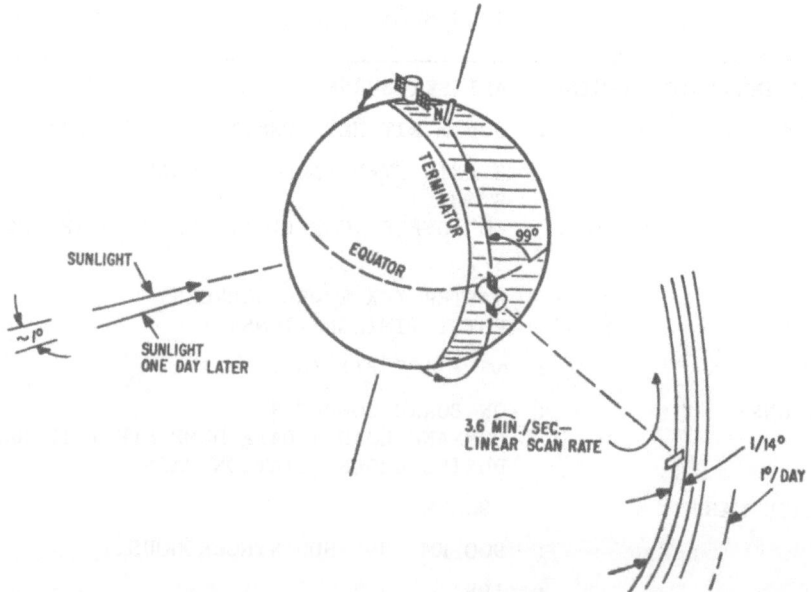

Fig. 4. Sun synchronous orbital configuration.

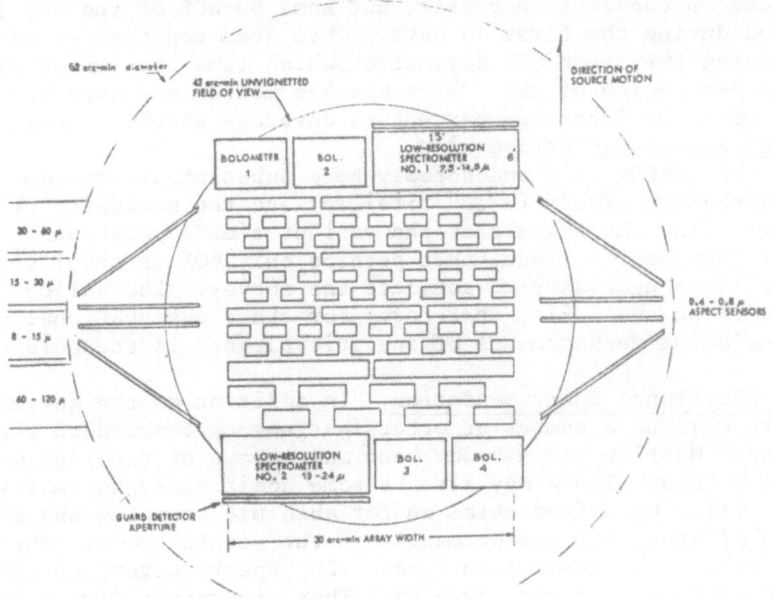

Fig. 5. Proposed focal plane layout for IRAS.

Estimates higher than 10^7 are not unreasonable. Unfortunately, along with sensitivity to sources of interest, will go sensitivity to Solar System debris, dust particles, etc., whose signals must be identified and rejected. There may also be some problems due to charged particle hits on detectors. Great care will be necessary in planning the survey, consequently, to ensure its accuracy and reliability down to a given sensitivity limit. It is planned that this will be achieved mainly by including substantial redundancy in the observations. That is, each point on the sky will be observed more than once and at a range of time separations sufficient to identify moving objects. The periods between successive observations will be spaced roughly on the scale 1 sec., 100 minutes, 15 days, and 6 months. The shortest period corresponds to the passage of the parallel strips of detectors across the sky while the longest is the time for the orbit plane to precess through 180° If required, the entire survey can be repeated during the second 6 months on orbit. The period of 100 minutes is one orbit and that of 15 days results from the detailed programme for executing the survey. This makes use both of the array width ($\simeq$ 30 arc min.) and the offset pointing capability of the satellite to obtain a faster sky coverage rate than the nominal sun synchronous rate. After each half orbit the satellite is manoeuvred over the poles in such a way that the next half orbit scanned is displaced by about half the array width rather than the distance appropriate to the orbital precession rate of 4 arcmin/orbit. Each point on the sky is still viewed on consecutive orbits, but some 50-60% of the sky is now covered during the first 15 days. This area can then be re-scanned during the second 15 days after which time the survey proceeds to a new region of sky. It takes 5-6 months to cover the whole sky with the increased percentage coverage steadily decreasing during each 15 day period.

Data processing will undoubtedly be a substantial task due to the high data rate itself (> 10^8 bits/day) and the necessity of cross correlating the signals at the various time separations.

It is foreseen, in fact, that perhaps only 60% of the 1 year lifetime will be necessary to complete the survey. The balance of the time will be available, therefore, for other observational programmes including measurements of specific targets in the pointed mode.

IRAS Additional Instrumentation. In addition to the survey array there will be a number of other instruments located in the focal plane. Mention has already been made above of the long wavelength bolometers. There may also be some additional photometry channels, utilizing differential and/or absolute chopping and smaller fields of view, for use primarily in the pointed mode. There will also be two low resolution ($\lambda/\Delta\lambda \simeq 20$) spectrometers operating in the ranges 7.5-14 μm and 13-24 μm. They are rather simple in concept, using prisms as the dispersive elements and the orbital motion for wavelength scanning. The addition of some spectral information, e.g. on the silicate features, may prove to be invaluable, however, for identifying sources detected in the survey. Unfortunately, it will not be possible to obtain spectral information on the fainter sources detected during the survey.

4.2 COBE (Cosmic Background Explorer)

This NASA satellite project is directed towards a survey of the sky for diffuse sources of infrared and microwave emission. Various anticipated components of this emission are shown in Fig.6. At wavelengths longer than a few hundred μm, the prime interest is the cosmic background radiation which is believed to be the fossil remnants of radiation which has been diluted and red-shifted by the expansion of the Universe since the big bang (Clegg and Longair, this volume). Various measurements have already confirmed the existence of this radiation and have established that its spectrum is approximately that of a black body at around 2.7 K. Significant improvements in these measurements can now only be obtained however by performing observations from above the atmosphere. It is proposed with COBE to measure the spectrum of this radiation between 300 μm and 3 mm with an accuracy of 1 part in 10^3 at the peak and to determine any large scale anisotropies to 1 part in 10^5. Two different scientific instruments will be provided for these observations. The spectrum measurement will be made with a cooled polarizing Michelson interferometer at a typical resolution around 1 cm^{-1} and at a maximum resolution of 0.06 cm^{-1}. This instrument

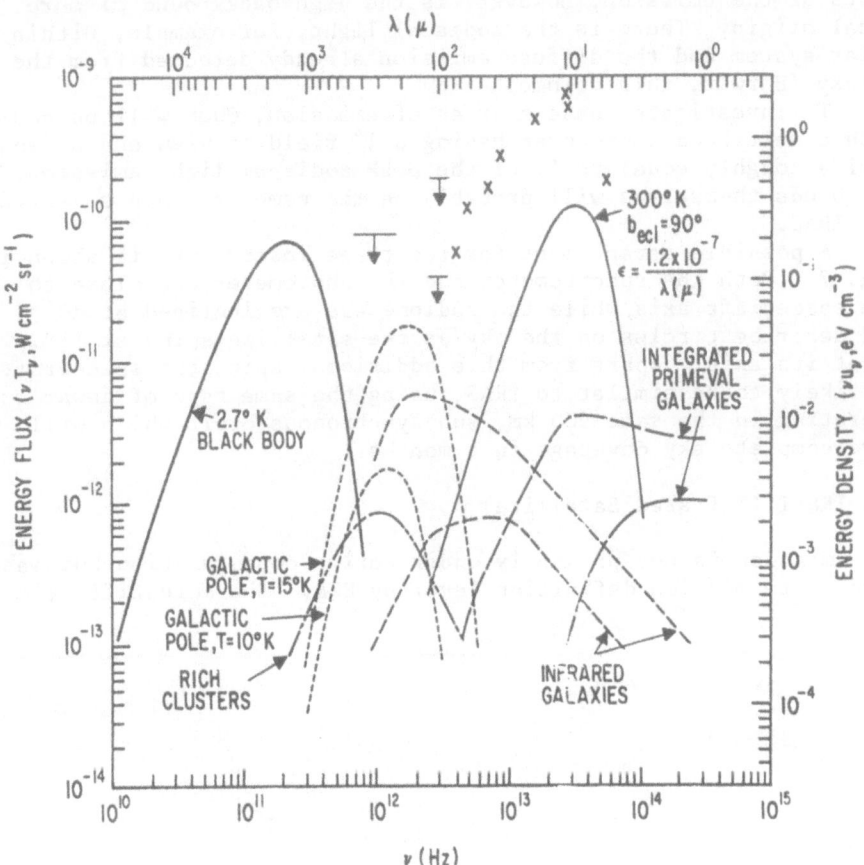

Fig. 6.   Sources of diffuse emission in the infrared and microwave
regions.

will be fed by a carefully designed horn which will define a 7°
field of view on the sky. It will be possible to periodically cover the horn with a cold black body to perform in flight calibration.
The isotropy measurements will be made in the microwave region using four differential radiometers at frequencies of 23.5, 31.4, 53
and 90 GHz. These will be uncooled instruments which switch between
7° fields located 60° apart on the sky. Each radiometer has two
channels measuring orthogonal planes of polarization.

   Moving downwards in wavelength into the infrared region there
are a number of possible contributors to a diffuse flux. Sources
of extragalactic origin may include the radiation from primeval
galaxies at z ∿ 10 which is sufficient to redshift the "visible"
spectrum into the infrared. There are also more evolved infrared
galaxies whose peak emission is at longer wavelengths. The

difficulty in extracting information on these extragalactic components of the emission, however, is the high background of more
local origin.   There is the zodiacal light, for example, within the
solar system and the diffuse emission already detected from the
Galaxy (Harwit, this volume).

To investigate these sources of emission, COBE will be equipped
with a multiband photometer having a 1° field of view and a sensitivity roughly equal to 1% of the peak zodiacal light emission.
The bands themselves will probably be the same as those selected
for IRAS.

A possible arrangement for the three instruments is shown in
Fig. 7.   Both the spectrometer and the photometer are close to
the spacecraft axis, while the radiometers are inclined at 30°
and describe circles on the sky as the satellite spins at 1 r.p.m.
about its axis.   Apart from this additional spin, the spacecraft
is likely to be similar to IRAS, using the same type of dewar and
operating in the same 900 km, sun synchronous orbit which will
give complete sky coverage in 6 months.

4.3 IRSAT (Infrared Satellite)

This mission is not presently under active consideration but was
studied to mission definition level by ESA.   The scientific aims

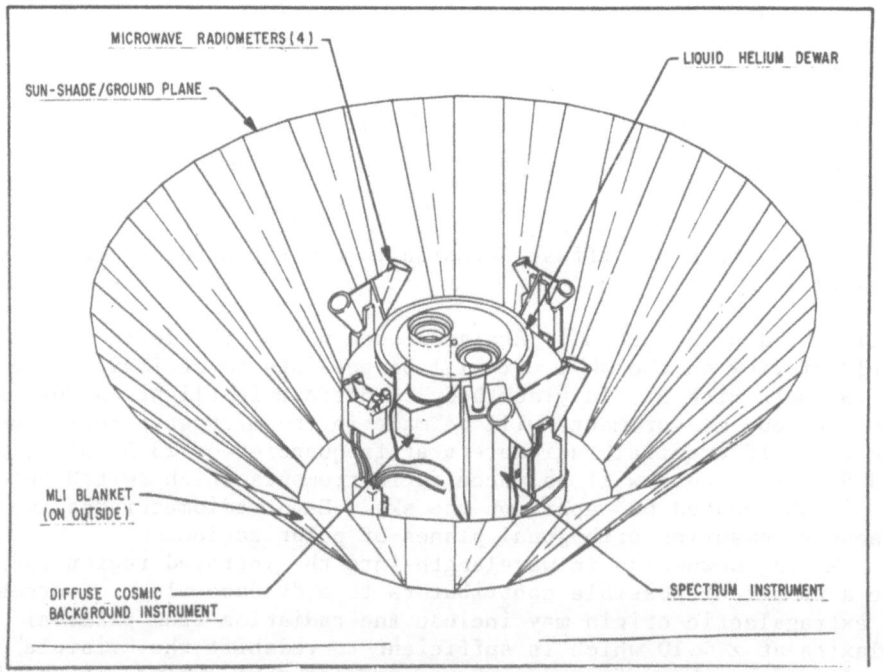

Fig. 7. Schematic of the COBE instrument configuration.

are similar to those for COBE for the cosmic background radiation.
Only a single instrument, a cooled polarizing interferometer, was
considered, however, and the whole payload was small enough for
launch by a Scout rocket. Its operation as a Spacelab experiment
and a Spacelab sub-satellite were also studied.

4.4 ST (Space Telescope)

The ST observatory is a large optical quality telscope offering
unprecedented pointing stability as well as freedom from atmospheric
absorption and seeing. Both on-orbit refurbishment and return
of the telescope to Earth will be possible using the Shuttle
during the planned 15 year life. Some of its major characteristics
are summarized in Table 4 and an exploded view showing the main
system elements is shown in Fig. 8.
    The telescope concept and performance have been driven by the
requirements of the ultraviolet and visible instruments to benefit
from the absence of seeing by utilizing the smallest possible field
stops.
    Although not originally foreseen, the possibility for infrared

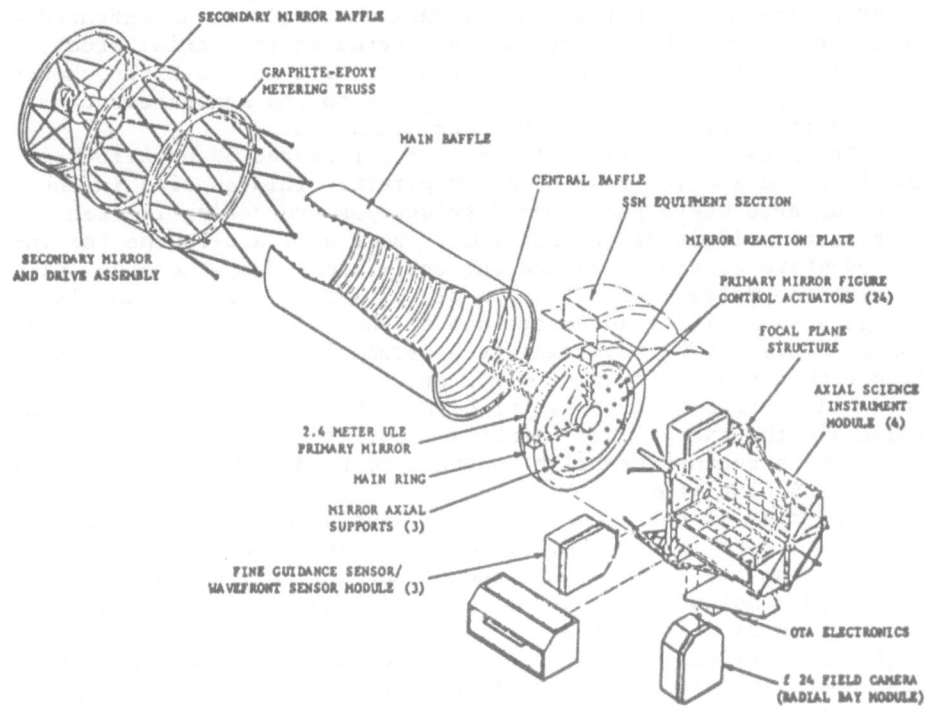

Fig. 8. System elements of one concept of the Space Telescope.
An infrared instrument could occupy one of the axial science
instrument modules.

Table 4

ST Major Characteristics

PRIMARY MIRROR	:	2.4 m, F/2.2 ULE 294 ± 1 K
OPTICAL SYSTEM	:	RITCHEY CHRETIEN F/24
FIELD OF VIEW	:	20'
FIGURE	:	$\lambda/20$ AT 6300 Å
POINTING STABILITY	:	0.006"
SPACECRAFT ALTITUDE	:	500 KM
LIFETIME	:	15 YEARS WITH REFURBISHMENT
OBSERVATION TIMES	:	<10 HRS.
NUMBER OF INSTRUMENTS:		∿5

observations has been taken up by NASA who included an infrared photometer among the instruments subjected to industrial study. The complement of focal plane instrument will not be known, however, until the proposals received in response to the Announcement of Opportunity (March, 1977) have been evaluated.

The great attraction of the ST for infrared observations is the improved spatial resolution it offers. This applies in the near infrared where groundbased telescopes are seeing limited (diffraction limit of the ST is 0.2" at 1 μm), and in the far infrared where observations are currently limited by the size (≤ 1 m) of airborne telescopes. The use of very small field stops also brings a sensitivity advantage due to the reduction in background noise. This may be important even around 10 μm where apertures larger than the diffraction limit are generally used on the ground to avoid increased sky noise problems. The absence of seeing in space and the exceptional pointing stability of the ST may also allow the structure of bright sources to be determined even below the diffraction limit.

The type of infrared photometer studied so far is fairly straightforward in concept. It operates over the entire range between 1 μm and 1 mm using a photoconductor detector out to about 25 μm and a bolometer at the longer wavelengths. A dichroic beamsplitter is used to divide the input beam into these two broad regions, and filter wheels in front of each detector are used to give approximately 10% spectral resolution for the overall system. There is also a rotating field stop wheel in front of each detector. A possible optical scheme for the photometer is shown in Fig. 9. The uncooled fore optics perform two functions. Firstly, as the ST has a fixed secondary mirror, they include a two position mirror

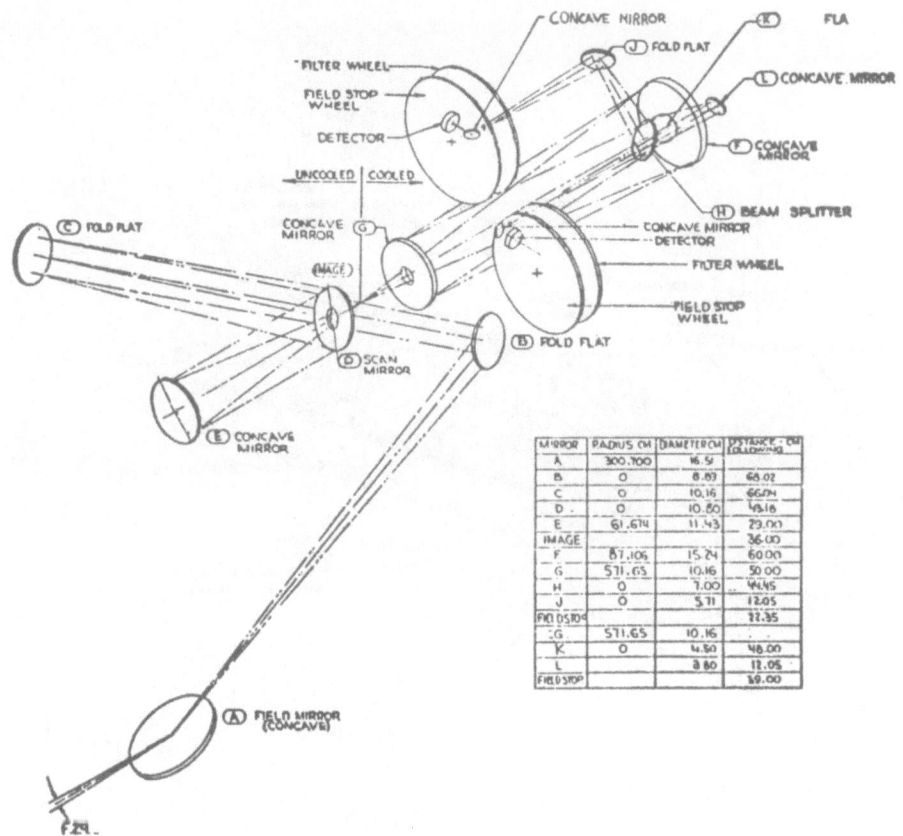

MIRROR	RADIUS CM	DIAMETER CM	DISTANCE CM FOLLOWING
A	300.700	16.51	
B	0	8.63	68.02
C	0	10.16	66.04
D	0	10.80	43.18
E	61.614	11.43	29.00
IMAGE			36.00
F	87.106	15.24	60.00
G	511.65	10.16	50.00
H	0	7.00	44.45
J	0	5.71	12.05
FIELD STOP			22.35
G	511.65	10.16	. .
K	0	4.50	48.00
L		3.60	12.05
FIELD STOP			19.00

Fig. 9. Possible optical schematic for a two channel infrared photometer for the Space Telescope.

(scan mirror) used for space chopping of the beam. Secondly, they increase the speed of the input beam from F/24 to F/4 to reduce the size of the aperture required to admit the beam into the cryo-genically cooled part of the instrument. The cooled optics per-form the various wavelength divisions and produce final focal ratios for the two channels which give acceptable physical sizes for the various field stops. A superfluid helium tank sized to fit one of the axial bays would provide cooling for around 1 year.

4.5 SIRTF (Shuttle Infrared Telescope Facility)

Telescope. SIRTF is a NASA facility instrument designed for re-peated flights on the Shuttle/Spacelab system over a number of years. Two views of the telescope are shown in Fig. 10. The major elements are the 1.16 m cryogenically cooled telescope, sun-shade, cooled instrument volume behind the primary mirror and

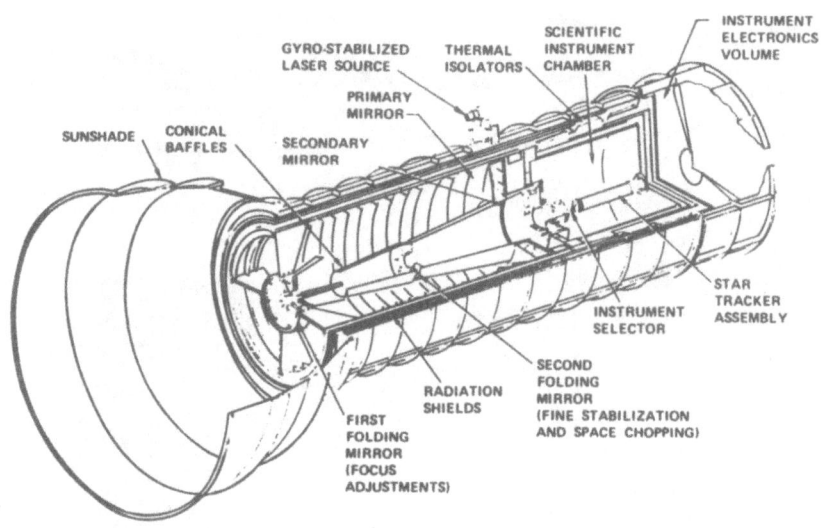

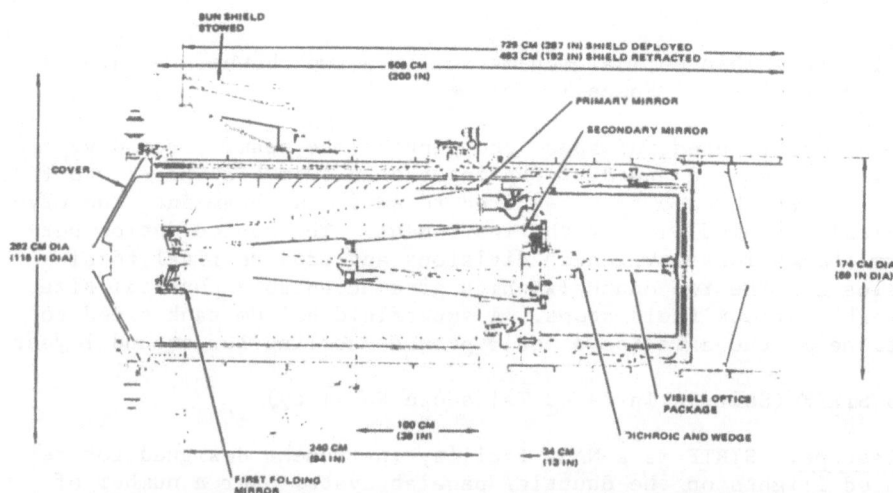

Fig. 10. Two cutaway views of the SIRTF telescope.

(not shown) the Instrument Pointing Subsystem (IPS). The IPS is
a general purpose stabilized platform capable of orienting a range
of payloads with arc second accuracy. It is being developed by
ESA as a Spacelab facility.

The telescope has beryllium optics and is cooled by supercritical helium gas (5 K, 5 atm.) stored in tanks attached to the
outer telescope structure. Temperatures below 5 K can be obtained
locally within the instrument volume by using Joule-Thompson expansion nozzles. The instrument volume can accommodate a number
of different scientific packages and a rotating selector directs
the beam as required during flight.

The optical scheme can be more easily understood by looking
at the simplified drawing on Fig. 11. The telescope configuration
is basically Gregorian but with two additional plane folding
mirrors. One advantage of this arrangement is that a cold, central
baffle can be used for stray light protection as shown. Also, the
second folding mirror can be used for space chopping without introducing extra optical aberration when it is tilted and without
causing a movement of the beam section on the primary mirror. This
mirror is additionally used for fine guidance on a star image under
control of the CCD star tracker located in the instrument volume.
The final pointing accuracy is then independent of any flexure
between the telescope itself and its outer case, and the image stability may be a fraction of an arc second. If no star bright
enough to close this loop is available then a gyrostabilized laser
on the outer case is used as an artificial star, and the gyro
drift is periodically corrected after integration on the brightest
field star available. Focussing of the telescope is achieved by
translating the first folding mirror along the telescope axis. The

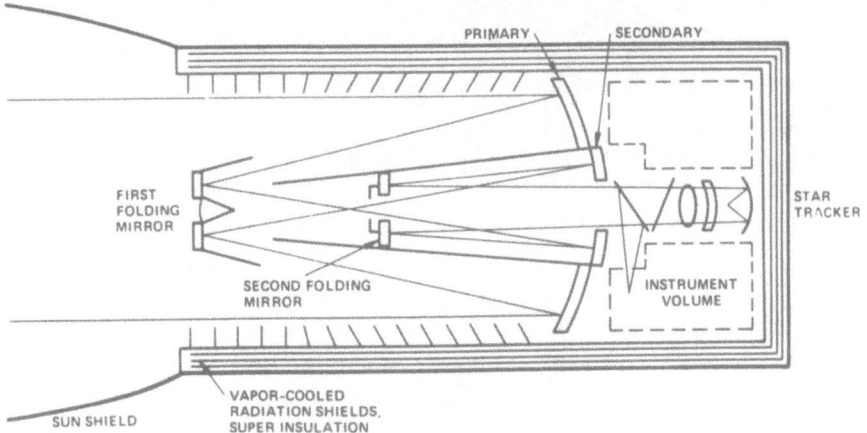

Fig. 11. Schematic of the SIRTF optical system. The telescope
configuration is Gregorian but is double folded by the addition
of the two plane mirrors.

image quality is sufficient for diffraction limited performance
down to a wavelength of 5 μm.

   The sensitivity achievable with SIRTF depends largely on the
radiative backgrounds experienced on-orbit. Those due to the
zodiacal light, telescope optics and molecular contamination are
summarized in Fig. 12. Superimposed on the background intensities
are curves corresponding to background limited NEP's for the reference condition of a 1 arc min. field of view and 10 μm bandpass. In the far infrared the limit will almost certainly be set
by the detectors themselves, particularly as the telescope temperature can be lowered if necessary. At mid and near infrared wavelengths the limit will be the zodiacal light, providing the molecular contamination does not exceed the limits shown. To ensure
that this is the case, great care will be exercised on-orbit to
minimize the ejection of molecules into the telescope environment.

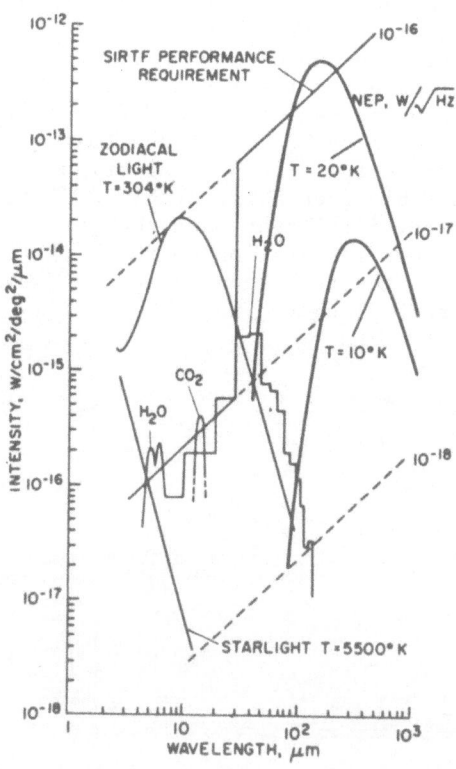

INFRARED BACKGROUND IN SHUTTLE ENVIRONMENT

Fig. 12. Summary of the expected radiation backgrounds in the
Shuttle environment and the corresponding background limited NEP's
of SIRTF in a 1 arc min. field of view and 10 μm bandwidth.

In particular, excess water from the fuel cells and evaporators will be stored in tanks rather than being ejected into space as normally planned. Also, the Shuttle thruster firings (RCS) will be kept to a strict minimum during observational sequences.

SIRTF Scientific Objectives. A complete list of scientific objectives for SIRTF would be very long indeed and would include a range of observations on essentially all known classes of objects in the Universe. It is of more interest here, therefore, to isolate a few of the areas in which its performance is unique amongst the various space observations being planned.

These derive primarily from its exceptional infrared sensitivity which can be achieved even with a relatively large optical throughput. It is ideally suited, therefore, to photometric observations of intrinsically faint and very distant objects and to total flux measurement of extended sources. A very exciting way of using this capability is clearly for "guided" searches for new types of infrared object. The ideal instrument for this purpose would be a photometric camera which utilizes a large area imaging detector. Such devices, based on the use of CCD's, can certainly be anticipated by the 1980's. It would then be possible to take "photographs" of limited areas of sky down to an extremely low flux level. Rather than an all sky survey, such as will have been completed by IRAS, therefore, one can make a very deep survey, i.e. to very large distances in a limited region of the sky. An exposure of around ten minutes, for example, would be sufficient to detect protogalaxies at a redshift $z \sim 10$ if their luminosities exceed 10^{12} L_O as expected on some models. Positive identification of candidate objects would, of course, involve further spectroscopic observations capable of detecting redshifted features such as the Lyman edge and "visible" spectral lines. Somewhat less speculative is the probability of detecting more evolved galaxies whose energy output is entirely in the infrared. Observations of visible galaxies have already revealed examples such as the Seyfert galaxy NGC 1068 which radiates more than 98% of its energy in the infrared. At the other end of the scale the sensitivity of SIRTF may be sufficient to reveal extremely cool, low mass stars within the Galaxy. Objects such as the postulated Brown and Pink dwarfs with masses below 10^{-2} M_O and temperatures less than 1000 K. These few examples illustrate the exploratory capability of SIRTF to open up new areas of study in the infrared. It should not be forgotten, however, that there are also many astrophysical problems associated with known objects which require high sensitivity observations. Quite close to home, for example, are the many unanswered questions concerning the composition and chemical history of the Solar System. SIRTF would be a very powerful tool in this area for spectroscopic studies aimed at determining the minerology of solid bodies such as asteroids and planetary satellites.

Finally, it should be remembered that the results of the IRAS survey will be extremely important in planning observations with all the telescopes which follow it.

Fig. 13. An artist's impression of LIRTS on Spacelab in its short
module configuration.

4.6 LIRTS (Large Infrared Telescope on Spacelab)

Telescope. This large, uncooled infrared telescope is under consideration by ESA as a possible Spacelab facility instrument. A phase A study of the telescope has been completed and more detailed studies of potential focal plane instrumentation will be carried out in the near future. The studies of SIRTF and LIRTS have proceeded almost in parallel, in fact, and they are seen to be complementary in their scientific objectives.

An artist's impression of LIRTS in orbit on Spacelab is shown in Fig. 13. It is seen here flying in the 'short module' configuration although 'pallet only' missions are also foreseen. More details of the telescope are visible on Fig. 14. In many respects it is identical with a groundbased telescope. The 2.8 m diameter F/2 primary mirror, however, is an intrinsically lightweight structure fabricated from ULE (ultra low expansion) glass. Considerable effort has also gone into the thermal and mechanical design of the structure to ensure that temperature fluctuations

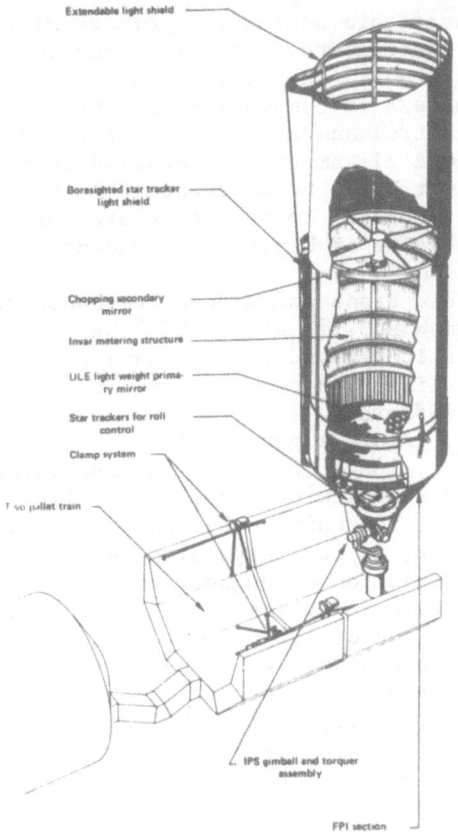

Fig. 14. Main elements of the LIRTS concept.

and gradients contribute no excess noise above the intrinsic photon noise from the two optical surfaces viewed by the detectors.
A long extendable shield is also provided to allow operation within
45° of the Sun and 30° of the Earth. The telescope optical configuration is Cassegrainian and the mirror surfaces themselves
are figured to give good optical performance over a large field
and when space chopping at the secondary mirror. Diffraction
limited performance is achieved throughout the far infrared over
a 10 arc min. field and with beam separations of up to 20 arc min.
Even at 10 μm it is possible to work at the diffraction limit up
to about 1 arc min. off axis and with beam separations of the same
order.

Pointing is by means of ESA's Instrument Pointing Subsystem
mentioned already in connection with SIRTF. The unusually compact
gimbal assembly has been designed specifically for zero-g operation and avoids the overall diameter restriction imposed by conventional 'outside ring' systems. A boresighted and two roll star
trackers as well as the gyro package are attached to the telescope
structure itself.

A large volume (~ 7 m^3) is available behind the primary mirror
for the location of focal plane instruments. This is shown in
some detail in Fig. 15. Also located in this volume is an alignment monitoring unit which includes a TV camera for viewing the
focal plane, sensors for focus control and a star simulator for
checking the star tracker/telescope alignment. A rotating beam
diverter mirror is provided to switch the beam between this unit
and the various focal plane instruments. These latter are currently
foreseen to be autonomous in the sense of having independent cooling
systems.

With diffraction limited field stops and a relative bandwidth
$\Delta\lambda/\lambda = 0.1$ the background limited N.E.P. is about 10^{-15} WHz$^{-1/2}$
around 20 μm and improves towards longer and shorter wavelengths.
The overall performance is still likely to be detector limited,
therefore, over much of the far infrared.

LIRTS Scientific Objectives. Again a general list of detailed
scientific objectives would be very long. LIRTS is an observatory
type facility capable of accommodating a wide range of focal plane
instruments which can be changed or refurbished between flights.
Its characteristic feature, however, is its large diameter and
correspondingly higher spatial resolution throughout the far infrared than both existing airborne systems and the planned cold telescopes in space. This feature will be utilized in many programmes
to map out source structures within extended complexes aimed at a
better understanding of the evolution, fragmentation and collapse
of clouds within the Galaxy. It will also be possible to determine
the far infrared brightness distribution of many external galaxies.

Spatial resolution is not exclusively associated with mapping,
however. Sensible spectroscopy also requires the use of small field
stops to isolate the physical region of interest in complex regions
where source confusion is high. Many spectroscopic techniques

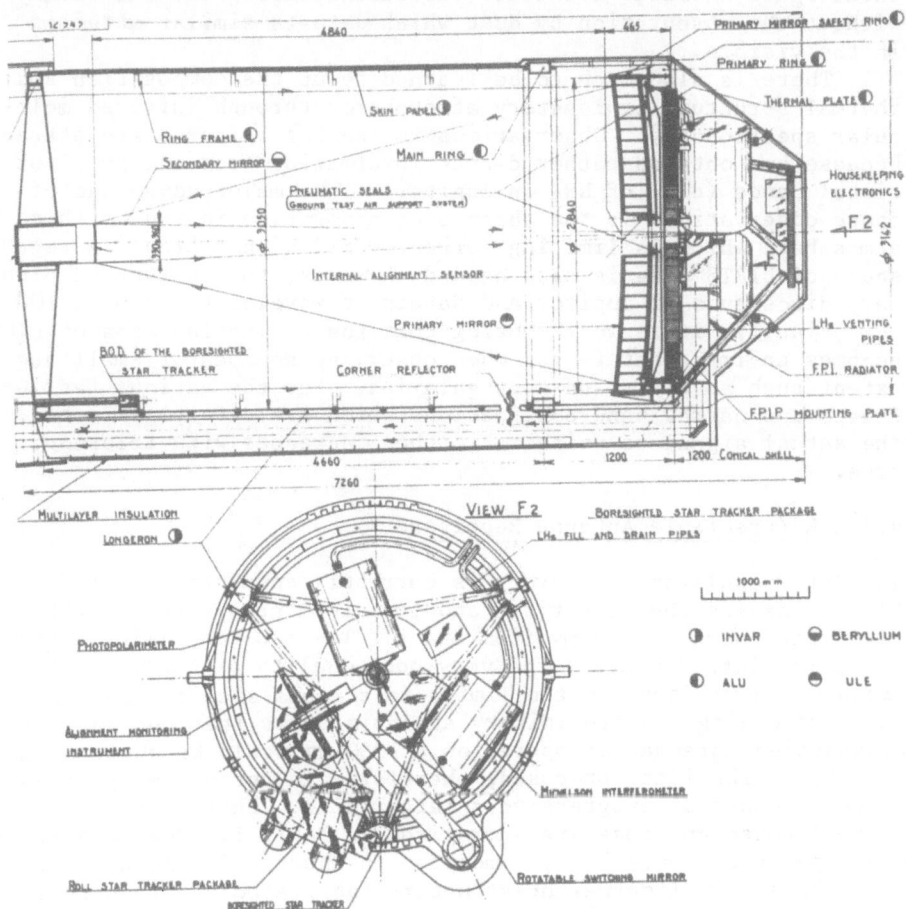

Fig. 15. LIRTS telescope and focal plane layout. Two foçal plane instruments are shown but more can be installed if required.

are also not limited by the telescope photon noise, and hence the spatial resolution and collecting area advantages of a large warm telescope can be gained without sacrificing sensitivity. High resolution observations in single spectral lines such as the important 28 μm H_2 line, for example, with a Fabry Perot interferometer would still be detector limited. Another very important example is heterodyne spectroscopy which will be used for molecular line investigations of the chemical composition, physical conditions and dynamics within cold regions of the interstellar medium and particularly in regions of star formation. Similar information in hotter, ionised regions can be obtained through atomic fine structure line observations. It will be possible, for example, to

investigate chemical abundance variations throughout the Galaxy
without the obscuration by dust which defeats similar attempts
in the visible.

There is also much to be learned about the composition and
thermal structure of planetary atmospheres through infrared molecular spectroscopy. Observations in the far infrared are attractive
because one obtains rather direct information from the pure rotational bands (e.g. of NH_3 in Jupiter). The major advantage of
space observations is the absence of absorption in the Earth's
atmosphere which is limiting current work in this area. A telescope of LIRTS size is also needed, however, to resolve the planetary discs even of Jupiter and Saturn at wavelengths around 100 μm.

Finally, it is worth noting that the collecting area of LIRTS
becomes an important factor when observing sources of small angular
extent such as stars and most galaxies. For the various far infrared and narrow band applications which are detector limited,
the actual signal to noise ratio then increases with telescope
area.

4.7 EAR (Erectable Antenna Receivers ?)

The Jet Propulsion Laboratory is currently conducting a study for
NASA to assess the scientific and technical prospects for submillimeter wave astronomy from space. The two compelling arguments
for going into space are the very poor quality of the available
groundbased windows and the limits to antenna size set by gravity
and windloading. A preliminary analysis shows that the largest
groundbased antenna for operation at 300 μm would be about 10 m in
diameter. The ideal proposed, therefore, is for a family of Shuttle
borne antennas of progressively larger size. Characteristics of
some of these antennas are summarized in Table 5. The largest single dish which could be accommodated on Shuttle is around 4 m. This
could be a very lightweight structure fabricated of graphite epoxy.
Preferably, however, the first generation antenna flown would be deployable and have a diameter of around 10 m. It could be formed
from 7 hexagonal panels each of 3.5 m in diameter which would be
stowed in a folded configuration and deployed on-orbit as shown in
Fig. 16. The largest antenna of this type which could be accommodated would have deployed diameter of 24 m and would consist of 37
hexagonal elements.

An alternative approach, which may lead eventually to antennas
considerably larger than 30 m is based on electrostatically control-
ling a precision reflector surface made of a thin elastic conducting
sheet. The shape is measured by means of a scanning optical system
and deviations corrected by depositing charge with a scanning electron gun. This possibility is seen to be a longer term prospect,
however, compared with the essentially available technology required
for the first generation deployable antennas.

The high resolution spectral line observations planned with
these antennas will use receivers based on essentially classical

Table 5

Characteristics of Possible Space Antennas

	Space EAR-1	Space EAR-2	Space EAR-3
Antenna Diameter	4.2 m	10 m	30 m
Antenna Type	fixed parabola	deployable	deployable
Maximum Frequency (GHz)	600	800	1000
Required Surface Accuracy	30 μm	25 μm	18 μm
Minimum Beamwidth	27 arc sec	8.6 arc sec	2.1 arc sec
Required Pointing Accuracy	5 arc sec	2 arc sec	0.4 arc sec
Maximum Gain (db)	88	98	110
Estimated Availability of Antenna Structure	1979	1982	1985

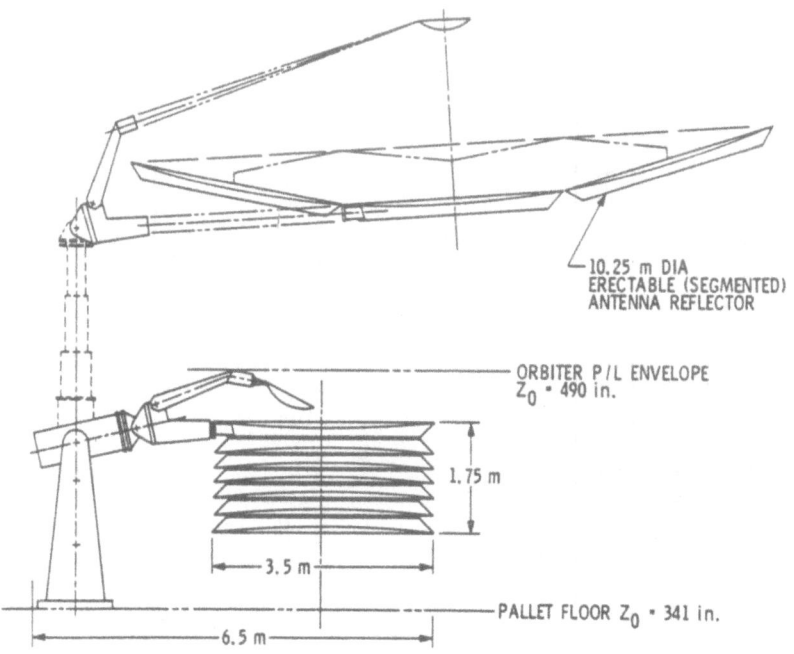

10.25 m DIA
ERECTABLE (SEGMENTED)
ANTENNA REFLECTOR

ORBITER P/L ENVELOPE
Z_0 = 490 in.

1.75 m

3.5 m

PALLET FLOOR Z_0 = 341 in.

6.5 m

A 7-Panel Space-Erectable Antenna (Space EAR-2?)
in its Stowed and Deployed Configurations

Fig. 16.   Conceptual view of a 10 m erectable antenna for submillimetre wave astronomy.

radio techniques, but utilizing local oscillators (e.g. klystrons,
carcinotrons) and mixers (e.g. Schottky barrier diodes) appropriate
to the higher frequencies. These systems are still mainly development items and their performance is improving rapidly. The data in
Table 6, however, shows that satisfactory performance could be obtained even with current state of the art receivers. A sensitivity
of 0.2 K is possible at 1 km/sec resolution within 1 minute of integration time. It is also worth noting that, for equivalent signal noise ratio, 1 minute of integration on a 10 m space telescope
is equivalent to about 150 hours on a 1 m airborne telescope.
      Scientific interest in the submillimetre region derives prin-
cipally from the large number of light molecules such as OH, CH, NH_3,
H_2O, CO, H_2S, HCl, etc. whose rotational spectra fall in this range.
The resulting spectral lines provide extensive information on the
chemistry, physical conditions and dynamics in all the situations
where these molecules are formed.
      Important questions concerning the evolution and collapse of
interstellar molecular clouds may be answered, for example. Are
the chemical composition and evolutionary state related? Does
cooling via the molecular lines, and particularly in maser lines,
play an important role in the collapse? What are the dynamics of

Table 6

Projected Sensitivities Achievable in 1 Minute Using State of the Art
Receivers on Various Telescopes

Freq. GHz	T_{rec} (°K)	$\Delta\nu$(1 km/sec) (MHz)	T(1 km/sec) (°K)	T(1 GHz) (°K)	$S(\times 10^{26}$ wm^{-2} Hz) ($\Delta\nu = 1$ GHz)		
					4 m diam	10 m diam	30 m diam
300	3000	1	0.2	.02	4	.7	0.08
600	5000	2	0.2	.03	.6	1	0.1
1000	6000	3.3	0.2	.04	9	1.4	0.2

the collapse.

It will also be of interest to determine the molecular composition of the Solar System and to compare that of relatively unmodified components such as comets with the protostellar cloud compositions derived elsewhere in the Galaxy.

There is also the possibility within both the Galaxy and external galaxies of utilizing the molecular clouds as probes of large scale dynamics and to map out spiral density waves and investigate the role of shock wave effects in star formation.

Pursuing a somewhat different line, it has been suggested that it may be useful to search for small scale anisotropies in the cosmic background radiation and also to probe the possible existence of hot intergalactic gas which may be revealed through Compton scattering of the 3K photons.

4.8 Small Cold Telescopes on Spacelab

As mentioned earlier, a number of small telescopes have been proposed for the second Spacelab flight in 1980. These are more limited instruments than the facilities discussed above, proposed by groups of collaborating laboratories. Final results of the selection procedure are not yet known, and it is not intended here to enter into descriptions of the various instruments. It is worth remarking, however, that in addition to their astronomical goals these instruments could yield valuable information on the quality of the Spacelab environment for infrared observations.

A national facility, GIRL (German Infrared Laboratory) is also being studied at present and could fly as early as 1982. This is a 0.4 m cooled telescope ($\sim$10 k) intended for both astronomical and atmospheric studies. The cryogenic system uses superfluid helium and up to five instruments are foreseen in the focal plane. These would include a detector array, a photometer/polarimeter and various medium and high resolution spectrometers.

5. CONCLUSIONS

There is no shortage of scientific objectives for infrared telescopes in space. They range from the purely exploratory to the extremely detailed study of objects ranging from the Solar System out to distant galaxies. Even before the sky has been comprehensively surveyed it is possible to anticipate the large return of information simply from answering questions about known objects whose further study with existing facilities is now limited. It is perhaps over optimistic to expect that all the instruments covered in this review will be provided - at least in the short term. Eventually, however, there will be the need to provide facilities possibly even more diverse than those already proposed if infrared astronomy is to be pursued in a comprehensive way. In the now relatively short term we can already look forward to

IRAS, however, and confidently expect that the unexpected will be
the dominant feature of its mission.

ACKNOWLEDGEMENTS

I am grateful for helpful discussions with many people connected
with the various projects and wish particularly to thank Dr. R.J.
van Duinen, Dr. S. Gulkis, Mr. J. de Koomen, Dr. D. Lemke, Dr.
C.J. Mather and Mr. L. Young for communicating information on
recent developments and for providing visual material. My thanks
also to Professor G. Setti and Dr. G.G. Fazio for inviting me to
present this review in Erice and for encouraging me to write it up
afterwards.

REFERENCES

Baluteau, J.P., Marten, A., Bussoletti, E., Anderegg, M., Beckman,
J.E., Moorwood, A.F.M. and Coron, N., 1977, Infrared Physics 17,
283.
Gillett, F.C., 'A 1-meter Cyrogenic Telescope for the Space Shut-
tle', p. 195.
Heinsheimer, T.F., Maran, S.P. and Sweeney, 1977, 'Satellite Obser-
vations of New Infrared Sources'. Preprint of paper prepared for
presentation to COSPAR, session on "Late Significant Results",
Working Group III, Panel 3A. Tel Aviv, June 1977.
Maran, S.P. et al., 'Long-Term Monitoring of Stellar Sources from
Earth Orbit', (Abstract) p. 35.
Moorwood, A.F.M.,'LIRTS: A Large Telescope for Spacelab', p. 207.
Neugebauer, G., 'Use of the Large Space Telescope for Infrared
Observations', p. 195.
Price, S.D. and Walker, R.G., 1977, 'Results from the Air Force
Geophysics Laboratory Survey Catalog'. Infrared and Submillimeter
Astronomy. Proceedings of a COSPAR Symposium held in Philadelphia,
June 1976 (G.G. Fazio, editor). Astrophysics and Space Science
Library, Vol. 63, D. Reidel Publishing Co., 1977.
Van Duinen, R.J., 'Infrared Astronomical Satellite (IRAS)', p. 177.

Other sources of information consulted during the preparation of
this review are acknowledged and listed below.

IRAS: Report of the Joint Scientific Mission Definition Team for
an Infrared Astronomical Satellite, May 1976. IRAS Project
Description (Report R-PM-76-10) Oct. 1976.

COBE: Mather, J.C., Report to the Space Science Board (Jan. 1977).

IRSAT: Report on the Mission Definition Study (ESA, DP/PS(76)9),
April, 1976.

ST: Report by the Instrument Definition Team on Final Instrument Definition (no ref.).

Infrared Photometer Final Report prepared for ITEK (subcontract No. 8237-AOOO2) by Lockheed Missles and Space Company Inc., August 1975.

Phase B Definition Study of the Large Space Telescope Infrared Photometer Instrument. Final Report prepared for the Perking Elmer Corporation (contract 25140-PT) by Ball Brothers Research Corporation, August 1975.

SIRTF: Shuttle Infrared Telescope Facility (SIRTF) Preliminary Design Study Final Report. Prepared for NASA Ames Research Center by Hughes Aircraft Company, August 1976.

Preliminary Reports on SIRTF Science prepared by members of the FIRST Instrument Definition Team.

LIRTS: Report on the Phase A Study (ESA, DP/PS(76)18), May 1976. Phase A Final Report (Matra No. 60/181), June 1976.

EAR: Gulkis, S., Kuiper, T.B.H. and Swanson, P.N., Report on Scientific and Technical Prospects for Submillimetre Wavelength Radio Astronomy from Space, February 1977.

GIRL: Lemke, D., Klipping, G. and Römsich, N., Liquid Helium Cooled Infrared Telescope for Astronomical and Atmospherical Measurements from Spacelab. Preprint 1977.

SEMINARS

FAR INFRARED EMISSION OF MOLECULAR CLOUDS AND STAR FORMATION
IN THE GALAXY

Charles Ryter

Division de la Physique, Service d'Electronique Physique,
SES, Centre d'Etudes Nucléaires de Saclay, B.P. 1,
91190 Gif-sur-Yvette, France

1. FAR INFRARED EMISSION ASSOCIATED WITH MOLECULAR CLOUDS

Basic reasoning: Start from a purely observational point of view.
Some HII region – giant molecular cloud complexes are observed to
be far IR sources. The flux received in a 0.5° beam is larger than
that received in a 0.25° beam, showing that the source (say at
∿5 kpc away) is ∿50 pc in size, typical for a molecular cloud, and
too large for an HII region. An estimate of the dust temperature
is obtained from 2 color photometry (75-95 μm band and 115-196 μm
band). The column density of the cloud is derived from ^{13}CO observations. The IR brightness, I_{IR}, is related to the column density,
N_H, by

$$I_{IR} = L_{IR}^{H} N_H / 4\pi \qquad (1.1)$$

which defines an observed luminosity, L_{IR}^{H}, normalized per hydrogen
atom. The results of the analysis of nine clouds are given in
Table 1. Relevant data are taken from Olthof (1974, 1975), and
Wilson et al. (1974). See Ryter and Puget (1977) for details. It
is found that

$$L_{IR}^{H} \sim 2 \times 10^{-30} \quad W(H \text{ atom})^{-1} \qquad (1.2)$$

When corrected for the contribution of the hot HII region, the
color temperature is of the order $T_{dust} \sim 25$ K, in excellent
agreement with the value deduced from standard dust properties,
and from the column density.

G. Setti and G. G. Fazio (eds.), Infrared Astronomy, 319-326.

Table 1

Far Infrared Emission of Molecular Clouds

Object	Distance (pc)	Column Density $(10^{22}cm^{-2})$	Minimum Extent (0)	$F_{114-196}$	F_{71-95}	L_{IR}^H	T_{dust}
				$(10^{-13}Wcm^{-2})$		$(10^{-30}W)$	(K)
NGC 6334	700	11	0.4	2.75	4.8	2.8-4.3	35
M 8	1000	2.5		0.65	1.2	2.4	37
W 31	4500	13	0.5	0.95	1.3	0.73	32
W 33	4600	23	0.3	1.25	1.65	0.54	31
M 17	2100	10	0.5	2.75	5.8	1.5	38
M 16	2500	5.8	0.2	0.95	1.3	1.6-4.0	32
W 41	6500	14	0.5	0.95	0.85	0.54	28
W 43	7000	14	0.3	1.25	1.65	0.8-2.4	31
W 44	2000	20	0.4	0.95	0.65	2.2-3.5	25
Average						2.1±1.3	32

2. GALACTIC FAR INFRARED DIFFUSE EMISSION

An application of the results in Table 1 to all the molecular clouds, would lead one to expect a galactic diffuse (but patchy) emission to be observable. This is indeed the case, and results of observations are summarized in Figure 1. The values of L_{IR}^H are summarized in Table 2 where it clearly appears that, at least within $|\ell| \lesssim 45°$, the far IR source function is the same as that found for molecular clouds mentioned in Table 1.

Table 2

Summary of the Far Infrared Emission, L_{IR}^H, of the Gas-and-Dust Mixture as Deduced from Available Observations

(ℓ)	$10^{30}L_{IR}^H$ W(H atom)$^{-1}$	
9 clouds	2.1 ± 1.3	Table 1
2.5	2.4	Pipher (1971)
350-30	2	Low et al. (1977) (no detailed analysis completed)
28	4.5	Rouan et al. (1977)
42-46	1.8	Serra et al. (1977)
55	1	ibid.

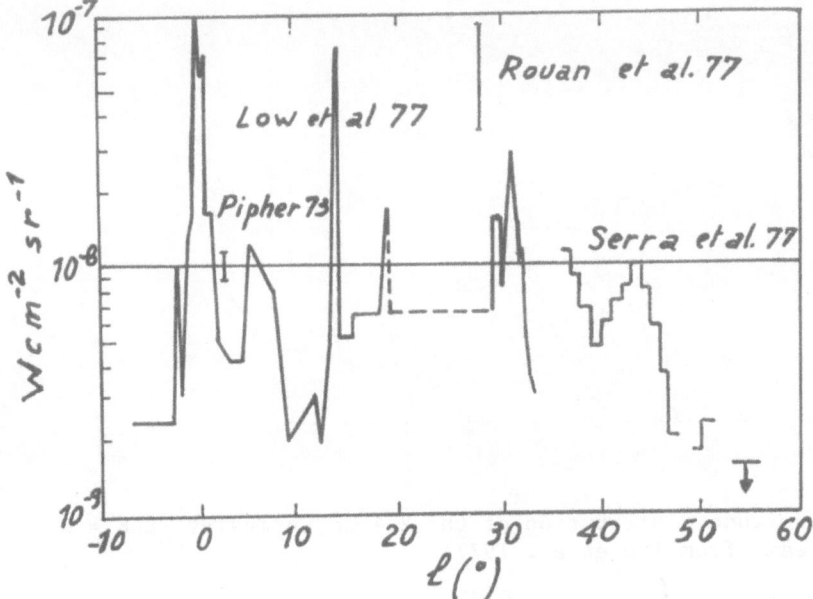

Fig. 1. Summary of observations of a "diffuse" far IR emission of the galactic plane.

3. 2.4 μm EMISSION OF THE MILKY WAY

A mapping of the Milky Way at 2.4 μm is now available between $\ell = 350°$ and $\ell = 60°$ (Hayakawa et al. 1976; Okuda et al. 1977a; Ito et al. 1977). See figure 2 for the galactic longitude profile in the strip $b = \pm 1°$ and note the humps at $\ell \approx 30°$ and at $\ell \approx 45°$. They are interpreted as arm components superimposed on a disc component (Hayakawa et al. 1977, Serra and Puget 1977).

3.1 Contribution of Disc Stars

Define the contribution at $\lambda = 2.4$ μm of a star of spectral type Sp (including luminosity class), effective temperature, T(sp), and luminosity, L(sp), as

$$\xi(Sp) = \frac{B_{2.4}\{T(Sp)\}}{\int_0^\infty B_\lambda\{T(Sp)\}d\lambda} = \frac{B_{2.4}\{T(Sp)\}}{L(Sp)} \tag{3.1}$$

where B_λ is the Planck function. With $\rho(\tilde{\omega},z)$ being the density of disc stars in the Galaxy and $\rho(10,0) = \rho_\Theta$ its value in the solar vicinity, the 2.4 μm volume luminosity is then given by

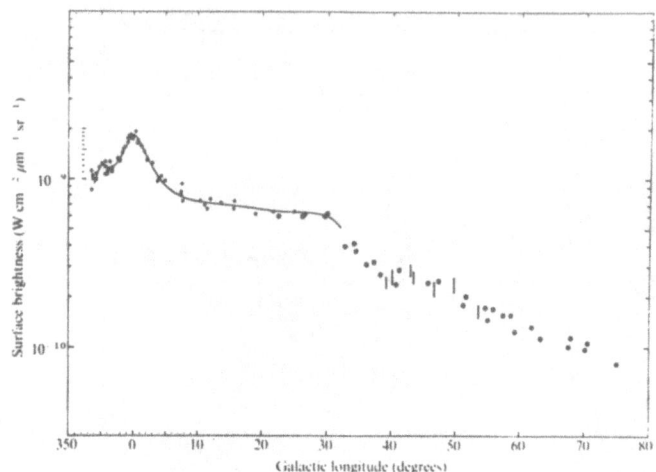

Fig.2. Longitude distribution of the 2.4 μm intensity at b = 0 in
2° x 2° beam (from Ito et al. 1977).

$$\mathcal{L}_{2.4}(\tilde{\omega},z) = \int L(Sp)\xi(Sp)\frac{\rho(\tilde{\omega},z)}{\rho_\theta}\phi_\theta(Sp)d(Sp) \qquad (3.2)$$

where $\phi_\theta(Sp)$ is the number density of stars of spectral type Sp in
the solar neighborhood. When absorption is taken into account (radial distribution of H_2, thickness of layer z_{eq} = 117 pc, N_H/A_v =
2 x 10^{21} cm^{-2}mag^{-1}, $A_{2.4}$ = 0.08 A_v), the Schmidt (1965) model for
the Galaxy gives the dashed line in Fig. 3. The agreement is good
at $\ell \approx 0$ and $\ell \approx 60°$, but the excesses at $\ell \approx 30°$ and $\ell \approx 45°$ are
obvious.

3.2 Contribution of Spiral Arms

The sharpness of the features implies relatively young objects.
Build a simple galaxy model made up of 3 rings with appropriate
width, $\Delta\omega$, thickness 300 pc, and volume luminosity, $\mathcal{L}_{2.4}$, as summarized in Table 3. The result of the calculation is shown in
Fig. 3 and gives a very good fit to the observations.

3.3 Some Implications

$\mathcal{L}_{2.4}^{arm}$ is attributed to red giants produced by stars earlier than
Sp $\approx$ F, for lifetimes on the main sequence to be consistent with
the "arm" characteristic. By postulating a steady star formation
rate over the last 2 x 10^8 yr, it is possible to deduce the total
present luminosity (visible,UV) of the stars in the 5 kpc ring from
the 2.4 μm luminosity. Since $\xi^{MS}(Sp)$ is negligible, the ratio of
the energy released by a star on the main sequence (luminosity L^{MS}
(Sp), lifetime $\tau^{MS}(Sp)$) to that released in the 2.4 μm band by the

Table 3

Adopted simplified model of the Galaxy: circular rings at $\tilde{\omega}$ = 5, 6.7 and 7.7 kpc + disc (from Serra and Puget 1977).

$\tilde{\omega}$ (kpc)	$\Delta\tilde{\omega}$ (kpc)	arm 2.4 (10^{25} W pc^{-3} μm^{-1})	disc 2.4
5	0.75	4.1	0.72
6.7	0.3	1.0	0.44
7.7	0.3	0.9	0.33
10		0	0.13

same star when it has moved into the red giant branch, is obviously given by

$$\frac{L^{MS}(Sp)\tau^{MS}(Sp)}{L^{RG}_{2.4}\tau^{RG}} = \frac{L^{MS}(Sp)\tau^{MS}(Sp)}{L^{RG}\tau^{RG}}\frac{1}{\xi^{RG}} = \{\xi^{RG}X(Sp)\}^{-1} \quad (3.3)$$

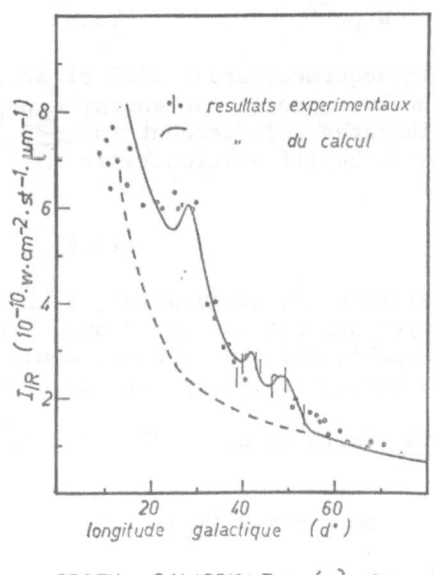

PROFIL GALACTIQUE (λ = 2,4 μm)

Fig. 3. The 2.4 μm intensity from the luminosity function in the solar neighborhood (dashed line), and the 2.4 μm intensity deduced from the complete model (solid line) of Table 3 (from Serra and Puget, 1977).

Table 4

Some values of the ratio of the total energy released on the main sequence by a
star of spectral type Sp, to the energy released per μm in the 2.4 μm band by the
same star during the red giant phase. {Data are taken from Iben (1965)}.

Sp on M.S.	M/M_0	L^{MS}/L_0	$\tau_{MS}(10^6 \text{yr})$	L^{RG}/L_0	$\tau_{RG}(10^6 \text{yr})$	$\{\xi^{RG}X(Sp)\}^{-1}$
B7	5	700	65	1000	7.5	15 μm
A3	3	100	222	200	70	28 μm
A5	2.25	30	480	100	38	60 μm

Some values of $\{\xi^{RG}X(Sp)\}^{-1}$ have been computed from eqs. (3.1) and
(3.3) based on the evolutionary tracks of Iben (1965). They are
given in column 7 of Table 4. We adopt the average value

$$< \xi^{RG}X(Sp) > = 1/30 \quad \mu m^{-1}, \tag{3.4}$$

which is appropriate for galactic open clusters and should not be
too sensitive to the initial mass function (I.M.F.). Since in the
steady state the number density of stars (MS or RG) is directly
proportional to their lifetimes, the total luminosity per unit volume in the 5 kpc ring becomes

$$\mathcal{L}_{tot}^{5 \text{ kpc}} = 30 \, \mathcal{L}_{2.4} = 1.2 \times 10^{27} \text{ W pc}^{-3}. \tag{3.5}$$

Star formation rate. On the main sequence, about 1/10 of star
mass is converted into He. In the nuclear process an energy equal
to 7/1000 the rest mass energy is liberated. To account for $\mathcal{L}_{tot}^{5 \text{ kpc}}$
the star formation rate (S.F.R.) has to be (if stationary in
time)

$$dM/dt = 9 \times 10^{-8} \text{ M}_\Theta \text{yr}^{-1} \text{pc}^{-2}, \tag{3.6}$$

(z_{eq} = 300pc has to be used for consistency of computation, even
if stars indeed form in a thinner layer and diffuse later on). Present surface gas density is $\sigma = 12$ M$_\Theta$pc^{-2}, and the lifetime would
be approximately 1.5×10^8 yr. Thus stationarity may not strictly
apply.

Comparison with Lα-photon counts. Smith et al. (1977) obtain
a star formation rate

$$dM/dt = 1.5 \times 10^{-8} \text{ M}_\Theta \text{yr}^{-1} \text{pc}^{-2} \quad (100 > M > 0.1 M_\Theta) \tag{3.7}$$

at a distance $\tilde{\omega} \sim 5$ kpc. Two effects may collaborate to account
for the factor of 6 between eqs. (3.6) and (3.7): (i) The S.F.R.
decreased by ~ 6 in the last $\sim 2 \times 10^8$yr and/or (ii) the I.M.F. is
not constant in the Galaxy and steepens at smaller $\tilde{\omega}$. A summary
of characteristics at $\tilde{\omega} = 5$ kpc and $\tilde{\omega} = 10$ kpc is given in Table 5.

Table 5

Summary of Luminosities, Densities, and Star Formation Rates in
the Solar Vicinity and in the 5 kpc Ring

	Sun	5 kpc	Ratio	
Disc Stars $(M_\odot pc^{-2})$	133	374	~ 2.8	Smith et al. 1977
Gas $(M_\odot pc^{-2})$	2.5	11.7	4.7	Gordon, Burton 1976
ΣN_c (Ly-continuum) $(10^{44} s^{-1} pc^{-2})$	0.41	2.3	5.6	Smith et al. 1977
S.N.R. ; P.S.R.			~ 5	
dM/dt (S.F.R.) $(10^{-9} M_\odot yr^{-1} pc^{-2})$	2.5	$15^\dagger$	6	Smith et al. 1977 $100 > M/M_\odot > 0.1$
$\mathcal{L}_{2.4}^{disc}$ $(10^{24} W\ pc^{-3} \mu m^{-1})$	1.3	7.2	5.5	Predicted from Schmidt model
$\mathcal{L}_{2.4}^{disc + arm}$ $(10^{24} W\ pc^{-3} \mu m^{-1})$	1.3	41	30	Observed (this work)
dM/dt (S.F.R.) $(10^{-2} M_\odot yr^{-1} pc^{-2})$	2.5	$\lesssim 90^{\dagger\dagger}$	$\lesssim 35$	$100 > M/M_\odot > 0.1$

$\dagger$ in the last 10^5 years (lifetime of HII region).
$\dagger\dagger$ if S.F.R. was steady in the last $\sim 2 \times 10^8$ yr.

4. CONNECTIONS BETWEEN 2.4 μm AND FAR INFRARED GALACTIC EMISSION

At 5 kpc, eq. (3.5) yields for the total luminosity normalized per
H atom of the interstellar medium

$$< L_{tot}^H > \approx 10 \times 10^{-30}\ W(H\ atom)^{-1}, \tag{4.1}$$

i.e. about 5 times the value $L_{IR}^H \sim 2 \times 10^{-30}\ W(H\ atom)^{-1}$ (Table 2).
It means that about 1/6 of the total luminosity of the 5 kpc ring
is radiated in the far infrared, i.e. that about 17 percent of visible and UV light is degraded in the interstellar medium. This figure is not unreasonable if the M.S. stars are located in the vicinity of the molecular clouds from which they are born, quite a likely

hypothesis.

The two independent approaches, i.e. the study of the far infrared emission of the dust mixed with the gas, and that of the 2.4 μm of the stars (on the red giant branch) lead consistently to the conclusions that in the 5 kpc ring, (i) the star formation rate per surface density unit of gas is larger by a factor $\sim$ 5 compared to the solar vicinity, and (ii) the birth rate of O stars appears to be deficient compared to less massive stars, provided the star formation rate did not dramatically slow down in the last $\sim 10^8$ years.

Acknowledgements: This work was done in collaboration with Drs. G.Serra and J.L.Puget. I am indebted to Professor G.Setti for his careful reviewing of the manuscript.

REFERENCES

Gordon, M.A., and Burton, W.B., 1976, Astrophys. J. 208, 346.
Hayakawa, S., Ito, K., Matsumoto, T., Ono, T., and Uyama, K., 1976, Nature 261, 29.
Hayakawa, S., Ito, K., Matsumoto, T., and Uyama, K., 1977, Astron. and Astrophys. 58, 325.
Iben, I., 1967, Ann. Rev. Astron. and Astrophys. 5, 571.
Ito, K., Matsumoto, T., and Uyama, K., 1977, Nature 265, 517.
Low, F.J., Kurtz, R.F., Poteet, W.M., and Nashimura, T., 1977, Astrophys. J. (Letters) 214, L 115.
Okuda, H., Maihara, T., Oda, N., and Sugiyama, T., 1977, Nature 265, 515.
Olthof, H., 1974, Astron. and Astrophys. 33, 471.
Olthof, H., 1975, Thesis, University of Groningen, The Netherlands.
Pipher, J., 1973, I.A.U. Symp. 52, 559 (Dordrecht: Reidel).
Rouan, D., Lena, P.J., Puget, J.L., de Boer, K.S., and Wijnbergen, J.J., 1977, Astrophys. J. (Letters) 213, L 35.
Ryter, C., and Puget, J.L., 1977, Astrophys. J. 215, 775.
Schmidt, M., 1965, in Galactic Structure, ed. by A. Blaauw and M. Schmidt, p. 513 (Chicago: University of Chicago Press).
Serra, G., 1977, Thesis in preparation.
Serra, G., Puget, J.L., and Ryter, C., 1977, Symp. Recent Results in Infrared Astronomy, NASA-AMES Jan. 13-15, to be published.
Serra, G., and Puget, J.L., 1977, Meeting of the French Physical Society, Poitiers, June 27-July 1.
Serra, G., Puget, J.L., Ryter, C., and Wijnbergen, J.J., 1977, Astrophys. J. (Letters) (to be published).
Smith, L.F., Bierman, P., and Mezger, P.G., 1977, Astron. and Astrophys. (in press).
Wilson, W.J., Schwartz, P.R., Epstein, E.E., Johnson, W.A., Etcheverry, R.D., Mori, T.T., Berry, G.G., and Dyson, H.B., 1974, Astrophys. J. 191, 357.

SHOCK INDUCED STAR FORMATION

Bruce G. Elmegreen

Harvard University, Cambridge, Massachusetts

1. OVERVIEW

Star formation seems to occur after all or part of an interstellar
cloud becomes gravitationally unstable. This instability may arise
in a variety of circumstances which can be classified broadly into
two categories. Some gravitational instabilities seem to proceed
"spontaneously" in a dense cloud, as a natural part of the cloud's
evolution: For example, a cloud may form in a state of high internal energy, either thermal, magnetic or turbulent in origin and
may become gradually unstable as this energy dissipates and the
cloud cools radiatively. Similarly, some clouds may undergo molecular or grain-molecular reactions, thereby changing the balance
between heating and cooling and eventually leading to gravitationally unstable regions.

The second manner in which interstellar gas can become gravitationally unstable is by "stimulated" collapse. This will occur
if some process external to a cloud can act on the cloud and alter
its state of gravitational stability. For example, an existing
stable cloud can be compressed as a whole (with or without shocks)
until it exceeds some critical density. Changes in heating and
cooling processes at this larger density may then lead to gravitational instabilities, or the cloud may find itself immediately
unstable in a dynamic sense. A similar result is obtained if only
part of a stable cloud is compressed and "forced" into a state of
gravitational instability. Of course, the possibility that a
cloud forms from lower density gas and is gravitationally unstable
immediately upon formation also may be included in either the
spontaneous or stimulated categories, depending on the manner in
which the low density medium condenses.

Following this initial phase, when interstellar matter is

327

G. Setti and G. G. Fazio (eds.), Infrared Astronomy, 327–333.

somehow put in a state of gravitational instability, there begins
a second phase in the process of star formation which concerns
the initial collapse of the cloud. Problems relating to the time
scale of the collapse, the unstable fragment masses, the geometry
and manner of early collapse will arise in this second phase. The
final collapse to a star is regarded here as a third phase. The
formation of individual stars, binary stars, accretion disks, etc.
becomes important here.

Within this broad framework of problems relating to star
formation, the concept of shock induced star formation (SISF)
emerges as one possible means of initiating gravitational instabilities. The initial collapse phase (Phase II above) may also
be unique to SISF because of the special geometry that is involved.
Details in the final stages of star formation are probably not
significantly different for shocked and unshocked regions.

2. BASICS OF SISF

Three conditions must be satisfied for a shock to induce significant star formation. First of all, the shock must exist for a time
that is long enough for gravitational collapse to become significant. Otherwise the compressed gas will expand back to its
original state after the shock disappears. Secondly, there should
be a favorable orientation of the ambient (preshock) magnetic
field with respect to the direction of shock propagation. This
allows the possibility of large compression. A third condition
for generating significant star forming activity is that there
must be sufficient gas available for compression, so that an isothermal shock can form and eventually accumulate a large amount
of matter.

With these intuitive conditions, we may immediately compare
the extent and magnitude of star formation that may result from
shocks of various types. For example, an HII region may expand
into a molecular cloud for a time at least as long as the main
sequence lifetime of an O star, some 2 to 3 million years. The
resulting shocks will have velocities that are typically 1 to
10 km/s, comparable to the expected Alfvén velocity in the clouds.
This coincidence in velocity allows the field to be relatively
straight and unperturbed in the part of expansion that proceeds
parallel to $\vec{B}$, while expansion perpendicular to $\vec{B}$ will have relatively low compression and may even occur without severe shocking.
Evidently, all three conditions of ample time, ample accumulation
and favorable field orientation may be satisfied near some expanding HII regions, and SISF may be expected to proceed under
typical conditions.

Supernova (SN) shocks are relatively short lived, ($\lesssim 10^5$
years), depending on the preshock density. If high post shock
accumulation is achieved by a SN shock advancing into a molecular
cloud, then the shock duration will be even shorter. Swept up

shells in model calculations also maintain significant magnetic
pressure and the field lines tend to lie in the plane of the
shock. This is because the SN pressure is typically much larger
than the preshock magnetic pressure (or shock velocity >> preshock
Alfvén velocity). Thus SN induced star formation may not be expected to occur readily in the swept up shell. Perhaps existing
clouds that are engulfed by the high pressure of a SNR are better
candidates for regions of SISF.

Shocks may also occur in a collision between two clouds. The
duration of the collision may be several million years and the
amount of material that is available for compression may be large
for large clouds. However, not all collisions may be efficient
in forming stars, because the field orientations in the clouds
will, in general, be random. A collision that compresses the
field may transfer momentum into the clouds with magnetic field
pressure alone and only weak compression may result. It is also
unclear at the present time whether or not clouds collide often
enough to make this process significant. The only region where
collision induced star formation has been suggested is NGC 1333
(Loren, 1976) and this is at the periphery of the Per OB II association. It is possible that an expanding HII region induced the
collision of two initially separate clouds or that the current
appearance of two colliding clouds is due to a single cloud with
one part shocked and moving into the other part.

It is conceivable that star formation could also occur in
compressed material that has been swept up by a strong stellar
wind. The lifetime of the resulting bubble, the magnetic field
orientation in the compressed matter and the amount of material
that is available for accumulation should be as favorable for
star formation in this case as in the case of an expanding HII
region.

Of course, the details of any gravitational instability
must be considered individually for each astrophysical situation.
Nevertheless, the possibility of star formation in various interstellar shock processes may be expected to depend on the three
basic qualities discussed here: the duration of the shock, the
magnetic field orientation relative to the shock and the amount
of gas available for accumulation.

3. SOME DETAILS OF THE COLLAPSE

A detailed numerical calculation of the dispersion relation for
gravitational collapse in a pressurized, plane parallel, isothermal
layer is now available (Elmegreen and Elmegreen, 1978). This has
direct application to the thin dense layer behind an isothermal
shock. The principal results are summarized here.

We introduce the quantity

$$A \equiv \left(1 - \frac{P}{\rho_{oo}c^2} \right)^{1/2} = \left(\frac{\pi G \sigma^2 / 2}{P + \pi G \sigma^2 / 2} \right)^{1/2} , \tag{3.1}$$

where P is the shock driving pressure, ρ_{oo} the density in the mid-
plane of the layer, c the sound velocity, G the gravitational constant and σ the column density in the layer, measured perpendicular
to the plane, in gm cm^{-2}. Perturbations of the form exp (iωt + ikx)
are introduced to the equilibrium distribution and ω is determined
as a function of k for appropriate boundary conditions of the
layer.

In the three dimensional collpase that results in Jeans'
criterion, we have $\omega^2 = -4\pi G\rho + c^2 k^2$. In the limit of infinite
wavelengths (and therefore masses) the collapse rate goes to a
constant, $(4\pi G\rho)^{1/2}$. For collapse in a plane, however, we expect
that ω will go to zero as k goes to zero. This follows simply
from the scaling approximation, $\omega^2 \sim v^2/L^2 \sim GM(L)/L^3$ for free
fall velocity v and fragment size L. For three dimensions, M(L) $\sim$
$\sim \rho L^3$ and ω^2 becomes a nonzero constant, independent of L. For
two dimensions, M(L) $\sim \sigma L^2$ and ω^2 decreases as L^{-1} for large fragments. We find from our calculation that the value of ω^2 at
(Ak) $\sim$ 0 is

$$\omega^2 = -Akc(8\pi G\rho_{oo})^{1/2} \qquad (3.2)$$

As k increases from zero, $(-\omega^2)$ increases and reaches a
maximum before decreasing back to zero and to negative values
(i.e., ω^2 positive). The second zero crossing marks the transition
from unstable (small k) to stable (large k) fragments. The
corresponding wavenumber determines the minimum unstable mass.
The wavenumber at the maximum growth rate determines the mass of
the fragments which dominate the collapse.

The qualitative nature of this isothermal collapse may be
understood by comparing the mass in a collapsing fragment to the
mass of a pressurized, isothermal sphere that is at the threshold
of gravitational collapse (Spitzer, 1968), i.e.

$$M_{sph} = \frac{1.18 \ c^4}{G^{3/2} \ P^{1/2}} \qquad (3.3)$$

We assume that the mass in a post shock fragment is equal to the
mass in a cylinder of diamter $\lambda/2$ for wavelength λ:

$$M_{cyl} = \frac{\pi}{4} \left(\frac{\lambda}{2}\right)^2 \sigma . \qquad (3.4)$$

We then obtain

$$\frac{M_{cyl}}{M_{sph}} = 0.83 \ \frac{A(1-A^2)^{1/2}}{\nu^2} \qquad (3.5)$$

after normalizing the wavenumber to $\nu = kc(2\pi G\rho_{oo})^{-1/2}$. We find
from our calculations that ν varies roughly in inverse proportion
to A, for a given value of ω. Thus M_{cyl}/M_{sph} varies with A roughly
as $A^3(1-A^2)^{1/2}$, which has a maximum for some A between 0 and 1.
From the calculation we determine that

M_{cyl}(minimum unstable mass) $< 0.36\ M_{sph}$.   (3.6)

Thus the mass of the smallest unstable fragment is stable after its collapse to a sphere regardless of A. Similarly, the mass of the most rapidly growing element satisfies M_{cyl}(maximum growth rate) $< M_{sph}$ if A < 0.54, and M_{cyl}(MGR) $> M_{sph}$ if A > 0.54. This condition on A corresponds to $\sigma > 0.9(P/\pi G)^{1/2}$. Thus, the dominant unstable mass, i.e. the one which collapses most rapidly, is also stable after it contracts to a sphere unless the layer has acquired a sufficiently large column density. After the layer is large, the most rapidly collapsing mass is unstable as an isothermal sphere.

The tendency for small unstable disks to stabilize as spheres may be understood as follows. Consider a sphere that is initially stable in the sense that its mass is less than M_{sph} for some external pressure. Now stretch the sphere to a disk and release it. The disk will return to a spherical shape and to its original state of stability (after several damped oscillations). In the same way, a sufficiently small fragment may be unstable as a thin disk but stable as a sphere.

Thus gravitational instabilities in plane parallel layers will not always lead to complete collapse. The isothermal calculation reported here suggests that significant collapse will occur after the post shock gas reaches a column density which satisfies $\sigma > > 0.9(P/\pi G)^{1/2}$ for shock driving pressure P. This inequality, or one like it, may eventually be a useful criterion to determine the onset of rapid star formation behind a shock. However, the dynamical details of the collapse, the way in which heating and cooling changes with density, magnetic diffusion and other complications prohibit the derivation of any universal criterion for star formation behind shocks at this time.

Sufficiently large masses may be unstable as spheres for any σ, but these large masses will grow slowly. It is useful to determine a maximum mass, M_{MAX}, such that the collapse rate of this mass, ω, is greater than $1/\tau$ for some time scale τ characteristic of the shock. This is the time during which post shock material could have collapsed. We obtain from our calculation for $\nu A \leq 0.1$,

$$M_{MAX} \cong 0.6\pi^5 G^2 \tau^4 \sigma^3$$

$$= 1\ M_{\odot}\ (\tau/10^5\ \text{years})^4\ (\sigma/0.028\ \text{gm cm}^{-2})^3$$ (3.7)

where $\sigma = 0.028$ gm cm^{-2} for an application over 10 pc of preshock density $n(H_2) = 200$ cm^{-3}. Masses larger than M_{MAX} will not be significantly unstable in times less than τ. The fact that a time dependent upper limit to the collapsing mass exists at all is a result of the 2-dimensional geometry of post shock layers, for which $\omega \to 0$ as $k \to 0$. We obtain the interesting result from our isothermal calculation that M_{MAX} is independent of c and P!

We may summarize these results as follows: (1) There will be

a delay in the time of active star formation which is in addition
to a collapse time or a main sequence turn-on time. This results
from the requirement that the column density in the layer must
first increase to a sufficiently large value before gravitational
instabilities can lead to complete collapse.

One observable consequence of this is related to the possibility that star formation may proceed as a chain reaction, involving successive generations of young stars and shocks. In that
case, the centers of star forming activity will be separated in
age and position by an amount depending on the preshock density,
the shock driving pressure, etc.

We have also found that the maximum mass of a fragment that
can collapse in a given time, which is determined by the relevant
timescale or age of the shock, depends sensitively on that time
and on the column density of matter that has accumulated behind
the shock. This maximum mass is relatively insensitive to rms
motions in the shocked gas.

The observational significance of this was anticipated by the
intuitive arguments in §II: Expanding HII regions (for which
$\tau \sim 10^6$ years and $\sigma \sim 0.1$ gm cm^{-2}) can create large self-gravita-
ting fragments compared to those created by supernova shocks
(where $\tau \sim 10^5$ years). This will be true even if σ is the same
in each case. The material that accumulates behind short-lived
shocks may eventually form stars after the shock has disappeared
(perhaps by a spontaneous process), or it may only form small
clouds, Bok globules, etc.

4. MAGNETIC FIELDS

Magnetic fields, $\vec{B}$, in the primordial cloud provide an asymmetric
source of internal pressure. This can influence the nature and
compression of the shock, depending on the direction of the shock
velocity relative to $\vec{B}$. Magnetic support against gravitational
collapse will be reduced in the part of a shock that propagates
parallel to $\vec{B}$. This follows from the variation of gravitational
energy density in an unstable mass, $\sim GM/R^4 \propto \rho^{3/2}$, compared to the
energy density of the field, $B^2/8\pi$. For compression parallel
to the field, $\vec{B}$ does not change and the ratio of gravitational
to magnetic energy density increases as $\rho^{3/2}$. For compression
perpendicular to $\vec{B}$, B varies as ρ and the same ratio is proportional to $\rho^{-1/2}$. Thus, magnetic support against gravitational
collapse will be enhanced in the part of the shock that propagates perpendicular to $\vec{B}$.

The observational significance of this is that the site of
active star formation should be in that part of the shock that
propagates parallel to $\vec{B}$. Subgroups of the Orion OB association
tend to be aligned in a near-linear sequence. This could be a
result of successive generations of SISF in a uniform cloud in
which the magnetic field orientation is also fairly uniform.

A second property of magnetic fields may be important for SISF. This concerns the angular momentum per unit mass of gravitationally unstable elements: the angular momentum per gram will be reduced in gas that is compressed along a magnetic field. This is a result of an increase in the gaseous density under conditions where the angular velocity due to galactic rotation is unchanged. The angular momentum lost by the compressed gas will be transferred to remote parts of the same cloud or to other clouds by the magnetic field. Thus the characteristics of star or cluster formation that depend on the removal of angular momentum should be enhanced in regions where compression has occurred along a magnetic field.

5. OBSERVATIONS OF SISF

To prove that stars in a particular region formed as a result of gravitational instabilities behind a shock, it is necessary to establish both a positional and temporal correlation between the stars and the suspected shock. This suggests a useful guideline for the observation of SISF.

Look for the formation of massive stars! Only massive stars turn on fast enough (as masers, IR sources or compact continuum sources) to appear before the shock disappears. Low mass stars may form in the same compressed gas, but because of their relatively long, pre-main sequence lifetime, they may only appear after evidence for the shock has become obscured. It is another matter altogether whether or not shocked gas preferentially forms massive stars. Indeed if this is true, then observations of star formation in regions where shocks still exist are possible, and are probably already available (see Elmegreen and Lada, 1977).

A spatial correlation between young stars and regions of suspected shock activity is not enough to prove that the shock triggered the star formation. A shock may simply clear away obscuration from a star-forming region that existed in a molecular cloud prior to the passage of the shock. Only if the ages of the stars can be determined, and then shown to be identical to the age of the shock, can some relationship between the shock and star formation be inferred. This age determination may be difficult for low mass stars.

REFERENCES

Elmegreen B.G., and Lada, D.J., 1977, Astrophys. J. <u>214</u>, 725.
Elmegreen, B.G., and Elmegreen, D.M., 1978, Astrophys. J. <u>220</u>, in press.
Loren, R.B., 1976, Astrophys. J. <u>209</u>, 466.

FOUR-COLOR INFRARED BOLOMETER SYSTEM FOR ONE-METER TELESCOPE

V. Daneu[+], C. Maxson, G. Peres, S. Serio[†], G. Vaiana[†]

Osservatorio Astronomico di Palermo and Center for
Astrophysics: Harvard College Observatory – Smithsonian
Observatory, Cambridge, Mass.

1. INTRODUCTION

As a part of the collaborative program between the University of
Palermo and the Center for Astrophysics (HCO/SAO) a far infrared
photometer has been designed, fabricated and tested by the Univ-
sity of Palermo to be used as a focal plane instrument on the Center of Astrophysics/University of Arizona balloon-borne 102 cm
telescope (Fazio et al. 1974).

The instrument uses four Gallium-doped Ge bolometers operating
at 1.8 K in a liquid-helium Dewar. The optical scheme is illustrated in Fig. 1a, which summarizes the layout of the components
mounted·on the Cu bottom plate of the Dewar. The incoming telescope beam (f/13.5 from the chopping secondary of the one-meter
Cassegrain) is focused on the Dewar entrance aperture, four milli-
meters in diameter, which limits a one-arc minute diameter field
of view. The beam is then split by an arrangement of Reststrahlen
plates and mirrors, and the four beams so obtained are individually
refocused, with f/1 optics, on four bolometer elements. Because
of the frequency dependence of reflection and transmission for
the crystals used, each beam leaving the Reststrahlen assembly
contains a well-defined wavelength band; the shortest wavelength
beam (the one undergoing transmission through all crystals) is
further passed through a multilayer interference filter to select
a narrow spectrum between 18 and 22μm. The other bands are centered
at 42, 70 and 140μm, respectively. Because the wavelength bands
are separated by amplitude division, the four beams coincide
geometrically and all correspond to the same resolution element

---

[+]Istituto di Elettrotecnica (Università di Palermo
[†]Istituto di Fisica (Università di Palermo)

G. Setti and G. G. Fazio (eds.), Infrared Astronomy, 335-344.

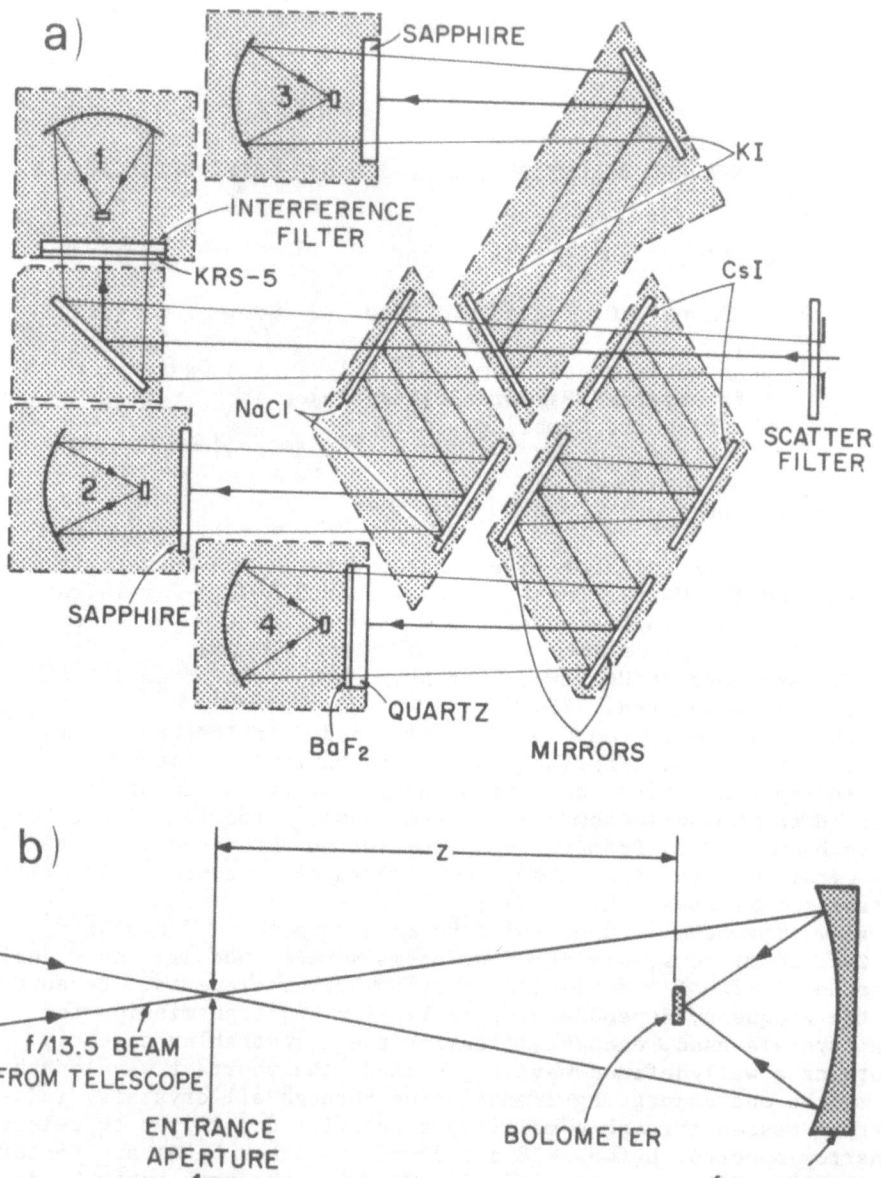

Fig. 1 a) Layout of optical system on Dewar cold plate; b) Unfolded optical scheme for 1 or 4 detectors.

in the telescope field of view.

The performance of the instrument has been found to be limited by the detectors, i.e. by their responsivity and Johnson noise;

background loading is unimportant, as can be expected given the relatively narrow band seen by any single bolometer.

2. OPTICS

The unfolded optical arrangement for one channel is shown in Fig. 1b. All bolometer elements are 0.4 x 0.4 mm in cross section. The spot size of the beam refocused on the bolometer is limited mainly by diffraction, being approximately 0.5 mm diameter for the long-wave (140μm) channel; spherical aberration of the f/1 mirror (silicon overcoated with Al) gives a disc of 0.12 mm diameter at the position of minimum confusion. To compensate for variations in the unfolded beam length z between the channels, the positions of the four focusing mirrors are individually adjusted while observing the image of the bolometer (illuminated with visible light) near the entrance aperture. This test is made with plane mirrors replacing the Reststrahlen crystals in the appropriate holders and allows very precise alignment of the four beams.

Since each Reststrahlen pair is supported individually and involves an even number of reflections, alignment is relatively tolerant of small displacements of the crystal holders; these can be checked outside the Dewar for parallelism using a He-Ne laser. Initial tests showed that alignment is not significantly changed when the system is brought to liquid-helium temperatures. The system has behaved very consistently since the initial assembly, and the individual focusing-mirror positions have never needed readjustment.

The materials employed for the Reststrahlen plates, as indicated in Fig. 1a are NaCl, KI and CsI in order of increasing wavelength of maximum reflectivity. Reflection and transmission data for these materials as a function of wavelength are given in Fig. 2a; 4 K curves have been used when available (Hadni et al., 1968; Mitsuishi et al., 1962). From these curves the overall transmission through the Reststrahlen plate assembly can be calculated for each beam. The resulting curves are shown in Fig. 2b. These curves show "leaks" of unwanted shortwave radiation[+] (visible and near-IR) due to the transparency of the Reststrahlen crystals in these regions. These leaks can give rise to ambiguity in the interpretation of the four-channel signals for sources which are not very cold; to eliminate them, various absorbing materials have been used as windows placed in front of the detector boxes (Fig.1). Further, a blocking filter for visible and near-IR light, consisting of 4 to 8μm diamond powder deposited on 1 mm thick polyethylene, has been mounted near the entrance

---

[+]The long-wave tail curve T_1 (20μm channel) in Fig. 2b comes from assuming that the interference filter (on a Ge substrate) becomes transparent at long wavelengths; a 0.5 mm thick KRS-5 filter is used to eliminate the tail.

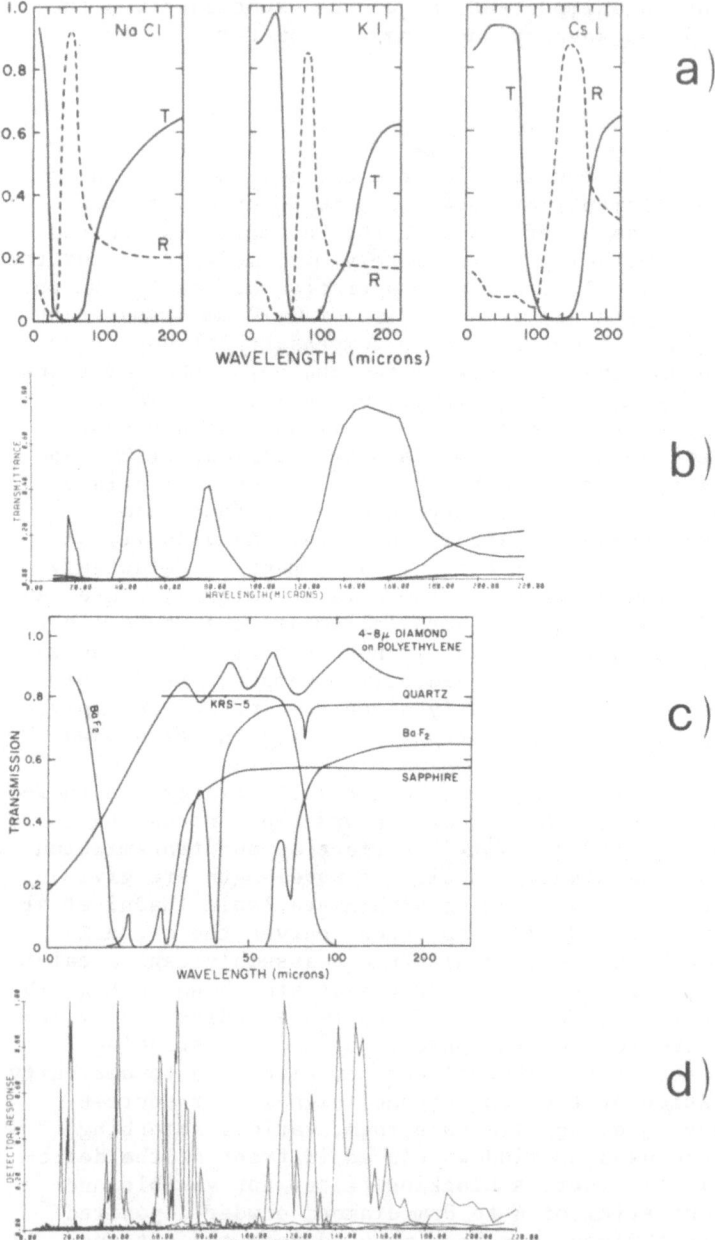

Fig. 2 a) Reflection and transmission curves for Reststrahlen crystals; b) Transmission spectra for the four channels as calculated from a); c) Transmission spectra of blocking filters shown in Fig. 1a (Armstrong, 1976); d) Measured spectral responses of the four channels (arbitrary units).

aperture on the cold Dewar plate. Transmission curves for absorption and scatter filters are shown in Fig. 2c (Armstrong, 1976).
     Calculations of background power from the telescope under
flight conditions (telescope at 220 K) give the levels shown in
Table 1 for the four detectors; for comparison, the background
power incident on each detector of the 1-color instrument used in
the 1-meter telescope is 1×10^{-8} W. These data may be difficult
to check directly; it is straightforward, however, to plot bolometer I-V curves (load curves) with the instrument "looking" at
surfaces of different emissivities. Such curves are shown in Fig.
3 for the system exposed to a 300 K blackbody and to a 300 K mirr-
or with an emissitivity of ∿0.05, respectively. It is clear that,
except possibly for channel 1, the effect of background radiation
on detector performance is negligible (see also Appendix).

3. ELECTRONICS

The bolometers are biased as indicated in Fig. 3; channels 1, 2
and 3 have a 3V bias applied through a 7MΩ resistor, while channel 4 has a 1.5V bias and a 3.5MΩ resistor. The bolometer noise
level is approximately 30 nV/Hz$^{\frac{1}{2}}$ at 16 Hz. Because the wire-
wound load resistors are mounted on the Dewar cold plate at 1.8 K,
their noise contribution is insignificant; this facilitates bias-
ing the bolometers, which would otherwise be more critically
dependent on choosing an operating point with a low dynamic impedance (see bolometer load curves) in order to short-circuit the
load resistor noise. In the present arrangement the detector
performance is nearly independent of bias point over a wide
range of bias battery voltage.
     A block diagram and schemantics for the preamplifiers are
shown in Fig. 4 for one of four identical circuits. The DC voltage across each bolometer element is monitored to provide information on the background level. The 16 Hz signal is AC coupled
into a low-noise JFET (selected Siliconix E230) whose output is
in turn fed to an operational amplifier connected as a current-
to-voltage converter. The gain of the first stage (JFET + Op Amp)
is 200; the circuit arrangement for this stage is similar to the
one described by Stefanovitch (1976). Two further stages follow,
each with a gain of 10; the first incorporates an active low-pass
filter with an 18 Hz cutoff frequency and an 18 dB/OCT roll-off.
The outputs from the two stages are both sent to the telemetry;
thus each detector has outputs at gains of 2000 and 20,000 for
the 16 Hz signal, and a gain of 8 for the DC level output. The
highest gain must bring the noise level above the telemetry resolution, which is ∿5mV; for an input noise of 30 nV/Hz$^{\frac{1}{2}}$, the
peak-to-peak noise at a gain of 20,000 is 14 mV over the amplifier
bandwidth of ∿16 Hz. The low-pass filter was included in the
design to prevent high-frequency microphonics pickup (mainly from
the chopping secondary drive) from cross-modulating the 16 Hz

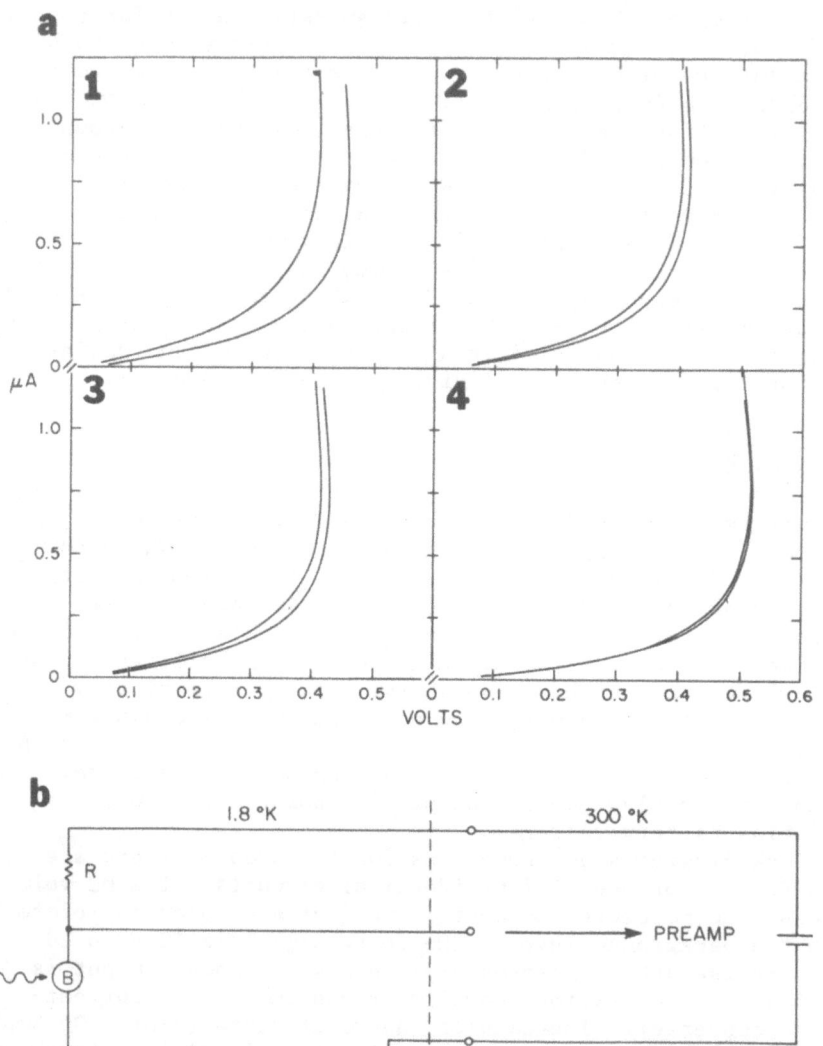

Fig. 3 a) Current–voltage charateristics (load curves) of the
four bolometers obtained by exposing the photometer to a 300 K
blackbody (upper curves) and a 300 K mirror (lower curves). b)
Bolometer bias circuit (1 of 4).

signal.
    The measured equivalent input noise of the preamplifier is
under 6 nV/Hz$^{\frac{1}{2}}$, so that the limiting sensitvity is determined by
detector noise (or possible detector microphonics).
    The preamplifiers are powered by individual ±9V alkaline
batteries; the current drain, ∿2 mA per channel, is low enough

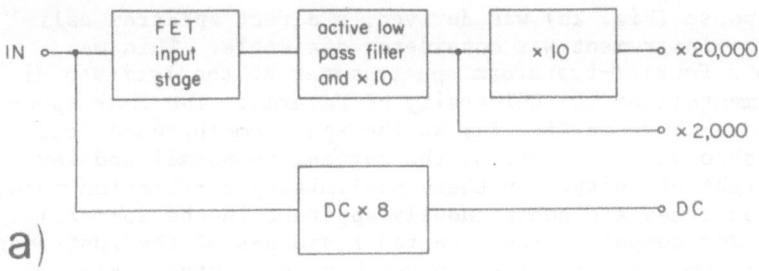

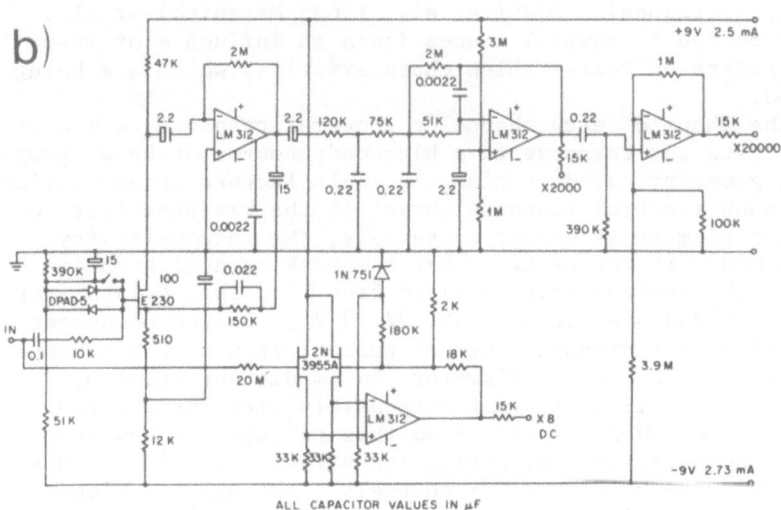

ALL CAPACITOR VALUES IN μF

Fig. 4 a) Bolck diagram of preamplifier. b) Circuit diagram of preamplifier.

to insure at least 48 hours continuous-duty lifetime. To maintain correct operating temperatures for batteries and circuits during flight, and to prevent hardening of O-ring seals, both the amplifier container and the Dewar body are heated with resistive pads to the payload 28V supply in series with thermostats adjusted to close at 0°C. The Dewar and the electronics box are enclosed in 2-inch thick styrofoam to insure thermal insulation. When energized, the heaters absorb a total of 37W; the duty cycle of the thermostats during simulated flight conditions is of the order 50%.

4. SYSTEM PERFORMANCE

Due to the uncertainties in the data from which the theoretical

spectral response (Fig. 2b) was derived, a direct spectral calibration of the instrument was considered desirable. This was
performed on a Fourier-transform spectrometer at the Istituto di
Fisica Sperimentale of the University of Palermo. The four spectral curves are shown in Fig. 2d; as the spectrometer used does
not give an absolute calibration, the curves are normalized for
a maximum height of unity. In these preliminary calibration runs,
H_2O absorption lines are conspicuously apparent in the spectral
responses[†]. For computing the expected responses of the instrument
during flight, smoothed response curves have been used. Fig. 2d
shows that the short- and long-wave leaks have largely been eliminated; the discrepancies between actual and calculated boundaries
of the spectral channels (Hadni et al., 1968; Mitsuishi et al.,
1962) could be due to several causes (such as influence of manu-
facturing process on Reststrahlen charateristics) which are being
investigated.

Given the shape of each channel's spectral response, a measur-
ment of the detector response to a blackbody source of known temperature and geometry can determine the scale factors (transmission
at peak of each spectral response curve) if the responsitivity of
the bolometer element is known; conversely, the responsitivity
can be determined if one assumes that the peak transmission is
equal to the theoretical value derived from Fig. 2b. A laboratory
calibration yielded signals 825, 55, 37, $9\mu V_{p-p}$ at the bolometer
leads, in order of increasing channel number, from a 16 Hz chopped
source at 550 K, 0.2 inch in diameter, 80 cm distant from the
Dewar window. Overall instrument responsitivities for the four
channels can be derived directly from this calibration and the
smoothed transmission curves, giving the values 2.4×10^4, 2.0×10^4, 2.9×10^4, 3.6×10^4 V/W for channels 1 through 4. Given
the bolometer noise of 30 nV/Hz$^{\frac{1}{2}}$, if one assumes an integration
time of 1/3 sec. (i.e. a 1 Hz signal bandwidth, compatible with
the present scanning speed of the one-meter telescope), it is
possible to estimate the minimum spectral irradiance required
for a signal-to-noise ratio of 1; the results are 29, 96, 102,
124 Jansky for channels 1 through 4.

Acknowledgements: Mark Stier of HCO participated in the preliminary design of the Dewar optics and filter assembly. Bolometers
were purchased from Infrared Laboratories, Inc., Tucson, Arizona.
The work was supported by CNR and CRRN (Italy). Partial support
by NASA Grant NSG 7176 is also acknowledged.

[†]These may be due to residual water vapor in the spectrometer
or to ice condensation inside the instrument at the time of the
test. Final testing is presently being conducted.

Table 1

Detector No.	1	2	3	4
220° K background power (Watts) integrated over				
I Main Pass Bands 16 → 23µm	3×10^{-8}			
35 → 65µm		2.1×10^{-8}		
65 → 110µm			2.5×10^{-9}	
100 → 220µm				1.9×10^{-9}
II Un-wanted Short Wave 10 → 20µm		8.7×10^{-10}		
10 → 30µm			4.4×10^{-9}	
10 → 100µm				9.5×10^{-9}
III Un-wanted Long Wave 150 → 220µm	1.2×10^{-10}	1.2×10^{-11}	1.2×10^{-11}	

APPENDIX

Background loading can increase the detector noise because of statistical fluctuation in the number of photons per second constituting the background flux. The most severely loaded bolometer is D_1, with an expected theoretical background power of $P=3 \times 10^{-8}$ W (Table 1). We assume all photons to have a wavelength $\lambda = 20\mu m$, and use the analogue to the Shot Effect formula $I_{N(RMS)} = (2Iq)^{\frac{1}{2}}$ where I is the relevant current and q is the charge; the noise power density due to background fluctuations is then given by

$$P_N = (2Ph\nu)^{\frac{1}{2}} = 2.4 \times 10^{-14} \text{ W/Hz}^{\frac{1}{2}} \qquad (A1)$$

The noise voltage due to P_N is $P_N \times S$, S being the detector responsivity (V/W). The worst case assumption used here is that S equals the electrical value (from the load curves), $S = 2 \times 10^6$ V/W. This gives a noise voltage due to background fluctuations

$$V_{NB} = S P_N = 48 \text{ nV/Hz}^{\frac{1}{2}} \qquad (A2)$$

slightly higher than the detector Johnson noise. As the actual responsitivity is much lower (or, equivalently, since the measured optical peak transmission is much lower than 1), this contribution to detector noise is insignificant.

References

Armstrong, K., 1976, private communication.
Fazio, G.G., Kleinmann, D.E., Noyes, R.W., Wright, E.L., and Low, F.J., 1974, Symp. on Telescope Systems for the Balloon-Borne Research, NASA Ames Research Center, Calif., NASA TM X-62, 397, p. 38.
Hadni, A., Claudel, J., Morlot, G. and Strimer, P., 1968, Applied Optics 7, 161.
Mitsuishi, A., Yamada, Y. and Yoshinaga, H., 1962, Journal of the Optical Soc. of America 52, 14.
Stefanovitch, D., 1976, Rev. Sci. Instrum. 41, No. 2.

SUBJECT INDEX

absorption lines   10
absorption at 9.7 µm   85ff
absorption at 3.07 µm   85
adiabatic fluctuations, 210, 212
 - non-linear development of   219
Airy disk 258
Airy function   255, 256, 260
antenna temperature   102, 185, 186
atmospheric emission   175, 234-236
 - emission lines   192, 193, 235-238
 - emission lines (atomic)   239
AOO235+164   167

background photon noise   245, 288, 289
background radiation, 173ff, 227, 228
 - from young galaxies   223, 224
 - in the infrared   173, 174, 177, 178, 297
 - in the visible   175, 176, 178
baryon pair production   205
beam cancellation   240, 245, 258, 288
Becklin-Neugebauer object   46, 68, 91, 142
BL Lac objects, properties of   166
 - infrared emission from   166
bolometer, 248
 - four-color   335ff
Bose-Einstein distribution 217
bremsstrahlung   9, 277
brightness temperature   102

"Cartwheel" ring galaxy   156
Cassegrain telescope configuration   308
charged coupled devices (CCD)   252
cloud fragmentation   143ff, 328, 329
clusters of galaxies, formation of   220
 - mean density and size   220
COBE satellite 216, 296-298
CO molecule column density   107, 108, 317